T0189677

Studies in Computational Intelligence

Volume 497

Series Editor

J. Kacprzyk, Warsaw, Poland

For further volumes:
http://www.springer.com/series/7092

Olivier Pivert · Sławomir Zadrożny
Editors

Flexible Approaches in Data, Information and Knowledge Management

 Springer

Editors
Olivier Pivert
IRISA/ENSSAT
Université de Rennes 1
Lannion Cedex
France

Sławomir Zadrożny
Systems Research Institute
Polish Academy of Sciences
Warsaw
Poland

ISSN 1860-949X
ISBN 978-3-319-34663-2
DOI 10.1007/978-3-319-00954-4
Springer Cham Heidelberg New York Dordrecht London

ISSN 1860-9503 (electronic)
ISBN 978-3-319-00954-4 (eBook)

Printed on acid-free paper

Springer is part of Springer Science+Business Media (www.springer.com)

Preface

1 Introduction

Nowadays, data management constitutes a blossoming domain that expands rapidly: the development of numeric applications and mobile devices generates a profusion of data in numerous activity areas. Data sources multiply at a sustained pace: social networks, sensor networks, user data, business data, web data. Data sets are getting huge, more and more open, often pervasive, and their exploitation becomes a major societal challenge.

In this context, research issues are numerous and diverse, in particular those concerning access to information: customization, preference queries, data summaries, database mining, heterogeneous and complex data handling, data integration, uncertain data management.

A crucial issue in database research these days is how to make systems more *flexible* and more *human-centric*. This implies, among other aspects, taking into account the preferences of the users as well as their context, to be able to deal with uncertain data, to exploit metadata such as ontologies in order to answer queries, to devise systems that exhibit a cooperative behavior, etc. Notice that similar problems arise in information retrieval as well. Besides classical database tools that need to be extended, knowledge discovery techniques can also be leveraged to make the content (and the structure) of a database more intelligible to users.

The application of fuzzy set theory to the database domain is an already old story that started in the late 1970s. Indeed, it appeared very early that fuzzy sets constitute an intuitive and powerful tool to model and handle gradual concepts in the context of databases (e.g., vague predicates involved in flexible queries, approximate functional dependencies, imprecise values, user profiles). Pioneering works are those by V. Tahani, H. Prade, P. Bosc, B. Buckles and F. Petry, M. Zemankova and A. Kandel, M. Umano, J. Kacprzyk, to cite a few.

Patrick Bosc played a leading role in this community from the very beginning. His position is confirmed with many publications, plenary Lectures at many prestigious conferences, but also with the response we have received while sending invitations to contribute to this volume dedicated to Patrick, to celebrate his retirement. Patrick, an expert in database theory, quickly recognized a potential of fuzzy set theory to make standard database concepts and tools closer to human way of thinking. One of his great achievements was the development of the SQLf language, which combined a highly popular query language with elements of fuzzy logic in a very comprehensive way, covering all the details of SQL.

The topics listed above (preference queries, uncertainty management, cooperative answering, etc.) are now recognized in mainstream database research, but there is still some effort to make in order to convince the researchers from that community that fuzzy sets are an appropriate tool for tackling these issues. This volume is hopefully a step toward that goal.

2 Structure of the Volume

2.1 Bipolar Preference Queries

Bipolarity refers to the propensity of the human mind to reason and make decisions on the basis of positive and negative affects. Positive information states what is possible, satisfactory, permitted, desired, or considered as being acceptable. On the other hand, negative statements express what is impossible, rejected, or forbidden.

The first part of the volume, devoted to bipolar preference queries, includes four chapters. The first, by D. Dubois and H. Prade, revisits the modeling of the connective *if possible* in requirements of the form "*A and if possible B.*" The authors mainly distinguish between two types of understanding: either (i) *A* and *B* are requirements of the same nature and are viewed as constraints with different levels of priority, or (ii) they are of different nature (only *A* induces constraint(s) and *B* is only used for breaking ties among items that are equally satisfying *A*). The authors show that the two views are related to different types of bipolarity, and discuss them in relation with possibilistic logic. The disjunctive dual of the first view ("*A or at least B*") is then presented in this logical setting.

The second chapter, by T. Matthé, J. Nielandt, S. Zadrożny, and G. De Tré, provides an overview of two approaches to bipolar preferences. Both approaches use pairs of satisfaction degrees as an underlying framework, but have different semantics, and hence lead to different operators for criteria evaluation, ranking, aggregation, etc.

In the third chapter, by L. Liétard, D. Rocacher, and N. Tamani, the authors also consider two different types of bipolar preferences: conjunctive ones (*if possible*) and disjunctive ones (*or else*) and show that both of them can be interpreted in a hierarchical way. They then introduce a general form of fuzzy bipolar conditions and extend the operators of relational algebra so as to give them a bipolar semantics and make them work on bipolar fuzzy relations.

In the fourth chapter, J. Kacprzyk and S. Zadrożny investigate different aspects and interpretations of bipolarity in database querying. The authors analyze various ways to deal with bipolarity, recast them in a unified perspective, and clarify them with respect to conceptual, algorithmic, and implementation related aspects.

See also the chapter by Buche et al. in the third part of the volume (cf. Sect. 2.3) that deals with bipolar queries to uncertain databases using ontologies.

2.2 Ontology-Based Data Access

The second part of the volume, made of two chapters, is devoted to Ontology Based Data Access (OBDA), whose aim is to use an ontology to mediate the access to data sources. The added value of OBDA, w.r.t. accessing a data source directly, is, on the one hand, that the ontology provides a semantic account of the information stored in the data source. On the other hand, the answer to user queries may be enriched by exploiting the constraints expressed by the ontology, thus overcoming incompleteness that may be present in the data.

The first chapter, by U. Straccia, discusses the problem of evaluating ranked top-k queries in the context of ontology mediated access over relational databases. An ontology layer is used to define the relevant abstract concepts and relations of the application domain, while facts with associated score are stored into a relational database. Queries are conjunctive queries with ranking aggregates and scoring functions. The results of a query may be ranked according to the score and the problem is to find efficiently the top-k ranked query answers.

The second chapter, by J. R. Campana, J. M. Medina, and M. A. Vila, presents a schema and a transformation algorithm to store OWL ontologies in Object Relational Databases. The database schema makes it possible to represent an ontology structure, while the transformation algorithm creates an appropriate schema to store its instances preserving all information. The approach presented enables using instance data of imprecise nature, mostly fuzzy numerical data.

2.3 Uncertain Databases

In the late 1970s, Database Researchers (notably E. F. Codd and W. Lipski) started investigating the issue of extending the relational database model so as to represent unknown (null) values. Since then, many authors have proposed diverse approaches to the modeling and handling of databases involving uncertain or incomplete data. In particular, the last two decades have witnessed an explosion of research on this topic, with many chapters devoted to probabilistic databases (and also a few dealing with possibilistic databases, notably by Bosc et al.).

This third part of the volume, devoted to uncertain databases, includes three chapters. The first, by T. Beaubouef and F. Petry, discusses rough set, fuzzy rough set, and intuitionistic rough set approaches and how to incorporate uncertainty management using them in the relational database model.

The second chapter, by P. Buche, S. Destercke, O. Haemmerlé, and R. Thomopoulos, proposes an approach to query a database where the user preferences can be bipolar (cf. Sect. 2.1 above) and the data stored in the database can be uncertain. Query results are then completely ordered with respect to these bipolar preferences, giving priority to constraints over wishes. Furthermore, the authors consider user preferences expressed on a domain of values which is not "flat," but contains values that are more specific than others according to the *kind of* relation.

The third chapter, by J. Pons, C. Billiet, O. Pons and G. De Tré, deals with imperfect data in temporal databases. The authors first provide an overview of the basic concepts and issues related to the modeling of time as such or in relational database models and the imperfections that may arise during or as a result of this modeling. Then, they present a novel technique for handling some of these imperfections.

2.4 Flexible Queries over Nonstandard Data

The notion of fuzziness in information retrieval has been long recognized. Indeed, broadly meant fuzziness may be existent in representing documents, in representing queries, in matching the representations of documents to the representations of queries, in evaluating the retrieved documents by users, and even in evaluation of the performance of the retrieval system. The first chapter of this fourth part of the volume, by G. Pasi, G. Bordogna, and G. Psaila, presents a unifying model of flexible queries with distinct semantics of search terms weights. When querying documents archives, there is often the need to specify importance weights of the search terms that define flexible selection conditions on documents representation. Several interpretations of the semantics of these weights have been proposed within distinct information retrieval models. In this contribution, the authors define a unifying model of information retrieval based on a vector p-norm, where importance weights with distinct semantics can be specified in flexible queries.

Social networks have rapidly become an important technology in our digital-based information intense world. In addition to enabling people from all over the world to connect with each other, they provide vast sources of information about the individual participants in the network. Each of these participants can view a kind of database containing information about themselves. Thus, a social network can be viewed as a kind of network of databases. The second chapter of this fourth part of the volume, by R. Yager, discusses fuzzy relationships and their role in modeling weighted social relational networks. The author describes how the idea of "computing with words" can provide a bridge between a network analyst's linguistic description of social network concepts and the formal model of the network. The idea of vector-valued nodes is introduced and then the basic elements of a technology of social network database theory are presented.

2.5 *Fuzzy Knowledge Discovery and Exploitation*

The paradigm of cooperative answering originated from the works concerning natural-language question-answering in the early 1980s. One of the aims of such works is to prevent systems from producing "there is no result" when a query fails. In the first chapter of this fifth part, G. Smits, O. Pivert, and A. Hadjali propose a unified framework for cooperative answering, that relies on a fuzzy-cardinality-based summary of the database. The authors show how this type of summary can be efficiently used to explain failing queries or to revise queries returning a plethoric answer set.

The last chapter, by M. Q. Flores, F. Del Razo, A. Laurent, and N. Sicard deals with the parallelization of fuzzy database mining algorithms. The authors discuss why a parallel approach is crucial to tackle the problem of scalability and optimal performance in the context of database mining. They then present parallel algorithms on multi-core architectures devoted to four knowledge discovery paradigms, namely fuzzy association rules, fuzzy clustering, fuzzy gradual dependencies, and fuzzy tree mining.

The editors wish to thank the following reviewers for their invaluable help: Troels Andreasen, Isabelle Bloch, Gloria Bordogna, Patrice Buche, Jesus Roque Campana, Davide Ciucci, Sébastien Destercke, Guy De Tré, Didier Dubois, Michel Grabisch, Allel Hadjali, Olgierd Hryniewicz, Hélène Jaudoin, Janusz Kacprzyk, Donald Kraft, Ludovic Liétard, Zongmin Ma, Christophe Marsala, Arnaud Martin, Adam Niewiadomski, Henri Prade, Grégory Smits, Umberto Straccia, Eulalia Szmidt, Laurent Ughetto, Peter Vojtas.

Olivier Pivert
Sławomir Zadrożny

Contents

Part I Bipolar Preference Queries

1 Modeling *"and if possible"* and *"or at least"*: Different
   Forms of Bipolarity in Flexible Querying 3
   Didier Dubois and Henri Prade

2 Constraint-Wish and Satisfied-Dissatisfied: An Overview
   of Two Approaches for Dealing with Bipolar Querying 21
   Tom Matthé, Joachim Nielandt, Sławomir Zadrożny
   and Guy De Tré

3 A Relational Algebra for Generalized Fuzzy
   Bipolar Conditions 45
   Ludovic Liétard, Daniel Rocacher and Nouredine Tamani

4 Bipolarity in Database Querying: Various Aspects
   and Interpretations.................................... 71
   Sławomir Zadrożny and Janusz Kacprzyk

Part II Ontology-based Data Access

5 On the Top-k Retrieval Problem for Ontology-Based
   Access to Databases 95
   Umberto Straccia

6 Semantic Data Management Using Fuzzy Relational
   Databases... 115
   Jesùs R. Campaña, Juan M. Medina and Maria A. Vila

Part III Uncertain Databases

**7 Information Systems Uncertainty Design and Implementation
Combining: Rough, Fuzzy, and Intuitionistic Approaches** 143
Theresa Beaubouef and Frederick Petry

**8 Flexible Bipolar Querying of Uncertain Data
Using an Ontology** 165
Patrice Buche, Sébastien Destercke, Valérie Guillard,
Ollivier Haemmerlé and Rallou Thomopoulos

**9 Aspects of Dealing with Imperfect Data
in Temporal Databases** 189
José Pons, Christophe Billiet, Olga Pons and Guy De Tré

Part IV Flexible Queries Over Nonstandard Data

**10 A Unifying Model of Flexible Queries with Distinct Semantics
of Search Term Weights** 223
Gloria Bordogna, Gabriella Pasi and Giuseppe Psaila

**11 Social Network Database Querying Based on Computing
with Words** 241
Ronald R. Yager

Part V Fuzzy Knowledge Discovery and Exploitation

12 Fuzzy Cardinalities as a Basis to Cooperative Answering 261
Grégory Smits, Olivier Pivert and Allel Hadjali

13 Scalability and Fuzzy Systems: What Parallelization Can Do 291
Malaquias Q. Flores, Federico Del Razo,
Anne Laurent and Nicolas Sicard

Part I
Bipolar Preference Queries

Chapter 1
Modeling *"and if possible"* and *"or at least"*: Different Forms of Bipolarity in Flexible Querying

Didier Dubois and Henri Prade

Abstract This research note revisits an important issue with respect to the representation of preference queries, namely the modeling of *"if possible"* in requirements of the form *"A and if possible B"*. We mainly distinguish between two types of understanding: either (i) A and B are requirements of the same nature and are viewed as constraints with different levels of priority, or (ii) they are of different nature (only A induces constraint(s) and B is only used for breaking ties among items that are equally satisfying A). We indicate that the two views are related to different types of bipolarity, and discuss them in relation with possibilistic logic. The disjunctive dual of the first view (*"A or at least B"*) is then presented in this logical setting. We also briefly mention the idea of an extension of the second view where B may refer both to bonus conditions or malus conditions that may increase or decrease respectively the interest in an item satisfying A.

1 Introduction

Preference and flexible queries have attracted a considerable interest in different circles and at different epochs among databases researchers [5, 6, 13, 14, 20, 29, 30, 32–35, 38, 39]. There may exist slightly different motivations for using flexible queries. One may want to introduce some tolerance by tacitly enlarging crisp queries to similar requested items. One may more often try to express preferences. The expectation is then to both (i) rank-order retrieved items according to the extent to which they are satisfactory, and (ii) try to avoid empty answers by not restricting the query

D. Dubois (✉) · H. Prade
IRIT, Université Paul Sabatier, Toulouse Cedex 09,
Toulouse 31062, France
e-mail: dubois@irit.fr

H. Prade
e-mail: prade@irit.fr

O. Pivert and S. Zadrożny (eds.), *Flexible Approaches in Data, Information and Knowledge Management*, Studies in Computational Intelligence 497,
DOI: 10.1007/978-3-319-00954-4_1, © Springer International Publishing Switzerland 2014

to the profile of the most preferred items (which may not exist in the database). The interested reader may consult [28] for a comparative overview of different approaches to preference queries in database systems, ranging from early mainstream databases proposals that distinguish between mandatory conditions and secondary conditions, or use similarity relations, to fuzzy set-based approaches involving gradual membership functions and priorities, to Pareto ordering-based preference models (where no commensurability hypothesis between the satisfaction degrees pertaining to the different attributes involved is needed), and finally to conditional ceteris paribus preferences (where the request may take the form of comparative preferences stated in specific contexts). More generally, a broad panorama of approaches to the representation and the handling of preferences in operations research, databases, and artificial intelligence can be found in [16].

In the following, we focus the discussion on issues related to the idea of bipolarity in the expression of preference queries, an idea that has been developed in Patrick Bosc's group [1, 7, 9, 31] in the recent years. The idea of bipolarity refers to the distinction between what is regarded as positive and what is regarded as negative, or in other words, between what is found satisfactory and what is rejected. Bipolarity may be encountered in preference (or in knowledge) representation under different forms [24]. One may use bipolar univariate scales ranging from what is completely bad to what is completely good. Another form of bipolarity takes place when items are judged according to two independent evaluations on unipolar scales, a positive one for grading what is in favor of the items, and a negative one for what is in disfavor of them. The evaluations may play a symmetric role or not. In the asymmetric case [25], the evaluations are not based on specifications of the same nature.

More specifically, we examine the problem of modeling requests looking for items that satisfy "*A and if possible B*". Such requirements have been considered early by database researchers in order to introduce some hierarchy between requested conditions [30]. In such a basic query, A stands for a (possibly compound) condition that should be imperatively satisfied. Thus, (at least) the items for which A is satisfied are "acceptable", but if they also satisfy B, they are considered as being "better" answers. Thus, the requirement "*A and if possible B*" has a bipolar flavor, since on the one hand the items not satisfying A are rejected, while on the other hand those satisfying A and B are positively favored among the items satisfying A.

This chapter is organized as follows. The next section recalls what the conjunctive condition "*A and if possible B*" may mean precisely, when A and B are of the same nature, i.e. play the role of constraints; we also study the disjunctive condition "*A or at least B*". In Sect. 3 we discuss another view of "*A and if possible B*" where B is no longer of the same nature as A, and is only expressing a wish that is used for breaking ties among items that equally satisfy A. This latter view appears to be more refined. Moreover, we briefly suggest that wishes may be graded on a bipolar univariate scale (rather than on a positive scale), thus allowing either a positive or a negative impact on the ranking of items satisfying a set of constraints equally. In Sect. 4 we relate the previous concerns to artificial intelligence works and more particularly to possibilistic logic, and enlarge the discussion from conjunctive and disjunctive forms to hybrid forms.

2 Modeling "*and if possible*" and "*or at least*" in Terms of Weighted Constraints

In the following, conditions "*A and if possible B*" and "*A or at least B*" are considered as expressing a hierarchy of constraints. We first examine the case where A and B are crisp, then we show how a set of nested conditions "A_n and possibly A_{n-1} and ... and possibly A_1", where the A_i's are crisp, is naturally associated with a fuzzy set. Lastly, the approach is extended to when A and B are fuzzy. Moreover, another view where "*if possible*" is understood as "*if consistent*" is also addressed.

2.1 "*If possible*": Crisp Case

Consider for the moment that A and B are crisp conditions that can be modeled by classical subsets. Thus A and B are sets of interpretations that correspond to configurations of attribute values describing potential items.

With a condition of the form "*A and if possible B*" where A and B are classical subsets we introduce a hierarchy between three sets of (potential) items, namely

  i. those that satisfy $A \cap B$, i.e. the preferred ones,
 ii. those in $A \cap \overline{B}$, which are still acceptable, and
iii. those in $\overline{A}$, which are fully rejected.

The condition "*A and if possible B*" could be stated otherwise as "*B provided that A*". It is clear that in this approach, the condition "*A and if possible B*" is equivalent to the condition "*A and if possible A $\cap$ B*".

We could then assume without loss of generality that $A \supseteq B$, since what is better should be normally inside what is acceptable. In case A and B are general unconstrained requests and if $B \cap \overline{A} \neq \emptyset$ (where $\overline{A}$ denotes the complement of A), it is thus possible to revise B into $B^{revised} = A \cap B$ in order to have $A \supseteq B^{revised}$ since we are only interested in items satisfying A. In case $B^{revised} = \emptyset$, no further discrimination can be made among the items satisfying A since *no* item satisfying A is fully satisfactory. If $B^{revised} = A$, no discrimination takes place either since *all* items satisfying A are fully satisfactory.

Then we may think of interpreting the condition "*A and if possible B*" as a pair of nested sets (A, B), with $A, B \neq \emptyset$, representing the *support* and the *core* of a *fuzzy set F* respectively, where for an item x with description $\partial(x)$ in terms of attribute values, we have the following membership grades: $F(\partial(x)) = 1$ if $\partial(x) \in A \cap B$; $1 > F(\partial(x)) > 0$ if $\partial(x) \in A \cap \overline{B}$; and $F(\partial(x)) = 0$ if $\partial(x) \notin A$. Thus, $F(\partial(x)) > 0$ iff $\partial(x) \in A$. For simplicity we write $F(x)$ instead of $F(\partial(x))$ in the following.

When $x \in A \cap \overline{B}$ we can prescribe some value $\lambda \in [0, 1]$ such that $F(x) = \lambda$. Then it can be checked that F is defined by

$$F(x) = \min(A(x), \max(\lambda, B(x)))\ \ \ \ \ \ \ \ \ \ \ \ \ (1)$$

where λ is a discounting factor (A has weight 1, and B has weight λ) and $A(\cdot)$, $B(\cdot)$ are characteristic functions of A and B. When $\lambda = 0$ (no discounting), we are back to a classical conjunction while if $\lambda = 1$ (total discounting of B) we get A, i.e., B is forgotten. This expression comes from the prioritized fusion of possibility distributions, and is a weighted conjunction first suggested by Dubois and Prade [18]. Interpreting $1 - \lambda$ as the degree of priority of B, it means A must be satisfied with full priority (1) and B with priority $1 - \lambda$, so $x \in A$ is satisfactory to degree λ if B is violated. This view has been exploited for the possibilistic approach to prioritized constraints [17]. The above equation can also be read as a formal translation of "A and if possible B" (where $\max(\lambda, B(y))$ can be seen as an implication connective $a \rightarrow b = \max(1 - a, b)$, viewing λ as the complement to 1 of the extent to which it is possible to consider B).

2.2 "Or at least": Crisp Case

Dual to the conjunctive "*and if possible*" conditions are disjunctive "*or at least*" conditions. Several authors [8, 31] write "or else" in place of "or at least", but we prefer this latter phrase which is better suggesting the idea of a hierarchy. It corresponds to the ordered disjunction in qualitative choice logic [12]. With a condition of the form "*A or at least B*" where A and B are classical subsets we introduce another hierarchy between three sets of (potential) items, namely

   i. those that satisfy A, which are the preferred ones,
  ii. those in $B \cap \overline{A}$, which are still acceptable, and
 iii. those in $\overline{A \cup B}$, which are fully rejected.

The condition "*A or at least B*" could be stated otherwise as "*B or better A*". It is clear that in this approach, the condition "*A or at least B*" is equivalent to the condition "*A or at least $A \cup B$*". Again, it is then possible to change B into $A \cup B$ without harming the condition, which means we could assume $A \subseteq B$ without loss of generality. Again we need to define a membership grade μ if $\partial(x) \in B \cap \overline{A}$. We obtain a fuzzy set G defined by:

$$G(x) = \max(A(x), \min(\mu, B(x)))\ \ \ \ \ \ \ \ \ \ \ \ (2)$$

so that μ is the degree of preference of B rather than its priority. It is clear that this is the so-called weighted disjunction [18], that can be read as a formal translation of "*B, or at least B*". Indeed, one may either insist that "*(at least) B and if possible B*" is required, or one may state that one would prefer A, or if not possible one would accept at least B by default. This expression entertains a close relationship with the weighted minimum.

Some properties are worth noticing. First, it is clear that *"A and if possible B"* is equivalent to *"A ∩ B or at least A"*, and *"A or at least B"* is equivalent to *"A ∪ B and if possible A"* (if $\mu = \lambda$). This is clear, from Eqs. (1) and (2), and mutual distributivity between union and intersection.

So, if $A \supseteq B$ it holds that *"A or at least B"* is equivalent to *"B and if possible A"* since then:

$$F(x) = \max(B(x), \min(\lambda, A(x))) = \min(A(x), \max(\lambda, B(x))) \qquad (3)$$

Moreover some De Morgan-type laws are valid as follows: *"not (A or at least B)"* means *"not A and if possible not B"* (if $\mu = 1 - \lambda$, using the negation $1 - (\cdot)$ for negating fuzzy set G in (2)). Indeed, $1 - \max(a, \min(1 - \lambda, b)) = \min(1 - a, \max(\lambda, 1 - b))$.

2.3 Fuzzy Case

We may more generally consider pairs (A, B) where A and B are normalized *fuzzy* sets, such that $\exists x, y, A(x) = B(y) = 1$. The above results and definitions carry over to this context using idempotent connectives such as the fuzzy weighted minimum and maximum [18]. Namely the fuzzy *and if possible* and *or at least* conditions yield fuzzy sets F and G defined by Eqs. (1) and (2) respectively, with all membership grades in the unit interval. The fact that we can restrict to nested fuzzy sets $A \supseteq B$, i.e., $\forall y\ A(y) \geq B(y)$ is due to the following equalities: $\forall \alpha, \beta, \lambda, \mu \in \mathbb{R}$,

$$\min(\alpha, \max(\beta, \lambda)) = \min(\alpha, \max(\min(\alpha, \beta), \lambda)) = \max(\min(\alpha, \beta), \min(\alpha, \lambda));$$
$$\max(\alpha, \min(\beta, \mu)) = \max(\alpha, \min(\max(\alpha, \beta), \mu)) = \min(\max(\alpha, \beta), \max(\alpha, \mu));$$

that ensure the validity of fuzzy extensions of Eqs. (1), and (2) and the equivalence between the following statements in the fuzzy case :

- *"A and if possible B"* $\iff$ *"A and if possible A ∩ B"* $\iff$ *"A ∩ B or at least A"*;
- *"A or at least B"* $\iff$ *"A or at least A ∪ B"* $\iff$ *"A ∪ B and if possible A"*.

In particular, if $A \supseteq B$, i.e., $\forall y\ A(y) \geq B(y)$, then we retrieve fuzzy extensions of Eq. (3), which is the median $med(A(x), \lambda, B(x))$. So, *"A and if possible B"* can be turned into *"B or at least A"* (under the condition $A \supseteq B$ of fuzzy inclusion). Note that if $B = A$, we get $F = A$ (whatever the value of λ), as expected since *"A and if possible A"* means *"A"* indeed, even if A is fuzzy.

Both *"and if possible"* conjunctions and *"or at least"* disjunctions are studied in detail by Bosc and Pivert [8] in a more general axiomatic setting.

2.4 "If possible" as "if consistent"

In the case of fuzzy bipolar queries "A *and if possible B*", the importance coefficient λ may reflect the consistency between A and B, interpreting the constraint as "A *and, if B is consistent with A, B*". Then, one may choose $\lambda = cons(A, B) = \sup_y \min(A(y), B(y))$, which estimates the consistency of A and B. This is what has been proposed in [19, 36] for the purpose of information fusion. It applies when one of the pieces of information is certain, while the other is a piece of default information. Expressions formally similar to Eqs. (1) and (2) then express the prioritized conjunction and disjunction of A with B (where A has priority over B):

$$(A \cap B)(y) = \min(A(y), \max(1 - cons(A, B), B(y))); \quad (4)$$

$$(A \cup B)(y) = \max(A(y), \min(cons(A, B), B(y))). \quad (5)$$

As can be seen, when $cons(A, B) = 1$, regular conjunctions and disjunctions are retrieved. When, $cons(A, B) = 0$, A is obtained in both cases. In flexible querying, the possible conflict between A and B does not come only from the specifications as in information fusion, but also from the fact that maybe there does not exist items in the database satisfying B (and then A). Note that $cons(A, B)$ accounts only for the extensions of A and B on an attribute domain. Taking the supremum on the items x's such as $y = \partial(x)$ may make $cons(A, B)$ smaller since there may exist values y of the attribute domain that are not reached by any item x in the database. The second equation is a prioritized disjunction that becomes all the more restrictive as B becomes incompatible with A, in which case only A remains. It means that if B is in slight conflict with A, then the objects for which B is true are less preferred, although not fully rejected. De Morgan laws hold between these prioritized conjunctions and disjunction. However, this approach becomes trivial if there is an inclusion relation between normalized A and B.

2.5 Nested "if possible" Conditions

Let a fuzzy set F be defined on a finite scale $\alpha_1 = 1 > \alpha_2 > \ldots > \alpha_n > \alpha_{n+1} = 0$, and consider its level cuts $F_{\alpha_i} = \{y | F(y) \geq \alpha_i\}$. Clearly, F_{α_1} is the core and F_{α_n} the support of F, and $F_{\alpha_n} \supseteq F_{\alpha_{n-1}} \cdots \supseteq F_{\alpha_1}$. The membership function of F is then obtained from its level cuts as

$$F(y) = \max_{i=1,n} \min(\alpha_i, F_{\alpha_i}(y)) = \min_{i=1,n} \max(\alpha_{i+1}, F_{\alpha_i}(y)) \quad (6)$$

where $F_{\alpha_i}(y) = 1$ if $y \in F_{\alpha_i}$ and $F_{\alpha_i}(y) = 0$ otherwise. The first equality is just Zadeh's representation of a fuzzy set in terms of its cuts [37]. It can be viewed as requesting "F_{α_1} *or at least* F_{α_2} *or* ... *or at least* F_{α_n}". The other one is the decomposition of a fuzzy constraint into crisp prioritized ones [17]. Then it is clear that such a fuzzy set can also be seen as a representation of "F_{α_n} and if possible $F_{\alpha_{n-1}}$ and ... and if possible F_{α_1}". Indeed an item x will be all the better as its description $\partial(x)$ has a greater degree of membership in F, i.e. according to Eq. (6), belongs to a level cut with a higher value of α_i, which means that more *"if possible"* conditions are satisfied by x.

When $n = 2$, we get

$$F(y) = \max(F_{\alpha_1}(y), \min(\alpha_2, F_{\alpha_2}(y))) = \min(F_{\alpha_2}(y), \max(\alpha_2, F_{\alpha_1}(y))) \quad (7)$$

which returns 1 if $y \in F_{\alpha_2} \cap F_{\alpha_1} = F_{\alpha_1}$, $1 > \alpha_2 > 0$ if $y \in F_{\alpha_2} \cap \overline{F_{\alpha_1}}$, and 0 if $\overline{F_{\alpha_2}}$, in agreement with the above representations of "A *and if possible* B" and "B *or at least* A" with $A = F_{\alpha_2}$ and $B = F_{\alpha_1}$.

3 Modeling *"and if possible"* in Terms of Constraints and Wishes

In [21, 23], we have proposed and advocated another view of "A *and if possible* B" where A is still a constraint, while B is only used for breaking ties between items having the same evaluation w.r.t. A.

3.1 Asymmetric Handling of Bipolar Queries

In this approach, the condition "A *and if possible* B" means that B can be used only if the condition A leaves indifferent options. Considering the pair of valuations $(A(x), B(x)) \in [0, 1]^2$ qualifying the merit of object x, the preference on the set of objects is defined by means of the lexicographic ranking of these pairs. Namely

$$(\alpha, \beta) \succ (\gamma, \delta) \iff \text{either } \alpha > \gamma \text{ or } (\alpha = \gamma > 0 \text{ and } \beta > \delta). \quad (8)$$

Note that the requirement (A, B) is not the same as $(A, A \cap B)$ since if $B(x_1) > A(x_1) > 0$, $B(x_2) > A(x_2) > 0$, and $A(x_1) = A(x_2)$, $B(x_1) > B(x_2)$, it is clear that x_1 is preferred to x_2 under request (A, B) while they are indifferent with $(A, A \cap B)$. However, objects x such that $A(x) = 0$ are rejected whatever the value of $B(x)$. Moreover, it is not clear how to model "B *or at least* A" in the lexicographic approach, unless we directly define it as "$A \cup B$ *and if possible* B", and compute a lexicographic ranking of the pairs $(\max(A(x), B(x)), B(x))$. But again, it is no longer equivalent to "B *or at least* A" using a lexicographic ranking of the pairs $(B(x), A(x))$. The latter option is proposed by Liétard et al. [31] (see also their chapter in this book), under the condition $A \supseteq B$. But while under this condition, "A *and if possible* B" is equivalent to "B *or at least* A" in the weighted constraint approach, as intuitively expected, this is no longer true in the lexicographic approach, since even if $A \supseteq B$, the lexicographic ranking of the pairs $(B(x), A(x))$ is not the same as lexicographic ranking of the pairs $(A(x), B(x))$.

Yet another idea for further study could be to replace $\cap$ and $\cup$ by leximin and leximax in the processing of $(A, A \cap B)$ and $(A \cup B, B)$.

It is important to notice that this approach where wishes are used for breaking ties between the items that satisfy the constraints only makes sense when the possible levels of satisfaction of the constraints belong to a discrete chain (as it is the case when dealing with nested requirements such as "A_n and if possible A_{n-1} and ... and if possible A_1" and the A_i's are crisp). In case of a continuum of objects valued on $[0,1]$, preferring, e.g., an item x such that $A(x) = 0.85$ and $B(x) = 0$ to an item x' such that $A(x) = 0.84$ and $B(x) = 1$ would sound highly debatable. Still, in practice, it would be possible to accommodate a request such as "a reasonably priced apartment, if possible close to the train station" with this approach, by discretizing the scale $[0,1]$ into a finite set of levels corresponding to significantly different prices, which requires a granulation step.

3.2 Comparing the Two Approaches

In the elementary case, where A and B are both crisp conditions, the views of B as a weighted constraint, as discussed in the previous section, or as a criterion for breaking ties lead to the same ranking of the considered items, namely first those that satisfy both A and B, then those that satisfy A without B, and finally those that do not satisfy A.

When A and B become fuzzy, the two views are no longer equivalent. This point can be checked by taking $A(y) = \max(A_1(y), \min(\alpha, A_2(y)))$ and $B(y) = \max(B_1(y), \min(\alpha', B_2(y)))$, the A_i's and B_j's being crisp subsets. The condition $A \supseteq B$ means here $A_1 \supseteq B_1$ $A_2 \supseteq B_2$ and $\alpha \geq \alpha'$ assuming $B_2 \cap A_2 \neq \emptyset$. Then, by applying Eq. (3), the view of A and B as constraints yields:

$$F(y) = \max(B(y), \min(\lambda, \max(A_1(y), \min(\alpha, A_2(y)))$$
$$= \max(B_1(y), \min(\alpha', B_2(y), \min(\lambda, A_1(y)), \min(\alpha, \lambda, A_2(y))))$$

Note that $A \supseteq B$ means here $A_2 \supseteq A_1 \supseteq B_1$, $A_2 \supseteq B_2 \supseteq B_1$ and when $\alpha' < \alpha < \lambda$, we get the following evaluations $F(\partial(x))$ for items x:

- 1 if $\partial(x) \in A_1 \cap B_1$
- λ if $\partial(x) \in A_1 \cap \overline{B_1}$
- α if $\partial(x) \in A_2 \cap \overline{A_1} \cap B_2$
- 0 if $\partial(x) \in \overline{A_2}$

The other view, where B is used to break ties among items that satisfy A to some extent, leads to rank-order the items in the following way:

(1st) those that satisfy A_1 (and thus A_2) and B_1
(2nd) those that satisfy A_1 (and A_2) and not B_1 but B_2,
(3rd) those that satisfy A_1 (and A_2) and not B_2 (and thus not B_1),
(4th) those that satisfy A_2 (but not A_1) and B_2,
(5th) those that satisfy A_2 but neither A_1 nor B_2,
(6th) those that do not satisfy A_2.

It is clear that this second view is more refined, since we now have 6 distinct layers of items in the above example, instead of 4 with the hierarchically organized constraints approach. Another more sophisticated example illustrating the difference between the two views can be found in [23, 25].

3.3 Positive and Negative Wishes

In often found examples of constraints with wishes (e.g., [21, 23]), wishes have a positive flavor: if a wish is satisfied, it provides a bonus in favor of the item satisfying it against other items that have similar levels of satisfaction with respect to constraints, but do not satisfy this wish. Still, there may exist a negative counterpart to wishes. For instance, a request such as "a reasonably priced apartment, if possible not on the groundfloor" may not just mean that "not being on the groundfloor" provides a bonus, but rather that "being on the groundfloor" has a negative flavor. In fact, one may have both positive and negative wishes, as in "a reasonably priced apartment, if possible close to the train station, and if possible not on the groundfloor".

This calls for the use of a bipolar univariate scale where a neutral level separates the positive grades from the negative ones, and where the positive wishes and the negative wishes are handled separately. Then, one may for instance give priority to negative wishes, and use the negative wishes for breaking ties between items having similar levels of satisfaction with respect to constraints, and then use positive wishes for breaking further ties if any left. One may also think of combining evaluations pertaining to the positive wishes and the negative wishes, thus introducing compensation between them.

Another simpler option, already discussed in [2], would be to rank-order the items on the basis of the constraints and then for a given constraint satisfaction level to use the number of wishes satisfied (for instance the positive ones) for each item, providing the user with arguments pro and con respectively corresponding to the positive and to the negative wishes fulfilled by the item.

4 Possibilistic Logic Modeling

In this section, we reexamine the modeling of *"and if possible"* in a logical setting and enlarge the discussion to non conjunctive queries. We consider different forms of queries asking for items satisfying conditions C_1, C_2, C_3, with the information that C_1 is more important than C_2, which is itself more important than C_3. Conditions are supposed to be binary. They are *not necessarily nested*. They may be logically independent or not. For the sake of simplicity, we use here three conditions only, but what follows would straightforwardly extend to n conditions. We denote $[C_i]$, $[C_i \wedge C_j]$, the set of items (if any) satisfying condition C_i, the set of items (if any) satisfying C_i and C_j, and so on. We first consider conjunctive queries.

4.1 Conjunctive Queries

Consider the query of the type "C_1 is required *and if possible C_2 also and if possible C_3 too*", with the following intended meaning ($\gg$ reads "is preferred to") in terms of items:

$$[C_1 \wedge C_2 \wedge C_3] \gg [C_1 \wedge C_2 \wedge \neg C_3] \gg [C_1 \wedge \neg C_2] \gg [\neg C_1] \qquad (9)$$

i.e., one prefers to have the three conditions satisfied rather than the two first ones only, which is itself better than having just the first condition satisfied (which in turn is better than not having even the first condition satisfied).

This may be described in possibilistic logic [22] (see Appendix) in different ways. First, it can be expressed as the conjunction of prioritized goals $\mathcal{C} = \{(C_1, \gamma_1), (C_2, \gamma_2), (C_3, \gamma_3)\}$ with $1 = \gamma_1 > \gamma_2 > \gamma_3 > 0$. Indeed, this possibilistic logic base is associated with the possibility distribution

$$\begin{aligned} \pi_{\mathcal{C}}(\omega) =\,& 1 \text{ if } \omega \in [C_1 \wedge C_2 \wedge C_3] \\ & 1 - \gamma_3 \text{ if } \omega \in [C_1 \wedge C_2 \wedge \neg C_3] \\ & 1 - \gamma_2 \text{ if } \omega \in [C_1 \wedge \neg C_2] \\ & 0 \text{ if } \omega \in [\neg C_1]. \end{aligned}$$

which fully agrees with the ordering (9).

Besides, in a logical encoding, a query such as "find the x's such that condition Q is true", i.e., $\exists x \, Q(x)$? is usually processed by refutation. Using a small old trick due to Green [27], it amounts to adding the formula(s) corresponding to $\neg Q(x) \vee answer(x)$, expressing that if item x satisfies condition Q it belongs to the answer, to the logical base describing the content of the database. It enables theorem-proving by resolution to be applied to question-answering. This idea extends to preference queries expressed in a possibilistic logic setting [10]. The expression of the query Q corresponding to the above set of prioritized goals is then of the form

$$\begin{aligned} Q =\,& \{(\neg C_1(x) \vee \neg C_2(x) \vee \neg C_3(x) \vee answer(x), 1), \\ & (\neg C_1(x) \vee \neg C_2(x) \vee answer(x), 1 - \gamma_3), \\ & (\neg C_1(x) \vee answer(x), 1 - \gamma_2)\}. \end{aligned}$$

where $1 > 1 - \gamma_3 > 1 - \gamma_2$. Then, the levels associated with the possibilistic logic formulas expressing the preference query are directly associated with the possibility levels of the possibility distribution π_C providing its semantics.

4.2 The Two Bipolar Approaches in Possibilistic Logic

Let us go back to the example considered in Sect. 3.2. We considered a request of the form *"A and if possible B"*, where both A and B correspond to sets of prioritized goals, namely and respectively:

$$A = \{(A_2, 1), (A_1, 1 - \alpha)\} \text{ with } 1 > 1 - \alpha > 0, \text{ and}$$
$$B = \{(B_2, 1), (B_1, 1 - \alpha')\} \text{ with } 1 > 1 - \alpha' > 0.$$

Remember we assumed $A_2 \supseteq A_1 \supseteq B_1$, $A_2 \supseteq B_2 \supseteq B_1$ and took $\alpha' < \alpha < \lambda$, with $B_2 \cap A_2 \neq \emptyset$. Then, when both A and B are viewed as constraints, with priority to the ones associated with A, the request *"A and if possible B"* translates into a *unique* set G of prioritized goals, where the goals in B are discounted by $1 - \lambda$:

$$G = \{(A_2, 1), (A_1, 1 - \alpha), (B_2, \min(1, 1 - \lambda)), (B_1, \min(1 - \alpha', 1 - \lambda))\}$$

This possibilistic logic base is associated with the possibility distribution

$$\begin{aligned} \pi_G(\omega) = &1 \text{ if } \omega \in [A_1 \wedge B_1] \\ &\lambda \text{ if } \omega \in [A_1 \wedge \neg B_1] \\ &\alpha \text{ if } \omega \in [A_2 \wedge \neg A_1 \wedge B_2] \\ &0 \text{ if } \omega \in [\neg A_2]. \end{aligned}$$

This corresponds exactly to the membership function of F in Sect. 3.2.

Let us now consider the second view where only A is regarded as a set of prioritized constraints, while B is a set of prioritized wishes. Now we keep A and B separate. Each interpretation ω is the associated with a pair of values: the first (resp. the second) value is equal to $1 - \gamma^*$ (resp. $1 - \delta^*$) where γ^* (resp. δ^*) is the priority of the formula violated by ω having the highest priority in A (resp. B). We obtain, the following *vector-valued* possibility distribution:

$$\begin{aligned} \pi_{(A,B)}(\omega) = &(1, 1) \text{ if } \omega \in [A_1 \wedge B_1] \\ &(1, \alpha') \text{ if } \omega \in [A_1 \wedge \neg B_1 \wedge B_2] \\ &(1, 0) \text{ iff } \omega \in [A_1 \wedge \neg B_2] \\ &(\alpha, \alpha') \text{ if } \omega \in [A_2 \wedge \neg A_1 \wedge B_2] \\ &(\alpha, 0) \text{ if } \omega \in [A_2 \wedge \neg A_1 \wedge \neg B_2] \\ &(0, 0) \text{ if } \omega \in [\neg A_2]. \end{aligned}$$

Note the lexicographic ordering of the evaluation vectors. This corresponds to the 6 layers of interpretations found in 3.2, and makes it clear that this second view is (trivially) more refined.

4.3 Disjunctive Queries

We now consider *disjunctive* queries of the form "at least C_3 is required, or better C_2, or still better C_1", as discussed in [10] and in Sect. 2.2. It can be equivalently stated starting with what is preferred: "C_1 is required with priority, or failing this C_2, or still failing this C_3". It has the following intended meaning in terms of interpretations:

$$[C_1] \gg [\neg C_1 \wedge C_2] \gg [\neg C_1 \wedge \neg C_2 \wedge C_3] \gg [\neg C_1 \wedge \neg C_2 \wedge \neg C_3]. \qquad (10)$$

As can be checked, it corresponds to the following possibilistic logic base representing a conjunction of prioritized goals:

$$\mathcal{D} = \{(C_1 \vee C_2 \vee C_3, 1), (C_1 \vee C_2, \gamma_2), (C_1, \gamma_3)\}.$$

(with $\gamma_1 = 1 > \gamma_2 > \gamma_3$) whose associated possibility distribution is

$$\begin{aligned}
\pi_{\mathcal{D}}(\omega) = &1 \text{ if } \omega \in [C_1] \\
&1 - \gamma_3 \text{ if } \omega \in [\neg C_1 \wedge C_2] \\
&1 - \gamma_2 \text{ if } \omega \in [\neg C_1 \wedge \neg C_2 \wedge C_3] \\
&0 \text{ if } \omega \in [\neg C_1 \wedge \neg C_2 \wedge \neg C_3],
\end{aligned}$$

which is clearly in agreement with the ordering (10). It can be also equivalently expressed in a question-answering perspective by the possibilistic logic base:

$$\begin{aligned}
\mathcal{Q}' = \{&(\neg C_1(x) \vee answer(x), 1), \\
&(\neg C_2(x) \vee answer(x), 1 - \gamma_3), \\
&(\neg C_3(x) \vee answer(x), 1 - \gamma_2)\}.
\end{aligned}$$

which states that if an item x satisfies C_1, then it belongs to the answer to degree 1, and if it satisfies C_2 (resp. C_3), then it belongs to the answer to a degree at least equal to $1 - \gamma_3$ (resp $1 - \gamma_2$).

Let us also explain the relation between the possibilistic representation and qualitative choice logic (QCL) [12]. Indeed QCL introduces a new connective denoted $\times$, where $C_1 \times C_2$ means "if possible C_1, but if C_1 is impossible then (at least) C_2". This corresponds to a disjunctive preference of the above type. Then, the query "C_1, *or at least* C_2, *or at least* C_3", which, as already explained, corresponds to stating that C_1 is fully satisfactory, C_2 instead is less satisfactory, and C_3 instead is still less satisfactory, can be directly represented in a *non classical* possibilistic logic (see Appendix and [3]) based on guaranteed possibility measures, rather than on necessity measures. Using the notation in the Appendix, the corresponding weighted base simply writes $\mathcal{D}_\Delta = \{[C_1, 1], [C_2, 1 - \gamma_3], [C_3, 1 - \gamma_2]\}$, which clearly echoes $\mathcal{Q}'$, and encodes the same possibility distributions on models as $\mathcal{D}$.

Note that in $\mathcal{Q}'$, as in $\mathcal{Q}$, the weights of the possibilistic logic formulas express a priority among the answers x that may be obtained. They may be also viewed as representing the levels of satisfaction of the answers obtained.

4.4 Relation Between Conjunctive and Disjunctive Queries

The linguistic expression of conjunctive queries may suggest that C_1, C_2, C_3 are logically independent conditions that one would like to cumulate, as in the query "I am looking for a reasonably priced hotel, *if possible* downtown, *and if possible* not far from the station", while in disjunctive queries one may think of C_3 as a relaxation of C_2, itself a relaxation of C_1. In fact there is no implicit limitation on the type of conditions involved in conjunctive or disjunctive queries. For instance, a conjunctive query such as "I am looking for a hotel less than 2 km from the beach, *if possible* less than 1 km from the beach, *and if possible* on the beach", corresponds to the idea of approximating a fuzzy requirement, such as "close to the beach" by three of its level cuts, which are then relaxation or strengthening of one another.

As noticed in [10] and in Sect. 2.3, there is a perfect duality between conjunctive and disjunctive queries. Indeed the disjunctive query "C_3 is required, or better C_2, or still better C_1" can be also equivalently expressed under the conjunctive form "C_1 or C_2 or C_3 is required *and if possible* C_1 or C_2, *and if possible* C_1". This can be checked by noticing that changing C_1 into $C_1 \vee C_2 \vee C_3$, C_2 into $C_1 \vee C_2$, and C_3 in C_1, (9) is changed into (10). Conversely, the conjunctive query "C_1 is required *and if possible* C_2 *and if possible* C_3" can be equivalently stated as the disjunctive query "C_1 is required, or better C_1 and C_2, or still better C_1 and C_2 and C_3". It can be checked that changing C_1 into $C_1 \wedge C_2 \wedge C_3$, C_2 into $C_1 \wedge C_2$ and C_3 in C_1, (10) is changed into (9). The duality between the two types of queries, laid bare in Sect. 2.3, can be checked as well on their respective possibilistic logic representations.

4.5 Hybrid Queries

When considering the two above types of queries, Bosc et al. [11], while discussing extended divisions for fuzzy queries (in a non logical setting), have also introduced a mutual refinement of both queries, called "full discrimination-based queries". It amounts to a lexicographic ordering of the different worlds (here $2^3 = 8$ with 3 conditions), under the tacit, default assumption that it is always better to have a condition fulfilled rather than not, even if a more important condition is not satisfied.

However, it is clear that sometimes satisfying an auxiliary condition while failing to satisfy the main condition may be of no interest, as in the example "I would like a coffee if possible with sugar", where having sugar or not, if no coffee is available, makes no difference. There are even situations, in case of a conditional preference, where it is worse to have C_2 satisfied than not when C_1 cannot be satisfied, as in the example "I would like a Ford car if possible black" (if one prefers any other color for non Ford cars).

Full discrimination-based queries are thus associated with the following preference ordering [10]:

$$[C_1 \wedge C_2 \wedge C_3] \gg [C_1 \wedge C_2 \wedge \neg C_3] \gg [C_1 \wedge \neg C_2 \wedge C_3]$$
$$\gg [C_1 \wedge \neg C_2 \wedge \neg C_3] \gg$$
$$[\neg C_1 \wedge C_2 \wedge C_3] \gg [\neg C_1 \wedge C_2 \wedge \neg C_3] \gg [\neg C_1 \wedge \neg C_2 \wedge C_3]$$
$$\gg [\neg C_1 \wedge \neg C_2 \wedge \neg C_3]$$

It can be checked that it can be encoded in possibilistic logic under the form (we only give the question-answering form here):

$$\mathcal{Q}'' = \{(\neg C_1(x) \vee \neg C_2(x) \vee \neg C_3(x) \vee answer(x), 1),$$
$$(\neg C_1(x) \vee \neg C_2(x) \vee answer(x), \alpha), (\neg C_1(x) \vee \neg C_3(x) \vee answer(x), \alpha'),$$
$$(\neg C_1(x) \vee answer(x), \alpha''), (\neg C_2(x) \vee \neg C_3(x) \vee answer(x), \beta),$$
$$(\neg C_2(x) \vee answer(x), \beta'), (\neg C_3(x) \vee answer(x), \gamma)\}$$

with $1 > \alpha > \alpha' > \alpha'' > \beta > \beta' > \gamma$.

Thus, possibilistic logic offers a convenient language for the representation of preferences and the encoding of preference queries; see also [28] in that respect. Conjunctive, disjunctive or hybrid queries can be also processed in the setting of answer set programming, as discussed in [15].

5 Concluding Remarks

This chapter presents a detailed study of an issue closely related to the ideas of gradualness and bipolarity [26], namely the representation of requirements of the form "*A and if possible B*" or of a related form. We have emphasized the existence of two different views, according to whether B plays, or does not play, the role of a constraint of the same kind as A. We have discussed the logical expression of such requirements, and provided a unified presentation of a fuzzy logic-based approach together with other approaches to preference queries.

We are happy to publish this discussion in a book edited in honor of Patrick Bosc, at the time of his retirement, on a topic to which he contributed much. This is an opportunity to celebrate Patrick as one of the very few researchers to believe in the interest of fuzzy logic methods in flexible querying and imperfect data modeling since the early eighties. He successfully managed to progressively build a high quality team around him. This group has gained international recognition under his leadership, and developed important and seminal contributions to the fuzzy databases field over three decades. Patrick can be proud of what he has achieved. He can be confident that his team, and other ones as well, will continue to develop valuable and original pieces of work in the same vein. Congratulations to him!

Appendix

In a propositional possibilistic logic base $\mathcal{B} = \{(p_i, \alpha_i) | i = 1, n\}$, each formula (p_i, α_i) is a pair made of a classical logic proposition p_i and a weight $\alpha_i \in [0, 1]$ expressing a certainty level. It is semantically associated with the possibility distribution

$$\pi_{\mathcal{B}}(\omega) = \min_{i=1,n} \pi_{(p_i, \alpha_i)}(\omega)$$

with $\pi_{(p_i, \alpha_i)}(\omega) = 1$ if $\omega \in [p_i]$, and $\pi_{(p_i, \alpha_i)}(\omega) = 1 - \alpha_i$ if $\omega \notin [p_i]$. Thus, $\pi_{\mathcal{B}}$ is obtained as the min-based conjunctive combination of the representations of each formula in $\mathcal{B}$. Note that the possibility $\pi_{\mathcal{B}}(\omega)$ that an interpretation ω violates p_i is upper bounded by $1 - \alpha_i$ and the necessity of p_i, $N([p_i]) = 1 - \Pi(\overline{[p_i]}) = 1 - \max_{\omega \notin [p_i]} \pi_{\mathcal{B}}(\omega) \geq \alpha_i$, i.e., p_i is certain at least at level α_i.

There exists another type of possibilistic base, denoted $\mathcal{P} = \{[q_j, \gamma_j] | j = 1, k\}$ sometimes called positive possibilistic logic base. Its semantics is also given by a possibility distribution, namely

$$\delta_{\mathcal{P}}(\omega) = \max_{j=1,k} \delta_{[q_j, \gamma_j]}(\omega)$$

with $\delta_{[q_j, \gamma_j]}(\omega) = 0$ if $\omega \notin [q_j]$, and $\delta_{[q_j, \gamma_j]}(\omega) = \gamma_j$ if $\omega \in [q_j]$. Note that $\delta_{\mathcal{P}}$ is obtained as the max-based *disjunctive* combination of the representation of each formula in $\mathcal{P}$. Thus, the guaranteed possibility measure of q_j, $\Delta([q_j]) = \min_{\omega \in [q_j]} \delta_{\mathcal{P}}(\omega)$ is lower bounded by γ_j. See [4] for a detailed introduction. Note that the same possibility distribution can be associated both to a possibilistic logic base with a reading in terms of necessity measures, and a possibilistic logic base with a reading in terms of guaranteed possibility measures.

References

1. Abbaci, K., Lemos, F., Hadjali, A., Grigori, D., Liétard, L., Rocacher, D., Bouzeghoub, M.: A bipolar approach to the handling of user preferences in business processes retrieval. In: Proceedings of the 14th International Conference on Information Processing and Management of Uncertainty in Knowledge-based Systems (IPMU'12), Catania, 9–13 July , 2012, to appear
2. Amgoud, L., Prade, H., Serrut, M.: Flexible querying with argued answers. In: Proceedings of the 14th IEEE International Conference on Fuzzy Systems (FUZZ'05), pp. 573–578 (2005)
3. Benferhat, S., Brewka, G., Le Berre, D.: On the relation between qualitative choice logic and possibilistic logic. In: Proceedings of the 10th International Conference on Information Processing and Management of Uncertainty in Knowledge-Based Systems (IPMU'04), pp. 951–957, Perugia, 4–9 July 2004
4. Benferhat, S., Dubois, D., Kaci, S., Prade, H.: Modeling positive and negative information in possibility theory. Int. J. Intell. Syst. **23**(10), 1094–1118 (2008)
5. Bosc, P., Pivert, O.: Some approaches for relational databases flexible querying. J. Intell. Inf. Syst. **1**(3/4), 323–354 (1992)
6. Bosc, P., Pivert, O.: SQLf: a relational database language for fuzzy querying. IEEE Trans. Fuzzy Syst. **3**, 1–17 (1995)
7. Bosc, P., Pivert, O.: On diverse approaches to bipolar division operators. Int. J. Intell. Syst. **26**(10), 911–929 (2011)
8. Bosc, P., Pivert, O.: On four noncommutative fuzzy connectives and their axiomatization. Fuzzy sets and Systems, to appear, 2012
9. Bosc, P., Pivert, O., Mokhtari, A., Liétard, L.: Extending relational algebra to handle bipolarity. In: Shin, S.Y., Ossowski, S., Schumacher, M., Palakal, M.J., Hung, C.-C. (eds.) Proceedings of the 2010 ACM Symposium on Applied Computing (SAC), Sierre, pp. 1718–1722, 22–26 March 2010
10. Bosc, P., Pivert, O., Prade, H.: A possibilistic logic view of preference queries to an uncertain database. In: Proceedings of the 19th IEEE International Conference on Fuzzy Systems (FUZZ-IEEE'10), Barcelona, pp. 1–6, 18–23 July 2010
11. Bosc, P., Pivert, O., Soufflet, O.: On three classes of division queries involving ordinal preferences. J. Intell. Inf. Syst. **37**(3), 315–331 (2011)
12. Brewka, G., Benferhat, S., Le Berre, D.: Qualitative choice logic. Artif. Intell. **157**(1–2), 203–237 (2004)
13. Chomicki, J.: Preference formulas in relational queries. ACM Trans. Database Syst. **28**, 1–40 (2003)
14. de Calmès, M., Dubois, D., Hüllermeier, E., Prade, H., Sèdes, F.: Flexibility and fuzzy case-based evaluation in querying: an illustration in an experimental setting. Int. J. Uncertainty Fuzziness Knowl. Based Syst. **11**(1), 43–66 (2003)
15. Confalonieri, R., Prade, H.: Encoding preference queries to an uncertain database in possibilistic answer set programming. In: Proceedings of the 14th International Conference on Information Processing and Management of Uncertainty in Knowledge-based Systems (IPMU'12), Catania, 9–13 July 2012, to appear
16. Domshlak, C., Hüllermeier, E., Kaci, S., Prade, H.: Preferences in AI: an overview. Artif. Intell. **175**(7–8), 1037–1052 (2011)
17. Dubois, D., Fargier, H., Prade, H.: Possibility theory in constraint satisfaction problems: handling priority, preference and uncertainty. Appl. Intell. **6**, 287–309 (1996)
18. Dubois, D., Prade, H.: Weighted minimum and maximum operations. Inform. Sci. **39**, 205–210 (1986)
19. Dubois, D., Prade, H.: Default reasoning and possibility theory. Artif. Intell. **35**(2), 243–257 (1988)
20. Dubois, D., Prade, H.: Using fuzzy sets in flexible querying: why and how? In: Andreasen, T., Christiansen, H., Larsen, H.L. (eds.) Flexible Query Answering Systems, pp. 45–60. Kluwer Academic Publishers, Dordrecht (1997)

21. Dubois, D., Prade, H.: Bipolarity in flexible querying. In: Andreasen, T., Motro, A., Christiansen, H.H., Larsen, L. (eds.) Proceedings of the 5th International Conference on Flexible Query Answering Systems (FQAS'02), pp. 174–182. Springer, LNCS 2522, Copenhagen, 27–29 Oct 2002

22. Dubois, D., Prade, H.: Possibilistic logic: a retrospective and prospective view. Fuzzy Sets Syst. **144**(1), 3–23 (2004)

23. Dubois, D., Prade, H.: Handling bipolar queries in fuzzy information processing. In: Galindo, J. (ed.) Handbook of Research on Fuzzy Information Processing in Databases, pp. 97–114. IGI Global, Hershey (2008)

24. Dubois, D., Prade, H.: An introduction to bipolar representations of information and preference. Int. J. Intell. Syst. **23**(8), 866–877 (2008)

25. Dubois, D., Prade, H.: An overview of the asymmetric bipolar representation of positive and negative information in possibility theory. Fuzzy Sets Syst. **160**(10), 1355–1366 (2009)

26. Dubois, D., Prade, H.: Gradualness, uncertainty and bipolarity: making sense of fuzzy sets. Fuzzy Sets Syst. **192**, 3–24 (2012)

27. Green, C.: Theorem-proving by resolution as a basis for question-anwering systems. In: Michie, D., Meltzer, B. (eds.) Machine Intelligence, vol. 4, pp. 183–205. Edinburgh University Press, Edinburgh (1969)

28. Hadjali, A., Kaci, S., Prade, H.: Database preference queries—a possibilistic logic approach with symbolic priorities. (Preliminary version in Proceedings of the 5th International Symposium on Foundations of Information and Knowledge Systems (FoIKS 2008), Pisa, 11–14 Feb (S. Hartmann, G. Kern-Isberner, eds.), Springer, LNCS 4932, 291–310, 2008). Annals of Mathematics and Artificial Intelligence **63**(3–4), 357–383 (2011)

29. Kiessling, W.: Foundations of preferences in database systems. In: Proceedings of the 28th International Conference on Very Large Data Bases (VLDBO02), pp. 311–322 (2002)

30. Lacroix, M., Lavency, P.: Preferences: putting more knowledge into queries. In: Proceedings of the 13th Conference on Very Large Data Bases (VLDBO87), pp. 217–225 (1987)

31. Liétard, L., Tamani, N., Rocacher, D.: Fuzzy bipolar conditions of type or else. In: Proceedings of the 20th IEEE International Conference on Fuzzy Systems (FUZZ-IEEE'11), pp. 2546–2551. Taipei, 27–30 June 2011

32. Lukasiewicz, T., Schellhase, J.: Variable-strength conditional preferences for ranking objects in ontologies. J. Web Semant. **5**(3), 180–194 (2007)

33. Motro, A.: A user interface to relational databases that permits vague queries. ACM Trans. Inf. Syst. **6**, 187–214 (1988)

34. Pivert, O., Bosc, P.: Fuzzy Preference Queries to Relational Databses. Imperial College Press, London (2012)

35. Tahani, V.: A conceptual framework for fuzzy query processing—a step toward very intelligent database systems. Inf. Process. Manage. **12**, 289–303 (1977)

36. Yager, R.R.: Non-monotonic set theoretic operations. Fuzzy Sets Syst. **42**, 173–190 (1991)

37. Zadeh, L.A.: Fuzzy sets. Inf. Control **8**, 338–353 (1965)

38. Zadrozny, S., Kacprzyk, J.: Bipolar queries using various interpretations of logical connectives. In: Melin, P., Castillo, O., Aguilar, L.T., Kacprzyk, J., Pedrycz, W. (eds.) Proceedings of the 12th International Fuzzy Systems Association World Congress on Foundations of Fuzzy Logic and Soft Computing(IFSA'07), Cancun, pp. 181–190. Springer Verlag, LNCS 4529, 18–21 June 2007

39. Zadrozny, S., De Tré, G., De Caluwe, R., Kacprzyk, J.: An overview of fuzzy approaches to flexible database querying. In: Erickson, J. (ed.) Database Technologies: Concepts, Methodologies, Tools, and Applications, 4 Vol., pp. 135–156. IGI Global, Barcelona (2009)

Chapter 2
Constraint-Wish and Satisfied-Dissatisfied: An Overview of Two Approaches for Dealing with Bipolar Querying

Tom Matthé, Joachim Nielandt, Sławomir Zadrożny and Guy De Tré

Abstract In recent years, there has been an increasing interest in dealing with user preferences in flexible database querying, expressing both positive and negative information in a heterogeneous way. This is what is usually referred to as bipolar database querying. Different frameworks have been introduced to deal with such bipolarity. In this chapter, an overview of two approaches is given. The first approach is based on mandatory and desired requirements. Hereby the complement of a mandatory requirement can be considered as a specification of what is not desired at all. So, mandatory requirements indirectly contribute to negative information (expressing what the user does not want to retrieve), whereas desired requirements can be seen as positive information (expressing what the user prefers to retrieve). The second approach is directly based on positive requirements (expressing what the user wants to retrieve), and negative requirements (expressing what the user does not want to retrieve). Both approaches use pairs of satisfaction degrees as the underlying framework but have different semantics, and thus also different operators for criteria evaluation, ranking, aggregation, etc.

T. Matthé · J. Nielandt · G. De Tré (✉)
Department of Telecommunications and Information Processing, Ghent University,
Sint-Pietersnieuwstraat 41, B-9000 Ghent, Belgium
e-mail: Tom.Matthe@UGent.be

J. Nielandt
e-mail: Joachim.Nielandt@UGent.be

G. De Tré
e-mail: Guy.DeTre@UGent.be

S. Zadrożny
Systems Research Institute, Polish Academy of Sciences, ul. Newelska 6,01-447 Warsaw, Poland
e-mail: Slawomir.Zadrozny@ibspan.waw.pl

O. Pivert and S. Zadrożny (eds.), *Flexible Approaches in Data, Information and Knowledge Management*, Studies in Computational Intelligence 497, DOI: 10.1007/978-3-319-00954-4_2, © Springer International Publishing Switzerland 2014

1 Introduction

In daily life, it can be observed that people, whilst communicating their preferences, tend to use vague or fuzzy terms in expressing their desires. A typical example is a recruitment office that, e.g., is searching for *young* people with a *high* score in math. A lot of research has been done to translate this 'fuzziness' to the domain of database querying, resulting in 'fuzzy' querying of regular databases, where the queries are composed of several 'fuzzy' query conditions, interconnected by logical connectives. Indeed, the main lines of research in this area include the study of modeling linguistic terms (like, e.g., *young* or *high*) in the specification of elementary query conditions using elements of fuzzy logic [38] and the enhancement of fuzzy query formalism with soft aggregation operators [6, 15, 22, 23]. Both linguistic terms and soft aggregations model user's preferences [4] and, as such, require a query satisfaction modeling framework that supports rank-ordering the records retrieved in response to a query according to the degree to which they satisfy all conditions imposed by the query. Usually, query satisfaction in 'fuzzy' querying of regular databases is modelled by associating a *satisfaction degrees* with each record in the answer set of the query. These satisfaction degrees take values in the unit interval [0, 1] and are computed during query processing. The value 0 means complete lack of satisfaction and implies that the associated record does not belong to the query's answer set. The value 1 expresses full satisfaction, while all other, intermediate, values denote partial query satisfaction. Records with a satisfaction degree s that is lower than a given threshold value δ, i.e., for which $s < \delta$, are usually discarded from the query answer set.

A more advanced aspect of specifying user preferences in database queries concerns the handling of *bipolarity*. Bipolarity hereby refers to the fact that users might distinguish between positive and negative aspects (or between constraints and wishes) while specifying their query preferences. Positive statements may be used to express what is possible, satisfactory, permitted, desired or acceptable, whereas negative statements may express what is impossible, unsatisfactory, not permitted, rejected, undesired or unacceptable. Likewise, constraints express what is accepted, whereas wishes are used to specify which of the accepted values are really desired by the user. Bipolarity is inherent to human communication and natural language and should hence be reflected and dealt with in any querying system that aims to support human interaction as adequate as possible.

For example, consider the specification of user preferences in the context of selecting a car, more specifically concerning the color of a car. A positive statement is 'I like black or dark blue cars', while 'I do not want a white car' is a negative statement. In terms of constraints and desires, similar preferences might be expressed by 'I want a dark colored car' and 'if possible, I really prefer a black or dark blue car'. Remark that often, negative conditions might be translated to constraints, while positive conditions might be seen as wishes.

Depending on the situation, it may be more natural for a user to use negative conditions or positive conditions. Sometimes one can use both positive and negative conditions at the same time. This is especially the case if the user does not have complete knowledge of the domain on which the criterion is specified, or if this domain is too large to completely specify the user's preferences for every value in the domain, as can for example be the case with available car colors.

In standard approaches to regular 'fuzzy' querying it is explicitly assumed that a record that satisfies a query condition to a degree s, at the same time dissatisfies it, i.e., satisfies its negation, to a degree $1 - s$. This assumption does not generally hold when dealing with bipolar query criteria specifications as positive and negative conditions comprising a query are assumed to be independent, i.e., may assume any value from the interval [0,1]. In such situations of *heterogeneous bipolarity*, a semantically richer query satisfaction modeling approach, which is more consistent with human reasoning and is able to model this bipolarity, is preferred.

In this chapter, two such approaches to bipolar database querying are discussed. On the one hand, the *constraint-wish* (or mandatory-desired) approach will be presented, used amongst others by Dubois and Prade [16, 17] and Bosc et al. [10, 26, 27, 29, 39], and on the other hand, the *satisfied-dissatisfied* (or positive-negative) approach will be discussed, used amongst others by Zadrożny et al. [47] and De Tré et al. [13, 30, 33]. For both approaches, an overview is given, which consecutively handles the semantics of the actual framework, the evaluation of query conditions within this framework, the ranking of query results and the aggregation of compound query conditions.

The remainder of this chapter has been organised as follows: first, some preliminaries on bipolar query conditions will be presented in Sect. 2, explaining the two approaches that will be discussed in this chapter in more detail, together with their semantics. The next Sect. 3 discusses the ranking of the results of a bipolar query. Next, in Sect. 4, different techniques to aggregate the results of multiple query conditions are presented. Finally, Sect. 5 states some conclusions.

2 Bipolar Query Conditions

Pioneering work in the area of heterogeneous bipolar database querying has been done in [25], which seems to be the first approach where a distinction has been made between mandatory query conditions and desired query conditions. As mentioned earlier, desired and mandatory conditions can be viewed as specifying positive and negative information, respectively. Indeed, the opposite of a mandatory condition specifies what must be rejected and thus what is considered as being negative with respect to the query result, whereas desired conditions specify what is considered as being positive.

Later on, this idea has been further developed and adapted to be used in 'fuzzy' querying techniques. The use of the twofold fuzzy sets (TFS) to represent a bipolar elementary query condition with respect to a given attribute A is reported, e.g., in [16, 17]. A twofold fuzzy set expresses which domain values are accepted by the user and which among these accepted values are really desired by her or him.

An alternative approach, based on the concept of an Atanassov (intuitionistic) fuzzy set (AFS) and departing from the specification of which values are desired and which values are undesired, is presented in [12, 30]. Both approaches have in common that they deal with bipolarity that is specified *inside* elementary query conditions, i.e., in the domain of an attribute three subsets, in general fuzzy, are distinguished: of positively, negatively and neutrally evaluated elements.

Other approaches study bipolarity that is specified *between* elementary query conditions, meaning that these conditions are assigned different semantics. In particular, a distinction can be made between mandatory and desired query conditions. These conditions can still contain vague terms modelled by fuzzy sets as in regular 'fuzzy' querying [4, 5, 7, 21, 22, 38, 46]. For example, in [48] an approach is presented where bipolar queries are represented as a special case of the fuzzy 'winnow' operator. Bipolarity is thus studied considering queries with preferences as in [25]. An alternative assumption that can be made, is considering queries that consist of a number of 'positive' and 'negative' elementary conditions [13, 33].

In this chapter, only bipolarity that is specified *inside* elementary query conditions will be considered. In the following subsections, two approaches will be discussed in more detail: the *constraint-wish* and the *satisfied-dissatisfied* approach.

2.1 Constraint-Wish Approach

Consider a universe of discourse U corresponding to the domain of an attribute in question. In the *constraint-wish* approach (referred to elsewhere in this volume also as 'required-desired semantics'), the bipolar query condition consists of two parts: a constraint C, which describes the set of acceptable values of U, and a wish W, which defines the set of wished-for (or desired) values of U. In general, the constraint and the wish are specified using fuzzy sets C and W, defined on U [44], identified by their respective membership functions μ_C and μ_W. Because it is not coherent to wish something that is rejected (where the rejected values are represented by the complement of the fuzzy set C), a consistency condition is imposed. Two forms of consistency conditions may be considered:

- *Strong consistency,*

$$\forall x \in U : \mu_C(x) < 1 \Rightarrow \mu_W(x) = 0. \tag{1}$$

In this case, the support of the wish W is required to be a subset of the core of the constraint C, which means that the wish can play any role in evaluating only those records which fully satisfy the constraint. The pair of fuzzy sets C and W then form a *twofold fuzzy set* [11].

- *Weak consistency,*

$$\forall x \in U : \mu_W(x) \leq \mu_C(x). \tag{2}$$

In this case, the wish is required to be more specific than the constraint what represents the fact that it is harder to satisfy a wish than to satisfy a constraint, but the wish can also play a role in evaluating the records which do not fully satisfy the constraint. The pair of fuzzy sets C and W then form an interval-valued fuzzy set (IVFS) [19, 20, 36, 45].

Because a twofold fuzzy set is formally a special case of an interval-valued fuzzy set, the bipolar query condition can in both cases be modelled by means of an IVFS, which is defined by

$$F = \{(x, [\mu_{F_*}(x), \mu_{F^*}(x)]) | (x \in U) \wedge (0 \leq \mu_{F_*}(x) \leq \mu_{F^*}(x) \leq 1)\}. \quad (3)$$

Thus, a bipolar query condition is modelled by means of an IVFS, where the upper membership function μ_{F^*} models the constraint, i.e., $\mu_{F^*} = \mu_C$, and the lower membership function μ_{F_*} models the wish, i.e., $\mu_{F_*} = \mu_W$.

An important feature of this semantics is that the wish plays somehow a secondary role in the query. A bipolar query condition in the constraint-wish approach should be interpreted as '*satisfy C and, if possible, satisfy W*' [16].

Summarising, in this approach the evaluation of a record R against an elementary bipolar query condition '*A IS F*' with F composed of a couple (C, W) of fuzzy sets C and W results in a pair of satisfaction degrees $(c(R), w(R)) \in [0, 1]^2$ such that

$$c(R) = \mu_C(R[A]) \quad (4)$$
$$w(R) = \mu_W(R[A]) \quad (5)$$

where $R[A]$ denotes the value of record R for attribute A.

2.2 Satisfied-Dissatisfied Approach

In the *satisfied-dissatisfied* approach, the bipolar query condition also consists of two parts. One part specifies the values of an attribute A which are positively evaluated by the user with respect to her or his preferences and, *independently*, another part specifies the values for A which are negatively evaluated by the user. A pair of fuzzy sets, F^+ and F^-, expressing the respective parts of the query condition may be treated as a bipolar extension to the concept of fuzzy set. *Atanassov (intuitionistic) Fuzzy Sets* (AFSs) [1] are an example of such an extension. An AFS F over a universe U is formally defined by

$$F = \{(x, \mu_F(x), \nu_F(x)) | (x \in U) \wedge (0 \leq \mu_F(x) + \nu_F(x) \leq 1)\}. \quad (6)$$

where $\mu_F : U \rightarrow [0, 1]$ and $\nu_F : U \rightarrow [0, 1]$ are respectively called the membership and non-membership degree functions and $0 \leq \mu_F(x) + \nu_F(x) \leq 1, \forall x \in U$ reflects the consistency condition of the AFS. In the context of database querying, this consistency condition can be interpreted as stating that the degree of non-preference $\nu_F(x)$ for a given value x can never be larger than the complement $1 - \mu_F(x)$ of the degree of preference for that value (or, equivalently, that the degree of preference $\mu_F(x)$ for a value x can never be larger than the complement $1 - \nu_F(x)$ of the degree of non-preference for that value.)

Formally, in their basic form, AFSs are operationally equivalent to IVFSs and thus may be also used to represent preferences in the constraint-wish approach, but their intended semantics is closer to the idea of the satisfied-dissatisfied approach. However, in the satisfied-dissatisfied approach the total independence of positive and negative condition is assumed and the AFS's consistency condition does not meet this assumption. Thus, in what follows we will use the concept *bipolar AFS* which follows the two membership functions structure of AFSs but drops the consistency condition. In this respect, the presented approach is similar to the neutrosophic logic [35, 37]. It is also similar to a fuzzy version of Belnap's logic [2], proposed by Öztürk and Tsoukias [34] and further developed by Turunen et al. [40]; cf. also a study on links between Belnap's logic and bipolarity by Konieczny et al. [24]. However, it should be stressed that the degrees of satisfaction and dissatisfaction do not have any epistemic flavour here, i.e., e.g., they do not form an interval containing a 'true' degree to which the user likes the given value of an attribute in question. Instead, these degrees respectively express the genuine liking and disliking of the value which are assumed to occur simultaneously and independently of each other.

In what follows, we will thus often adopt the notation μ_F and ν_F instead of, respectively μ_{F+} and μ_{F-}, while referring to the sets of positively and negatively evaluated values of an attribute under consideration.

We have thus a couple $(\mu_F(x), \nu_F(x))$ which is referred to as the *bipolar satisfaction degree* (BSD) and represents the suitability of $x \in dom_A$ with respect to a condition A IS F, where A is an attribute and F is a bipolar AFS representing preferences of the user. Now, the question is how these couples are to be processed, i.e., used to order the records in an answer to the query and aggregated with the couples related to other elementary conditions. We discuss these issues in the following sections. An elementary bipolar query condition 'A IS F' in the satisfied-dissatisfied approach should be interpreted as '*preferably satisfy F^+ and preferably do not satisfy F^-*' [30].

Summarising, in this approach the evaluation of a record R against an elementary bipolar query condition 'A IS F' with F a bipolar AFS characterized by a pair (μ, ν) of membership functions μ and ν results in a pair $(s(R), d(R)) \in [0, 1]^2$ of values, referred to as *satisfaction degree* (s) and *dissatisfaction degree* (d), jointly called a Bipolar Satisfaction Degree (BSD) [30], such that

$$s(R) = \mu_F(R[A]) \tag{7}$$
$$d(R) = \nu_F(R[A]) \tag{8}$$

where $R[A]$ is the value of record R for attribute A. The set of all possible BSDs will be denoted as $\tilde{\mathbb{B}}$.

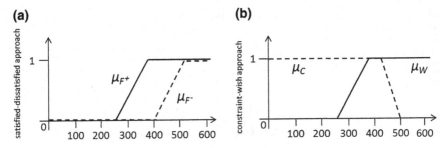

Fig. 1 Examples: **a** μ_{F^-}: 'too large', μ_{F^+}: 'large'; **b** μ_C: 'not too large', μ_W: 'large'

2.3 Examples

As an example of an elementary bipolar query condition, consider the case of a real estate application and a user who wants to find a suitable house to buy. An important criterion may be the size of the garden. The user may have a number of criteria in mind when judging which ranges of values of this attribute she or he prefers. For example, considering garden as a playground for children the user may use a positive unipolar scale to measure its suitability—the larger the size the better. On the other hand, taking into account the maintenance costs of the garden the user may use a negative unipolar scale—the larger the garden size the higher the costs.[1] Let us assume that the terms 'large' and 'too large', respectively, represent the preferences of the user along these two criteria and thus describe the sets F^+ and F^- of positive and negative parts of the bipolar condition.

Figure 1a shows how such preferences may be represented in the framework of the satisfied-dissatisfied approach. It is worth noticing that, for example, a garden size of 550 sq. m. is totally negatively evaluated from the point of view of the maintenance costs and, at the same time, totally positively evaluated from the point of view of fun for the children.

Looking for a counterpart in the constraint-wish approach we would like to interpret the positive condition as a wish and the negative condition as the complement of the constraint. However, this is not possible as the consistency condition implied by the semantics of the constraint-wish approach is not met, what is illustrated in Fig. 1b, i.e., there are some values x where, $\mu_W(x) > \mu_C(x)$.

In order to illustrate the constraint-wish approach at work let us assume the following scenario. The user may look for a 'large' garden but she or he would be most happy with a garden of size around 400–500 sq. m. Thus, the former may be interpreted as a constraint ('not large' garden is excluded) while the latter is just a desired size. Figure 2b shows an example of membership functions which may serve to represent such preferences in the framework of the constraint-wish approach.

---

[1] We are slightly simplifying the situation here as with respect to both criteria the user may have in mind two separate bipolar scales, but still it will result in sets of aggregated positively and negatively evaluated garden size values.

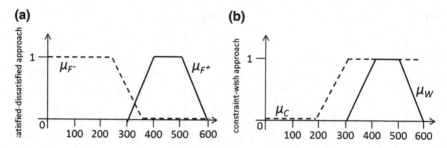

Fig. 2 Examples: **a** μ_{F-}: 'not large', μ_{F+}: 'around 400–500 sq. m.'; **b** μ_C: 'large', μ_W: 'around 400–500 sq. m'

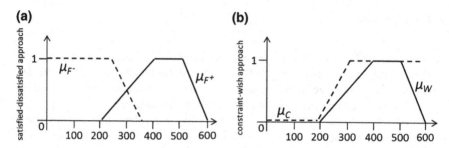

Fig. 3 Examples: **a** μ_{F-}: 'not large', μ_{F+}: 'around 400–500 sq. m. or slightly less'; **b** μ_C: 'large', μ_W: 'around 400–500 sq. m. or slightly less'

It should be stressed that the satisfied-dissatisfied approach is not suitable to represent such preferences. One can consider a kind of representation shown in Fig. 2a which is obtained by treating the wish and the constraint as, respectively, the positive evaluation and the complement of the negative evaluation. However, in the framework of the satisfied-dissatisfied approach Fig. 2a should be actually interpreted as representing the following preferences: the user has positive feelings about the garden size being ca. 400–500 sq. m. and does not like small gardens (more precisely: not large gardens).

Figure 3b shows a slightly different wish which may be expressed as 'preferred size of the garden is ca. 400–500 sq. m. or slightly less'. This case is still well suited to be represented in the constraint-wish based approach although only the weak consistency is preserved.

A kind of the counterpart of the above in the framework of the satisfied-dissatisfied approach, again in the spirit of Fig. 2a, is shown in Fig. 3a.

3 Ranking of Query Results

After evaluating a bipolar query condition for all potential query results, every result-ing record R_i will have an associated pair of calculated satisfaction degrees, either a pair $(c(R_i), w(R_i))$ or a BSD $(s(R_i), d(R_i))$. Now, we will deal with the question how the records should be ranked in the response to a query using these pairs of degrees.

3.1 Ranking in the Constraint-Wish Approach

In this approach it is assumed that constraints and wishes are not compensatory [10, 26], i.e., a higher satisfaction of a wish can not compensate a lower satisfaction of a constraint. Therefore, ranking is done primarily on the constraint satisfaction, and secondly, in case of ties, on the wish satisfaction. In general, one has [18]:

$$R_1 \succ R_2 \Leftrightarrow (c(R_1) > c(R_2)) \vee (c(R_1) = c(R_2) \wedge w(R_1) > w(R_2)) \qquad (9)$$

where $R_1 \succ R_2$ means that R_1 is preferred to R_2. Thus, this is the lexicographical ordering with respect to the pairs (c, w).

Another possibility is scalarization: a real function may be applied to the pairs $(c(R_i), w(R_i))$ and the records are then ranked according to the values obtained. Zadrożny and Kacprzyk [48], following Lacroix and Lavency [25], propose the aggre-gation of both degrees in the spirit of the 'and possibly' operator. In this approach the wish is taken into account only 'if possible', i.e., if its satisfaction does not interfere with the satisfaction of the constraint what is determined with respect to the content of the whole database. The same idea, applied in a different context, may be found in some earlier work of Bordogna and Pasi [3], Dubois and Prade [14] or Yager [41]. Recently, a lot of work has been done on the study of different interpretations of the 'and possibly' as well as its dual 'or at least' operators by Bosc, Pivert, Tamani, Hadjali (see, e.g., [8, 9, 28]) and by Dubois and Prade in this volume.

3.2 Ranking in the Satisfied-Dissatisfied Approach

Because, in the satisfied-dissatisfied approach, the satisfaction degree and the dis-satisfaction degree are assumed to be totally independent, both should have an equal impact on the ranking [30]. Naturally, the higher the satisfaction degree, the higher the ranking should be, and dually, the higher the dissatisfaction degree, the lower the ranking should be. A possible ranking function r for BSDs (s, d), with a com-plete symmetrical impact of both the satisfaction and dissatisfaction degrees, is the following:

$$r(s, d) = \frac{s + (1 - d)}{2}. \tag{10}$$

This ranking function produces values in [0, 1]. Three special cases can be distinguished:

- $r(s, d) = 1$: in this case it must be that $s = 1$ and $d = 0$, so this is the case of *full global satisfaction*.
- $r(s, d) = 0$: in this case it must be that $s = 0$ and $d = 1$, so this is the case of *full global dissatisfaction*.
- $r(s, d) = 0.5$: in this case it must be that $s = d$ and the ranking can be considered *neutral*. The condition is as satisfied as it is dissatisfied.

Remark that both degrees equally matter when ranking the records, as expected. For example, records for which the evaluation leads to a dissatisfaction degree $d = 1$, or dually a satisfaction degree of $s = 0$, should not a priori be excluded as being totally unsatisfactory. Indeed, e.g., the BSDs $(1, 1)$ and $(0, 0)$, although having $d = 1$ (respectively $s = 0$), both have neutral ranking ($r(s, d) = 0.5$) and are hence situated in the middle of the ranking spectrum.

Other ranking functions are also possible, e.g., assigning more importance to either the satisfaction degree or the dissatisfaction degree. In general, a suitable ranking function r for BSDs should meet the following minimal requirements:

1. $0 \leq r(x, y) \leq 1$, with (x, y) a BSD, i.e., $r : \tilde{\mathbb{B}} \to [0, 1]$.
2. $r(1, 0) = 1$, i.e., the BSD with full satisfaction and no dissatisfaction should be ranked the highest.
3. $r(0, 1) = 0$, i.e., the BSD with full dissatisfaction and no satisfaction should be ranked the lowest.
4. $\forall x, y \in [0, 1] : r(x, x) = r(y, y)$, i.e., for all BSDs with equal satisfaction degree and dissatisfaction degree, the ranking should also be equal. The reason for this requirement is that, ranking wise, it is impossible to make a sensible distinction between the cases of total indifference (i.e., BSD $(0, 0)$) and total conflict (i.e., BSD $(1, 1)$), and also all other intermediate cases where $s = d$ (i.e., BSD (x, x), $x \in [0, 1]$).
5. monotonicity: $r(x, y) \leq r(x + \varepsilon, y)$ and $r(x, y) \geq r(x, y + \varepsilon)$.

These minimal requirements eliminate the use of ranking functions which solely rank on either the satisfaction degree s or the dissatisfaction degree d, and use the other degree (d or s respectively) only as a 'tiebreaker', because they would violate the fourth requirement. Thus, for example, the lexicographical ordering, meaningfully used in the constraint-wish approach (see Eq. (9)), is not suitable for the satisfied-dissatisfied approach because of the assumed total independence between, and equally important role of, the satisfaction and dissatisfaction degrees.

A list of useful ranking functions for BSDs is listed below:

$$r_1 = \frac{s + (1 - d)}{2} \tag{11}$$

$$r_2 = \frac{s}{s + d} \tag{12}$$

$$r_3 = \frac{1 - d}{(1 - s) + (1 - d)} \tag{13}$$

$$r_4 = \frac{s}{s + d} \cdot \frac{1 - d}{(1 - s) + (1 - d)} \tag{14}$$

$$r_5 = \max\{0, s - d\} \tag{15}$$

$$r_6 = \min\{1 + s - d, 1\}. \tag{16}$$

Ranking function r_2 is discontinuous in BSD $(0, 0)$, r_3 is undefined for BSD $(1, 1)$, while r_4 is undefined for BSDs $(0, 0)$ and $(1, 1)$. More information on the behaviour and properties of these ranking functions can be found in [32].

3.3 Comparison and Discussion

The lexicographical ordering used in the constraint-wish approach makes wishes (positive information) rather secondary in comparison to the constraints (negative information), according to the assumed semantics. This may be, however, counter-intuitive in some cases. Let us consider two pairs of degrees $(c(R_1), w(R_1))$ and $(c(R_2), w(R_2))$ such that $c(R_1) = c(R_2) + \varepsilon$, while $w(R_1) = 0$ and $w(R_2) = 1$. In such a case R_1 will be ranked before R_2, even for ϵ very close to 0 what may be disputable. A possible escape is to assume a discrete scale for c's and w's with a small number of levels and to claim that the smallest difference in levels of the constraint satisfaction is large enough to justify its definite role in establishing the ranking of records whatever their satisfaction of wishes is.

A scalarization in the spirit of the 'and possibly' operator is an interesting option but it adopts a specific semantics of constraints and wishes.

Ranking in the satisfied-dissatisfied approach is based on the ranking of the BSDs. Due to their specific semantics and the total independence of the satisfaction and dissatisfaction degrees, BSDs can be ranked in different ways. A ranking function for BSDs should satisfy the requirements specified in Sect. 3.2. The selection of a ranking function depends on the requirements of the application.

- If it is necessary to assign an equal weight to $s(R)$ and $d(R)$, then the ranking function r_1 (cf. Eq. (11)) can be used. In this approach, query conditions are interpreted as preferences because records with $d(R) = 1$ or $s(R) = 0$ are not a priori excluded from the result, i.e., they do not necessarily result in a ranking value 0.
- If the non-zero satisfaction degree should be interpreted as an absolute requirement, i.e., if $s(R) = 0$ has to imply that the ranking value is 0, then the ranking function r_2 (cf. Eq. (12)) can be used.

- Dually, if avoiding the total dissatisfaction should be interpreted as an absolute requirement, i.e., if $d(R) = 1$ has to imply a ranking value 0, then the ranking function r_3 (cf. Eq. (13)) can be used.
- Ranking function r_4 (cf. Eq. (14)) can be used if both non-zero satisfaction degree is required and total dissatisfaction should be avoided.
- Finally, if the ranking should be based on the best of $s(R)$ and $d(R)$, then either ranking function r_5 (cf. Eq. (15)) or ranking function r_6 (cf. Eq. (16)) can be used. Hereby, $r_5 = 0$ if $s(R) \leq d(R)$ and $r_6 = 1$ if $s(R) \geq d(R)$.

It is worth noting that modelling bipolarity *inside* an elementary query condition using the constraint-wish approach (cf. Sect. 2.1) makes the ranking problem somehow trivial. Namely, it is easy to verify that due to the consistency condition, it is impossible to have two pairs of degrees $(c(R_1), w(R_1))$ and $(c(R_2), w(R_2))$ such that $c(R_1) < c(R_2)$ and at the same time $w(R_1) > w(R_2)$. This further justifies the primary role of the constraint satisfaction degree in the ranking process, as defined in (9). On the other hand, this is not the case in the satisfied-dissatisfied approach what makes room for more possible definitions of ranking.

4 Aggregation in Bipolar Query Processing

So far we have focused on bipolar queries comprising one elementary bipolar condition with respect to an attribute. In what follows we consider a compound bipolar query composed of many elementary bipolar conditions, possibly combined using explicit logical connectives of conjunction, disjunction and negation. The evaluation of an elementary bipolar query condition A IS F results in a pair of degrees (either $(c(R), w(R))$ or $(s(R), d(R))$) for every database record R. Now, consider the evaluation of an entire query composed of n elementary query conditions. First, for each relevant database record R, each elementary condition need to be evaluated resulting in n individual pairs of satisfaction degrees. Second, all these individual pairs must be *aggregated* to come up with a global result reflecting the extent to which R satisfies the entire bipolar query. The basic aggregation techniques in case of the constraint-wish approach and the satisfied-dissatisfied approach are presented in the two subsections below. A distinction has been made between techniques where the pairs of degrees are treated as a whole and techniques where these degrees are treated individually.

4.1 Aggregation in the Constraint-Wish Approach

Consider n elementary bipolar query conditions, the evaluation of which for a record R leads to a set of n pairs $(c_i(R), w_i(R))$, $i = 1, \ldots, n$. This set of n pairs needs to be aggregated to obtain the global satisfaction degree.

4.1.1 Treating $c(R)$ and $w(R)$ Individually

In this approach a bipolar query is meant as a list of elementary bipolar conditions and their conjunction is tacitly assumed. The $c_i(R)$'s and $w_i(R)$'s are separately aggregated [16, 17]. Both aggregations are guided appropriately by the semantics of the constraints and of the wishes. Namely, it is assumed that, if a record R does not satisfy a constraint then it should be rejected overall. Therefore, the degrees $c_i(R)$ are aggregated in a conjunctive way. On the other hand, if a record is desirable according to one wish then it is desirable overall. Therefore, the degrees $w_i(R)$ are aggregated in a disjunctive way. This then leads to a global pair $(c(R), w(R))$ expressing the satisfaction of the whole bipolar query by a record R:

$$(c(R), w(R)) = (\min_i c_i(R), \max_i w_i(R)). \tag{17}$$

Besides the minimum and maximum, other aggregation operators, based on triangular norms and co-norms, can also be used if a reinforcement effect is needed or desired.

Remark that, in general, this aggregation technique will not preserve consistency, i.e., it is possible that $w(R) > c(R)$. This can be solved by treating the 'global wish' not just as the mere disjunction of all wishes, but by also taking the conjunction of this disjunction with all the constraints [17]:

$$(c(R), w(R)) = (\min_i c_i(R), \min(\max_i w_i(R), \min_i c_i(R))). \tag{18}$$

4.1.2 Treating $(c(R), w(R))$ as a Whole

This approach, followed amongst others by Bosc et al. [10, 26], does not look at the $c_i(R)$'s and $w_i(R)$'s separately, but treats them as a whole. In contrast with Dubois and Prade, Bosc et al. consider both conjunction and disjunction of bipolar query conditions. As it is usually done in regular 'fuzzy' querying, conjunction is translated to a minimum operator and disjunction is translated to a maximum operator. In order to take the minimum or maximum, the set of $(c_i(R), w_i(R))$ pairs must be ordered. In this approach, a lexicographical ordering is assumed (see above, in Sect. 3.1) and the operators $lmin$ and $lmax$ are introduced as aggregation operators for respectively conjunction and disjunction of bipolar query conditions [10, 17, 26]. Let us assume that two elementary bipolar queries A_i IS F_i, $i = 1, 2$, result in two pairs of satisfaction degrees for a record R: $(c_i(R), w_i(R))$, $i = 1, 2$. Then, the pair of satisfaction degrees for the conjunction and disjunction of these elementary queries is defined as follows:

$$(c_{(A_1 \text{IS} F_1) \wedge (A_2 \text{IS} F_2)}(R), w_{(A_1 \text{IS} F_1) \wedge (A_2 \text{IS} F_2)}(R)) =$$
$$= lmin((c_1(R), w_1(R)), (c_2(R), w_2(R))) =$$
$$= \begin{cases} (c_1(R), w_1(R)) \text{ if } (c_1(R) < c_2(R)) \vee (c_1(R) = c_2(R) \wedge w_1(R) < w_2(R)) \\ (c_2(R), w_2(R)) \text{ otherwise} \end{cases}$$

$$(19)$$

$$(c_{(A_1 \text{IS} F_1) \vee (A_2 \text{IS} F_2)}(R), w_{(A_1 \text{IS} F_1) \vee (A_2 \text{IS} F_2)}(R)) =$$
$$= lmax((c_1(R), w_1(R)), (c_2(R), w_2(R))) =$$
$$= \begin{cases} (c_1(R), w_1(R)) \text{ if } (c_1(R) > c_2(R)) \vee (c_1(R) = c_2(R) \wedge w_1(R) > w_2(R)) \\ (c_2(R), w_2(R)) \text{ otherwise.} \end{cases}$$

$$(20)$$

Due to the associativity of the operators $lmin$ and $lmax$ formulas (19) and (20) may be easily extended to the case of a conjunction and disjunction, respectively, of n elementary bipolar queries.

By definition both $lmin$ and $lmax$ return one of the input pairs as the result. As all arguments are assumed to be consistent so is also the result of this type of aggregation.

4.2 Aggregation in the Satisfied-Dissatisfied Approach

Consider again n bipolar query conditions, evaluation of which for record R leads to a set $\{(s_i(R), d_i(R)), i = 1, \ldots, n\}$ of n BSDs. This set of n pairs needs to be aggregated to a BSD $(s(R), d(R))$ representing the global satisfaction and dissatisfaction when taking into account all imposed query conditions.

4.2.1 Treating s and d Individually

In this approach, as in the approach by Dubois and Prade, the BSDs are not aggregated as a whole but the lists of $s_i(R)$'s and $d_i(R)$'s are aggregated separately [30, 33]. But, unlike the Dubois and Prade approach, both conjunction and disjunction of bipolar query conditions are considered, as well as the negation. Moreover, this approach also allows to take into account weights to distinguish important from less important query conditions.

Because the bipolar query conditions in this approach are inspired by AFSs, the basic aggregation of BSDs (which are the result of the evaluation of such bipolar query conditions) is also inspired by the aggregation of AFSs. This means that the conjunction (respectively disjunction) of two BSDs is calculated in the same sense as the intersection (respectively union) of two AFSs. Moreover, these operations also coincide with those proposed in a continuous extension of Belnap's four-valued logic proposed by Öztürk and Tsoukiàs [34].

Non-Weighted Aggregation

Let us consider two elementary bipolar conditions: 'A_1 IS F_1' and 'A_2 IS F_2', and their conjunction and disjunction.

Conjunction.

The satisfaction and dissatisfaction degrees, i.e., a BSD, for the query '$(A_1$ IS $F_1) \wedge (A_2$ IS $F_2)$' is computed as follows:

$$(s_{(A_1 \text{ IS } F_1) \wedge (A_2 \text{ IS } F_2)}(R), d_{(A_1 \text{ IS } F_1) \wedge (A_2 \text{ IS } F_2)}(R)) =$$
$$= (\min(s_{A_1 \text{ IS } F_1}(R), s_{A_2 \text{ IS } F_2}(R)), \max(d_{A_1 \text{ IS } F_1}(R), d_{A_2 \text{ IS } F_2}(R))). \tag{21}$$

An intuitive justification for this formula is as follows:

- For the conjunction of two conditions to be satisfied, both conditions have to be satisfied. Therefore the minimum of both individual satisfaction degrees is taken as the satisfaction degree of their conjunction.
- For the conjunction to be dissatisfied, it is enough if one of them is dissatisfied. Therefore the maximum of both individual dissatisfaction degrees is taken as the dissatisfaction degree of their conjunction.

Besides the minimum and maximum, other aggregation operators based on triangular norms and co-norms can also be used if a reinforcement effect is needed or desired.

It should be noted that the formulas (17) and (21) although similar on the surface, are quite different. In both cases we have the minimum operator applied to the first components of the aggregated pairs and the maximum operator applied to the second components of these pairs. However, in the former case the minimum and maximum operators are applied to the complements of the negative evaluations and the positive evaluations, respectively, while in the latter case these are positive and negative evaluations, respectively.

Disjunction.

The satisfaction and dissatisfaction degrees, i.e., a BSD, for the query '$(A_1$ IS $F_1) \vee (A_2$ IS $F_2)$' is computed as follows:

$$(s_{(A_1 \text{ IS } F_1) \vee (A_2 \text{ IS } F_2)}(R), d_{(A_1 \text{ IS } F_1) \vee (A_2 \text{ IS } F_2)}(R)) =$$
$$= (\max(s_{A_1 \text{ IS } F_1}(R), s_{A_2 \text{ IS } F_2}(R)), \min(d_{A_1 \text{ IS } F_1}(R), d_{A_2 \text{ IS } F_2}(R))). \quad (22)$$

Similarly to the case of conjunction, an intuitive justification for this formula is as follows:

- For the disjunction of two conditions to be satisfied, it is enough for one of them to be satisfied. Therefore the maximum of both individual satisfaction degrees is taken as the satisfaction degree of their disjunction.
- For the disjunction of two conditions to be dissatisfied, both of them have to be dissatisfied. Therefore the minimum of both individual dissatisfaction degrees is taken as the dissatisfaction degree of their disjunction.

Negation.

The satisfaction and dissatisfaction degrees, i.e., a BSD, for the query '$\neg (A \text{ IS } F)$' is computed as follows:

$$(s_{\neg(A \text{ IS } F)}(R), d_{\neg(A \text{ IS } F)}(R)) = (d_{A \text{ IS } F}(R), s_{A \text{ IS } F}(R)). \quad (23)$$

The same effect of negation can also be achieved by swapping fuzzy sets of positively (F^+) and negatively (F^-) evaluated elements of dom_A composing F, $F = (F^+, F^-)$ in '$\neg (A \text{ IS } F)$', i.e., '$\neg(A \text{ IS } F)$' is thus equivalent to '$A \text{ IS } F'$', where $F' = (F^-, F^+)$.

Weighted Aggregation

When expressing queries (bipolar or not), one way to model the difference in importance between different elementary (bipolar) query conditions is by using weights. Also in the framework of BSDs, it is possible to deal with such weights [31]. The underlying aggregation operators are still appropriate basic aggregation operators, but a premodification step is performed on the elementary criteria evaluation results to take into account the impact of the weights. It is assumed that the importance of a condition, with respect to the final result, is linked with the condition itself, not with the degree to which the condition is satisfied. So weights $w_i \in [0, 1]$ can be attached to the individual elementary bipolar conditions. The semantics of the weights is as follows: $w_i = 1$ denotes that the condition is fully important, while $w_i = 0$ denotes that the condition is not important at all. Such a condition can be neglected (and hence should have no impact on the result). Conditions with intermediate weights should still be taken into account, but to a lesser extent than conditions with weight $w_i = 1$. In order to have an appropriate scaling, it is assumed that $\max_i w_i = 1$ [15].
   To reflect the impact of a weight on the evaluation of a condition, a premodification is performed on the initial BSDs, taking into account the weights. This means that,

before aggregating the individual BSDs, the impact of the weights on these BSDs is calculated first. Afterwards, the modified BSDs are aggregated using the regular aggregation techniques, as if they were regular, non-modified, BSDs. Let g be the operator that models this weight influence on the individual BSDs:

$$g : [0, 1] \times \tilde{\mathbb{B}} \rightarrow \tilde{\mathbb{B}} : (w, (s, d)) \mapsto g(w, (s, d)). \tag{24}$$

It has been shown that implication functions f_{im} and co-implication functions f_{im}^{co} can be used to model the impact of weights, where f_{im} and f_{im}^{co} are $[0, 1]$-valued extensions of Boolean implication and co-implication functions. As an example, consider the Kleene-Dienes implication and co-implication:

$$f_{im_{KD}}(x, y) = \max(1 - x, y)$$
$$f_{im_{KD}}^{co}(x, y) = \min(1 - x, y). \tag{25}$$

The impact of a weight on a BSD, in case of conjunction, can be defined as follows:

$$g^{\wedge} : [0, 1] \times \tilde{\mathbb{B}} \rightarrow \tilde{\mathbb{B}} : (w, (s, d)) \mapsto g^{\wedge}(w, (s, d)) = \left(s_{g^{\wedge}(w,(s,d))}, d_{g^{\wedge}(w,(s,d))}\right) \tag{26}$$

where

$$s_{g^{\wedge}(w,(s,d))} = f_{im}(w, s)$$
$$d_{g^{\wedge}(w,(s,d))} = f_{im}^{co}(1 - w, d).$$

As an example, consider the weight operator for conjunction based on the Kleene-Dienes implication:

$$g^{\wedge}(w, (s, d)) = (\max(1 - w, s), \min(w, d)). \tag{27}$$

Consider the basic conjunction operator $\wedge$ for BSDs, which is defined by

$$\wedge : (\tilde{\mathbb{B}})^2 \rightarrow \tilde{\mathbb{B}} : ((s_1, d_1), (s_2, d_2)) \mapsto (\min(s_1, s_2), \max(d_1, d_2)) \tag{28}$$

(cf. Eq. (21)). Using this definition and the definition of the weight impact operator $g^{\wedge}$, a definition of an extended operator for weighted conjunction $\wedge^w$ of BSDs can now be given as follows:

$$\wedge^w : ([0, 1] \times \tilde{\mathbb{B}})^2 \rightarrow \tilde{\mathbb{B}} \tag{29}$$
$$((w_1, (s_1, d_1)), (w_2, (s_2, d_2))) \mapsto g^{\wedge}(w_1, (s_1, d_1)) \wedge g^{\wedge}(w_2, (s_2, d_2)).$$

An extended operator for weighted disjunction can be defined analogously. Indeed, the impact of a weight on a BSD, in case of disjunction, can be defined by:

$$g^\vee : [0, 1] \times \tilde{\mathbb{B}} \to \tilde{\mathbb{B}} : (w, (s, d)) \mapsto g^\vee(w, (s, d)) = \left(s_{g^\vee(w,(s,d))}, d_{g^\vee(w,(s,d))}\right) \tag{30}$$

where

$$s_{g^\vee(w,(s,d))} = f_{im}^{co}(1 - w, s)$$
$$d_{g^\vee(w,(s,d))} = f_{im}(w, d).$$

Using the Kleene-Dienes implication, the following weight operator for disjunction is for example obtained:

$$g^\vee(w, (s, d)) = (\min(w, s), \max(1 - w, d)). \tag{31}$$

Consider the basic disjunction operator $\vee$ for BSDs, which is defined by

$$\vee : (\tilde{\mathbb{B}})^2 \to \tilde{\mathbb{B}} : ((s_1, d_1), (s_2, d_2)) \mapsto (\max(s_1, s_2), \min(d_1, d_2)) \tag{32}$$

(cf. Eq. (22)). Using this definition and the definition of the weight impact operator $g^\vee$, a definition of an extended operator for weighted conjunction $\vee^w$ of BSDs can then be given as follows:

$$\vee^w : ([0, 1] \times \tilde{\mathbb{B}})^2 \to \tilde{\mathbb{B}} \tag{33}$$
$$((w_1, (s_1, d_1)), (w_2, (s_2, d_2))) \mapsto g^\vee(w_1, (s_1, d_1)) \wedge g^\vee(w_2, (s_2, d_2)).$$

Averaging

Besides the basic aggregation operators based on the aggregation of AFSs, using triangular norms and co-norms, BSDs can also be aggregated using other operators, like averaging operators [31]. Some averaging operators that could be used are the arithmetic mean (AM), geometric mean (GM) or harmonic mean (HM). As an example, consider the traditional arithmetic mean:

$$AM(x_1, \ldots, x_n) = \frac{1}{n} \sum_{i=1}^{n} x_i. \tag{34}$$

Such averaging operators cannot be applied on BSDs as such, because a BSD consists of a pair of values. So again, the satisfaction degrees and dissatisfaction degrees need to be treated separately. An extended version of the above regular averaging operator can be defined, where this regular averaging operator is applied for the satisfaction degrees, and, separately, for the dissatisfaction degrees:

$$AM((s_1, d_1), \ldots, (s_n, d_n)) = \left(\frac{1}{n} \sum_{i=1}^{n} s_i, \frac{1}{n} \sum_{i=1}^{n} d_i\right). \tag{35}$$

A similar extension can be defined for other averaging operators (GM, HM, …).

Weighted Averaging

In the case of weighted averaging, it is again assumed that the importance of a condition, with respect to the final result, is linked with the condition itself, not with the degree to which the condition is satisfied. So weights $w_i \in [0, 1]$ can be connected with the individual bipolar conditions. Again, in order to have an appropriate scaling, it is assumed that $\max_i w_i = 1$. Weighted counterparts of the above averaging operators for BSDs (e.g., weighted arithmetic mean (AM^w), weighted geometric mean (GM^w), or weighted harmonic mean (HM^w)) can be used, where the satisfaction degrees on the one hand, and the dissatisfaction degrees on the other hand, are again aggregated separately using regular weighted averaging operators. As an example, consider the weighted arithmetic mean AM^w for BSDs:

$$AM^w : ([0, 1] \times \tilde{\mathbb{B}})^n \to \tilde{\mathbb{B}} \tag{36}$$

$$((w_1, (s_1, d_1)), \ldots, (w_n, (s_n, d_n))) \mapsto \left(\frac{\sum_{i=1}^n w_i \cdot s_i}{\sum_{i=1}^n w_i}, \frac{\sum_{i=1}^n w_i \cdot d_i}{\sum_{i=1}^n w_i} \right).$$

4.2.2 Treating (s, d) as a Whole

Aggregating BSDs as a whole can be done by using Ordered Weighted Averaging (OWA) operators for BSDs [31]. Ordered weighted averaging of BSDs can be based on the traditional OWA operators [42, 43] as done in the case of aggregating regular satisfaction degrees. The OWA operator of dimension n, i.e., accepting n arguments $x_1, \ldots, x_n$ is defined by:

$$OWA_W(x_1, \ldots, x_n) = \sum_{i=1}^n w_i \cdot x_i' \tag{37}$$

where x_i' is the ith largest value of $x_1, \ldots, x_n$ and $W = [w_1, \ldots, w_n]$; $\sum_{i=1}^n w_i = 1$ is a parameter of the OWA operator, referred to as the vector of weights.

This traditional OWA operator can also be extended to work with BSDs. To this aim, the BSDs are first rank ordered, for example by using one of the ranking functions presented in Sect. 3.2:

$$OWA_W : \mathbb{B}^n \to \mathbb{B} \tag{38}$$

$$((s_1, d_1), \ldots, (s_n, d_n)) \mapsto \left(\sum_{i=1}^n w_i \cdot s_i', \sum_{i=1}^n w_i \cdot d_i' \right)$$

where (s_i', d_i') is the ith largest BSD of $(s_1, d_1), \ldots, (s_n, d_n)$, according to the ranking function used.

Depending on the weight vector that is used, this extended OWA operator will behave differently (just like the regular OWA operator). In special cases, it can, e.g., act as a maximum function for BSDs ($w_1 = 1$, $w_i = 0$ for $i > 1$), a minimum function for BSDs ($w_n = 1$, $w_i = 0$ for $i < n$), or a median function for BSDs (for odd n: $w_{\lceil \frac{n}{2} \rceil} = 1$, $w_i = 0$ for $i \neq \lceil \frac{n}{2} \rceil$, where $\lceil \rceil$ denotes the ceiling function; for even n: $w_{\frac{n}{2}} = \frac{1}{2}$, $w_{\frac{n}{2}+1} = \frac{1}{2}$, $w_i = 0$ for $i \neq \frac{n}{2}$ and $i \neq \frac{n}{2} + 1$).

Remark that the exact behaviour of the maximum, minimum and median function for BSDs (and also for all other OWA operators) depends on the specific ranking function employed.

4.3 Comparison and Discussion

The aggregation of pairs of satisfaction degrees of elementary bipolar conditions should follow and reflect the semantics of the querying approach. This is why, for both approaches, specific aggregation techniques have been presented, hereby distinguishing techniques to aggregate both satisfaction degrees separately and to aggregate the satisfaction degree pairs as a whole.

In the constraint-wish approach, handling both satisfaction degrees separately boils down to treating all constraints together as a global constraint and treating all wishes together as a global wish, hereby preserving the applicable consistency condition, which requires some additional effort. Handling both satisfaction degrees as a whole boils down to lexicographical ordering. In both kinds of aggregation the semantics of constraints and wishes is retained.

In the satisfied-dissatisfied approach the satisfaction and dissatisfaction degrees are completely independent of each other. This characteristic offers more freedom to develop aggregation operators that treat both degrees separately.

- Basic aggregation operators, inspired by the aggregation of AFSs and based on the minimum triangular norm and maximum triangular co-norm, have been defined for non-weighted conjunction and disjunction. These operators retain the semantics of positive and negative information.
- To handle elementary query conditions of different importance, extended counterparts of the basic aggregation operators have been presented in literature. These operators use associated weights to model the relative importance of a query condition. First, the elementary conditions are evaluated as if there are no weights. Second, the impact of a weight on the evaluation of a condition is modelled in a premodification step using an implication and co-implication function.
- Instead of being based on a triangular norm and a triangular co-norm, an aggregation function can also be based on an averaging operator like the arithmetic, geometric or harmonic mean or on the weighted extension of such an averaging operator.

Choosing which aggregation operator to use depends on the requirements of the application. Aspects that may be considered in the selection of an adequate operator are: the need to better distinguish among the resulting records, the need for a reinforcement effect and the computation time.

BSDs can also be treated as a whole and aggregated based on their ranking. For that purpose, an OWA operator for BSDs has been presented in the literature. As is the case with regular OWA operators, the behaviour of the aggregation will then strongly depend on the used weight vector. Special cases are the minimum, maximum and median function for BSDs. Whether to use this kind of aggregation or not, and which weight vector should be chosen, again depend on the requirements of the application under consideration. Results obtained from an aggregation based on the ranking of BSDs are in general less informative to the user, because they do not provide independent information about the satisfaction of the positive and negative conditions in the user preferences. However, if a quantifier-based aggregation is required by the application, where at least (or at most) a specified (fuzzy) number of elementary conditions should be satisfied in order to satisfy the query, an OWA-based aggregation can be used.

5 Conclusions

In this chapter, an overview and comparison of two commonly known approaches to bipolar querying of databases have been presented: the constraint-wish approach and the satisfied-dissatisfied approach. The specification of bipolar query conditions and different aspects of query handling, including the evaluation of elementary conditions, their aggregation, as well as ranking of the query results have been described.

The constraint-wish approach has been specifically designed to cope with situations where user preferences express requirements—called constraints—which should be satisfied (at least to some extent) by the retrieved database records, and other, optional conditions—called wishes—which serve to distinguish among those records that satisfy the constraints to the same extent. Slightly different semantics is modelled by the 'and possibly' based approach to constraints and wishes, where the influence of the wishes on the results of a query depends on the existence of the records satisfying constraints and wishes at the same time.

The motivation for the satisfied-dissatisfied approach is to cope with user preferences that are composed of positive conditions—expressing what the user likes—and negative conditions—expressing what the user wants to avoid. The positive and negative conditions do not necessarily have to be complementary to each other.

Although both approaches result in pairs of satisfaction degrees (constraint satisfaction and wish satisfaction, or satisfaction degree and dissatisfaction degree), the semantics are quite different. In the constraint-wish approach, 'true' constraints, i.e., mandatory requirements, are treated as more important in a specific sense. In the satisfied-dissatisfied approach, the positive and negative requirements are considered in general as being equally important and independent. Due to this assumed independence, it is also possible to model inconsistent or conflicting situations in the satisfied-dissatisfied approach, which is not possible in the constraint-wish approach, where either strong or weak consistency must apply. Moreover, in the satisfied-dissatisfied approach, the set of operators that can be used for ranking or aggregating is more elaborate than in the constraint-wish approach (e.g., weighted aggregation operators). On the other hand, a complete '*bipolar*' relational algebra has been proposed for the constraint-wish approach, i.e., an extension of traditional relational algebra to handle bipolarity [10].

References

1. Atanassov, K.: Intuitionistic fuzzy sets. Fuzzy Sets Syst. **20**, 87–96 (1986)
2. Belnap, N.D.: Modern Uses of Multiple-Valued Logic, chap. A useful four-valued logic, pp. 8–37. Reidel, Dordrecht (1977)
3. Bordogna, G., Pasi, G.: Linguistic aggregation operators of selection criteria in fuzzy information retrieval. Int. J. Intell. Syst. **10**(2), 233–248 (1995)
4. Bosc, P., Kraft, D., Petry, F.: Fuzzy sets in database and information systems: status and opportunities. Fuzzy Sets Syst. **156**, 418–426 (2005)
5. Bosc, P., Pivert, O.: Some approaches for relational databases flexible querying. Int. J. Intell. Inf. Syst. **1**, 323–354 (1992)
6. Bosc, P., Pivert, O.: An approach for a hierarchical aggregation of fuzzy predicates. In: Proceedings of the 2nd IEEE International Conference on Fuzzy Systems (FUZZ-IEEE'93), pp. 1231–1236. San Francisco, USA (1993)
7. Bosc, P., Pivert, O.: Sqlf: a relational database language for fuzzy querying. IEEE Trans. Fuzzy Syst. **3**(1), 1–17 (1995)
8. Bosc, P., Pivert, O.: On three fuzzy connectives for flexible data retrieval and their axiomatization. In: Proceedings of the SAC'11 Conference, pp. 1114–1118. Taiwan (2011)
9. Bosc, P., Pivert, O.: On four noncommutative fuzzy connectives and their axiomatization. Fuzzy Sets Syst. **202**, 42–60 (2012)
10. Bosc, P., Pivert, O., Mokhtari, A., Liétard, L.: Extending relational algebra to handle bipolarity. In: Proceedings of the 2010 ACM Symposium on Applied Computing (SAC '10), pp. 1718–1722. Sierre, Switzerland (2010)
11. De Calmès, M., Dubois, D., Hüllermeier, E., Prade, H., Sèdes, F.: A fuzzy set approach to flexible case-based querying: methodology and experimentation. In: Proceedings of the 8th International Conference on Principles of Knowledge Representation and Reasoning (KR2002), pp. 449–458. Toulouse, France (2002)
12. De Tré, G., De Caluwe, R., Kacprzyk, J., Zadrożny, S.: On flexible querying via extensions to fuzzy sets. In: Proceedings of the EUSFLAT'05 and LFA'05 Joint Conference, pp. 1225–1230. Barcelona, Spain (2005)
13. De Tré, G., Zadrożny, S., Matthé, T., Kacprzyk, J., Bronselaer, A.: Dealing with positive and negative query criteria in fuzzy database querying : bipolar satisfaction degrees. In: Troels, A. (ed.) Lecture Notes in Computer Science, vol. 5822, pp. 593–604. Springer (2009)
14. Dubois, D., Prade, H.: Default reasoning and possibility theory. Artif. Intell. **35**(2), 243–257 (1988)
15. Dubois, D., Prade, H.: Using fuzzy sets in flexible querying: why and how?, pp. 45–60. Kluwer Academic Publishers, Norwell (1997)
16. Dubois, D., Prade, H.: Bipolarity in flexible querying. Lect. Notes Artif. Intell. **2522**, 174–182 (2002)

17. Dubois, D., Prade, H.: Handbook of Research on Fuzzy Information Processing in Databases, chap. Handling bipolar queries in Fuzzy Information Processing, pp. 97–114. Information Science Reference, New York (2008)
18. Dubois, D., Prade, H.: Gradualness, uncertainty and bipolarity: making sense of fuzzy sets. Fuzzy Sets Syst. **192**, 3–24 (2012)
19. Grattan-Guinness, I.: Fuzzy membership mapped onto intervals and many-valued quantities. Math. Logic Q. **22**(1), 149–160 (1976)
20. Jahn, K.U.: Intervall-wertige mengen. Mathematische Nachrichten **68**(1), 115–132 (1975)
21. Kacprzyk, J., Zadrożny, S.: Fuzziness in database management systems, chap. FQUERY for access: fuzzy querying for windows-based DBMS, pp. 415–433. Physica-Verlag, Heidelberg (1995)
22. Kacprzyk, J., Zadrożny, S., Ziólkowski, A.: Fquery iii+: a human-consistent database querying system based on fuzzy logic with linguistic quantifiers. Inf. Syst. **14**(6), 443–453 (1989)
23. Kacprzyk, J., Ziólkowski, A.: Database queries with fuzzy linguistic quantifiers. IEEE Trans. Syst. Man Cybern. **16**, 474–479 (1986)
24. Konieczny, S., Marquis, P., Besnard, P.: Bipolarity in bilattice logics. Int. J. Intell. Syst. **23**(10), 1046–1061 (2008)
25. Lacroix, M., Lavency, P.: Preferences: Putting more knowledge into queries. In: Proceedings of the VLDB'87 Conference, pp. 217–225. Brighton, UK (1987)
26. Liétard, L., Rocacher, D.: On the definition of extended norms and co-norms to aggregate fuzzy bipolar conditions. In: Proceedings of the 2009 IFSA/EUSFLAT Conference, pp. 513–518. Lisbon, Portugal (2009)
27. Liétard, L., Rocacher, D., Bosc, P.: On the extension of sql to fuzzy bipolar conditions. In: Proceedings of the 28th North American Information Processing Society Annual Conference (NAFIPS '09). Cincinnati, Ohio, USA (2009)
28. Liétard, L., Tamani, N., Rocacher, D.: Fuzzy bipolar conditions of type or else. In: Proceedings of the 2011 FUZZ-IEEE Conference, pp. 2546–2551. Taipei, Taiwan (2011)
29. Liétard, L., Tamani, N., Rocacher, D.: Linguistic quantifiers and bipolarity. In: Proceedings of the 2011 IFSA World Congress and the 2011 AFSS International Conference. Surabaya and Bali Island, Indonesia (2011)
30. Matthé, T., De Tré, G.: Bipolar query satisfaction using satisfaction and dissatisfaction degrees: bipolar satisfaction degrees. In: Proceedings of the ACM Symposium on Applied Computing (ACM SAC'09), pp. 1699–1703. Honolulu, Hawaii (2009)
31. Matthé, T., De Tré, G.: Weighted aggregation of bipolar satisfaction degrees. In: Proceedings of the 2011 IFSA World Congress and the 2011 AFSS International Conference. Surabaya and Bali Island, Indonesia (2011)
32. Matthé, T., De Tré, G.: Ranking of bipolar satisfaction degrees. In: Proceedings of the IPMU 2012 Conference on Communications in Computer and Information Sciences, vol. 298, pp. 461–470. Catania, Italy (2012)
33. Matthé, T., De Tré, G., Zadrożny, S., Kacprzyk, J., Bronselaer, A.: Bipolar database querying using bipolar satisfaction degrees. Int. J. Intell. Syst. **26**(10), 890–910 (2011)
34. Öztürk, M., Tsoukiàs, A.: Modelling uncertain positive and negative reasons in decision aiding. Decis. Support Syst. **43**(4), 1512–1526 (2007)
35. Rivieccio, U.: Neutrosophic logics: prospects and problems. Fuzzy Sets Syst. **159**(14), 1860–1868 (2008)
36. Sambuc, R.: Fonctions ϕ-floues. application à l'aide au diagnostic en pathologie thyroidienne. Ph.D. thesis, Université de Marseille, France (1975)
37. Smarandache, F.: A Unifying Field in Logics: Neutrosophic Logic. Neutrosophy, Neutrosophic Set, Neutrosophic Probability. American Research Press, Rehoboth (1999)
38. Tahani, V.: A conceptual framework for fuzzy query processing: a step toward very intelligent database systems. Inf. Process. Manage. **13**, 289–303 (1977)
39. Tamani, N., Liétard, L., Rocacher, D.: Bipolarity and the relational division. In: Proceedings of the 7th conference of the European Society for Fuzzy Logic and Technology (EUSFLAT-2011). Aix-les-Bains, France (2011)

40. Turunen, E., Öztürk, M., Tsoukiàs, A.: Paraconsistent semantics for Pavelka style fuzzy sentential logic. Fuzzy Sets Syst. **161**(14), 1926–1940 (2010)
41. Yager, R.: Fuzzy logic in the formulation of decision functions from linguistic specifications. Kybernetes **25**(4), 119–130 (1996)
42. Yager, R.R.: On ordered weighted averaging aggregation operators in multicriteria decision-making. IEEE Trans. Syst. Man Cybern. **18**(1), 183–190 (1988)
43. Yager, R.R., Kacprzyk, J.: The Ordered Weighted Averaging Operators : Theory and Applications. Kluwer Academic Publishers, Boston (1997)
44. Zadeh, L.A.: Fuzzy sets. Inf. Control. **8**(3), 338–353 (1965)
45. Zadeh, L.A.: The concept of a linguistic variable and its application to approximate reasoning— I. Inf. Sci. **8**(3), 199–249 (1975)
46. Zadrożny, S., De Tré, G., De Caluwe, R., Kacprzyk, J.: Handbook of Research on Fuzzy Information Processing in Databases, chap. An Overview of Fuzzy Approaches to Flexible Database Querying, pp. 34–54. Information Science Reference, New York (2008)
47. Zadrożny, S., De Tré, G., Kacprzyk, J.: Remarks on various aspects of bipolarity in database querying. In: Proceedings of the 2010 International Workshop on Database and Expert Systems Applications Proceedings (DEXA '10), pp. 323–327. Bilbao, Spain (2010)
48. Zadrożny, S., Kacprzyk, J.: Bipolar queries and queries with preferences. In: Proceedings of the DEXA'06 Conference, pp. 415–419. Kraków, Poland (2006)

Chapter 3
A Relational Algebra for Generalized Fuzzy Bipolar Conditions

Ludovic Liétard, Daniel Rocacher and Nouredine Tamani

Abstract Flexible querying of regular databases consists in expressing user's preferences (fuzzy conditions) inside queries instead of Boolean requirements. Fuzzy bipolar conditions are particular cases of fuzzy conditions which are made of two components, a mandatory fuzzy condition and an optional fuzzy condition. They define two different types of complex preferences which can be either of a conjunctive nature or of a disjunctive nature (both of them being interpreted in a hierarchical way). The first case leads to define fuzzy bipolar conditions of type *and if possible*, the second case leads to define fuzzy bipolar conditions of type *or else*. This chapter shows that a general form of fuzzy bipolar conditions having a hierarchical interpretation can be considered since these two forms are compatible. As a consequence, fuzzy bipolar conditions of both types can be used together in a single bipolar query and all the algebraic operators are extended to this generalization. The particular case (non algebraic) of the use of linguistic quantifiers is also studied.

1 Introduction

We consider the relational model of data and the integration of user's preferences inside queries which defines the flexible querying of relational databases (since the expression of user's preferences introduces a kind of flexibility). In this context, atomic conditions express preferences instead of Boolean requirements (as it is the

L. Liétard (✉)
IRISA/IUT/University Rennes 1, Rue Edouard Branly, BP 80519, 22305 Lannion Cedex, France
e-mail: ludovic.lietard@univ-rennes1.fr

D. Rocacher · N. Tamani
IRISA/ENSSAT/University Rennes 1, rue de Kerampont, BP 80518, 22305 Lannion
Cedex, France
e-mail: rocacher@enssat.fr

N. Tamani
e-mail: tamani@enssat.fr

O. Pivert and S. Zadrożny (eds.), *Flexible Approaches in Data, Information and Knowledge Management*, Studies in Computational Intelligence 497, DOI: 10.1007/978-3-319-00954-4_3, © Springer International Publishing Switzerland 2014

case for the ordinary querying) and a set of ranked answers from the most to the least preferred is obtained (instead of a set of indistinguishable answers). It has been shown that the fuzzy sets theory [15] provides a general framework for the definition and the interpretation of conditions expressing preferences. Atomic conditions are defined by fuzzy sets (to define fuzzy predicates also called fuzzy conditions) and the relational algebra has been extended to such conditions to propose the SQLf language [1, 10].

It is also possible to consider fuzzy bipolar conditions to model complex preferences [6, 8]. In this context, a fuzzy bipolar condition is made of two fuzzy predicates (two poles), the first one expressing a constraint to define the elements to be retrieved, the other one expressing a more restrictive attitude to define the best elements (among the ones satisfying the constraint). Obviously, the negation of a constraint is a set of values to be rejected. The advantage of this type of condition can be illustrated by the querying of a database containing cars to be sold by a company. A fuzzy bipolar condition can be useful to take into consideration two aspects: the requirement of the client (a red car with a *low* price) and those of the company (to sell in priority the *oldest* cars). A car which does not satisfy the conditions of the client is rejected, in other words the requirements of the client are defining a constraint. The best cars being the ones satisfying the two requirements, the conjunction of conditions stated by the client and the company is a wish (an *old* red car with a *low* price).

These two poles can been interpreted in a conjunctive way to define fuzzy bipolar conditions of type "and if possible" (fuzzy and-if-possible-bipolar conditions) and the algebraic operators (selection, projection, join, union, intersection) have been extended to this type of fuzzy bipolar conditions [3, 10]. These two poles can also be interpreted in a disjunctive way to define fuzzy bipolar conditions of type "or else" (fuzzy or-else-bipolar conditions)[11].

In this chapter, we aim at defining the basis of a general algebra extended to fuzzy bipolarity to express complex flexible queries. It is worth mentioning that this new type of flexible querying is a generalization of the one based on simple fuzzy predicates (non bipolar i.e. SQLf) which is also a generalization of Boolean querying (SQL). More precisely, we show that the two interpretations of fuzzy bipolar conditions can be generalized in an unique framework and we define the extension of the algebraic operators to this framework. The particular case of linguistic quantifiers is also studied since they provide a very powerful (non algebraic) aggregation of fuzzy bipolar conditions.

The remainder of this chapter is organized as follows. Section 2 introduces the fuzzy bipolar conditions of type "and if possible" and of type "or else". These two types of fuzzy bipolar conditions are not equivalent and can be interpreted in a hierarchical way. The differences between these two types are stressed and it is shown in Sect. 3 that they are mutually compatible (in particular the negation of a fuzzy bipolar condition of type "and if possible" is a fuzzy bipolar condition of type "or else" and vice versa) and can be generalized. The extension of the algebraic operators (complement, intersection, union, Cartesian product, selection, join, difference, projection) to this generalization is introduced in Sect. 4. Section 5 deals with the extension of linguistic quantifiers which allow to represent an intermediate attitude between the conjunction and the disjunction. Section 6 recalls our contribution and draws some lines for future works.

2 Fuzzy Bipolar Conditions

A bipolar condition is an association of a negative condition (negative pole) and positive condition (positive pole). In this chapter, a bipolar condition is made of two conditions defined on the same universe: (i) a constraint c, which describes the set of acceptable elements, (ii) a wish w which defines the set of desired or wished elements. The negation of c is the set of rejected elements since it describes non-acceptable elements. It is not coherent to wish a rejected element, consequently the following property of coherence holds: $w \subseteq c$.

In addition, condition c is mandatory since an element which does not satisfy c is rejected; $\neg c$ is then considered as the negative pole of the bipolar condition. Condition w is optional because its non-satisfaction does not automatically mean the rejection; w is then considered as the positive pole of the bipolar condition. Being of different nature, we propose not to allow the aggregation of these two poles because they convey different semantics. More precisely, one (the constraint c) is used to reject elements while it is not the case for the other (the wish w). As a consequence, we think it is important to keep these two poles separately because their aggregation leads to loose their different semantics. However, some authors consider that these two poles can be aggregated, as Dubois and Prade [5–7], de Tré et al. [4] and Zadrozny and Kacprzyk [17, 18].

If c and w are boolean conditions, the satisfaction with respect to a bipolar condition is a couple from $\{0, 1\}^2$. If c and w are fuzzy conditions defined on the same universe U (U can be a domain of an attribute or a Cartesian product of domains of attributes), the property of coherence becomes: $\forall u \in U, \mu_w(u) \leq \mu_c(u)$ (it is the inclusion in the sense of Zadeh [15]). The satisfaction with respect to a bipolar condition is then a couple of degrees from the unit interval $[0, 1]^2$.

These two poles can been interpreted in a conjunctive way to define fuzzy bipolar conditions of type "and if possible" (called fuzzy and-if-possible-bipolar conditions) introduced in Sect. 2.1 and in a disjunctive way to define fuzzy bipolar conditions of type "or else" (called fuzzy or-else-bipolar conditions) introduced in Sect. 2.2.

2.1 Fuzzy and-if-possible-Bipolar Conditions

A bipolar condition of type "and if possible" (a fuzzy and-if-possible-bipolar condition) is denoted (c, w) and means, *"to satisfy c and if possible to satisfy w"*. When c and w are Boolean and when querying a database with such a condition, tuples satisfying the constraint and the wish and tuples satisfying only the constraint are returned (tuples satisfying the constraint and the wish are ranked before the tuples satisfying only the constraint). When querying a relation R with a fuzzy and-if-possible-bipolar condition, each tuple t from R is then attached with a pair of grades denoted $(\mu_c(t), \mu_w(t))$ that expresses the degrees of its satisfaction respectively to the constraint c and the wish w (and a so-called fuzzy and-if-possible-bipolar relation is obtained). A tuple t is then denoted $(\mu_c(t), \mu_w(t))/t$. Any tuple t such that $\mu_c(t) = 0$ does not appear in the fuzzy and-if-possible-bipolar relation.

In such a context, since c is the most important pole, tuples can be ranked using the lexicographical order. In other words, tuple t_1 is preferred to tuple t_2 if and only if:

$$\mu_c(t_1) > \mu_c(t_2) \text{ or } ((\mu_c(t_1) = \mu_c(t_2)) \wedge (\mu_w(t_1) > \mu_w(t_2))),$$

which is denoted $(\mu_c(t_1), \mu_w(t_1)) \succ (\mu_c(t_2), \mu_w(t_2))$.
We note $(\mu_c(t_1), \mu_w(t_1)) \succeq (\mu_c(t_2), \mu_w(t_2))$ when $(\mu_c(t_1), \mu_w(t_1)) \succ (\mu_c(t_2), \mu_w(t_2))$ or $(\mu_c(t_1), \mu_w(t_1)) = (\mu_c(t_2), \mu_w(t_2))$.

In this case, the satisfaction with respect to the constraint is firstly used to discriminate between answers (the constraint being the most important pole). The satisfaction with respect to the wish being less important, it can only be used to discriminate between answers having the same evaluation with respect to the constraint (thus we obtain a hierarchical interpretation of c and w). A total order (lexicographical order) is then obtained on μ_c and μ_w (with $(1, 1)$ as the greatest element and $(0, 0)$ as the least element).

A fuzzy bipolar condition of type "and if possible" is a generalization of a fuzzy predicate. More precisely, when defining a fuzzy predicate, the set of rejected elements is the complement of the set of desired elements. As a consequence, a fuzzy predicate C can be rewritten "C and if possible C" (denoted (C,C)).

2.2 Fuzzy or-else-Bipolar Conditions

A bipolar condition of type "or else" (a fuzzy or-else-bipolar condition) is denoted $[w, c]$ and means, "to satisfy w or else to satisfy c" [11], where w and c are respectively a wish and a constraint. When querying a relation R with a fuzzy or-else-bipolar condition, each tuple t from R is then attached with a pair of grades denoted $[\mu_w(t), \mu_c(t)]$ that expresses the degrees of its satisfaction respectively to the wish w and to the constraint c (and a so-called fuzzy or-else-bipolar relation is obtained). A tuple t is then denoted $[\mu_w(t), \mu_c(t)]/t$ and any tuple t such that $\mu_c(t) = 0$ does not appear in the fuzzy or-else-bipolar relation.

As for bipolar conditions of type "and if possible", since condition w represents the most important pole, tuples can be ranked using the lexicographical order. In other words, tuple t_1 is preferred to tuple t_2 if and only if:

$$\mu_w(t_1) > \mu_w(t_2) \text{ or} (\mu_w(t_1) = \mu_w(t_2)) \wedge (\mu_c(t_1) > \mu_c(t_2)),$$

which is denoted $[\mu_w(t_1), \mu_c(t_1)] \succ [\mu_w(t_2), \mu_c(t_2)]$.
We denote $[\mu_w(t_1), \mu_c(t_1)] \succeq [\mu_w(t_2), \mu_c(t_2)]$ when $[\mu_w(t_1), \mu_c(t_1)] \succ [\mu_w(t_2), \mu_c(t_2)]$ or $[\mu_w(t_1), \mu_c(t_1)] = [\mu_w(t_2), \mu_c(t_2)]$.

In this case, the satisfaction with respect to the wish is firstly used to discriminate between answers (the wish being the most important pole). The satisfaction with respect to the constraint being less important, it can only be used to discriminate between answers having the same evaluation with respect to the wish. A total order is then obtained on μ_w and μ_c (with $[1, 1]$ as the greatest element and $[0, 0]$ as the least element).

Fuzzy bipolar conditions of type "or else" are a generalization of fuzzy predicate since a fuzzy predicate C can be rewritten "C or else C" ($[C,C]$). In this case, the set of rejected elements is the complement of the set of desired elements.

3 A Generalization for Fuzzy Bipolar Conditions

Section 3.1 recalls the main features of the two kinds of fuzzy bipolar conditions and it is shown that they can be unified within a general form. An operator of conjunction (*lexmin*) and of disjunction (*lexmax*) are also proposed. The next Sect. 3.2 defines a negation operator for this general form of bipolar fuzzy conditions. Section 3.3 shows that the *lexmin* and the *lexmax* operators satisfy the De Morgan's laws.

3.1 A General Form for Fuzzy Bipolar Conditions

Bipolar conditions of the form "c and if possible w" (denoted (c, w)) can be defined with the following properties:

1. $\neg c$ corresponds to the rejection (c denotes acceptable elements),
2. w corresponds to the optimal values,
3. The acceptability condition c is more important than the optimality (the condition w),
4. The set of optimal values is included in the set of acceptable values ($w \subseteq c$).

The property 3 means that the non-rejection (or the acceptability) is more important than the optimality, therefore, the lexicographical order can be used to rank between elements.

Similarly, a fuzzy bipolar conditions of the form "e or else f", denoted $[e, f]$, satisfies the following properties:

1. $\neg f$ corresponds to the rejection (f denotes acceptable elements),
2. e corresponds to the optimal values,
3. The optimality condition e is more important than the acceptability condition f,

Table 1 Example of behavior
of bipolar conditions

	Formalism (c, w)	
	$\mu_c(x_i)$	$\mu_w(x_i)$
t_1	1	1
t_2	0.8	0.2
t_3	0.7	0.5
t_4	0.4	0.3
	Formalism $[w, c]$	
	$\mu_w(x_i)$	$\mu_c(x_i)$
t_1	1	1
t_3	0.5	0.7
t_4	0.3	0.4
t_2	0.2	0.8

4. The set of optimal values is included in the set of acceptable values ($e \subseteq f$):
$\forall x, \mu_e(x) \leq \mu_f(x)$.

We notice strong similarities between these two formalisms, since in both cases
we have:

- c and f express the acceptable values (or non rejected values) (cf. point 1),
- $\neg c$ and $\neg f$ correspond to the discarded values (cf. point 1),
- w and e express the perfect or optimal values (cf. point 2),
- the set of perfect values is included in the set of acceptable values (cf. point 4).

Moreover, these two conditions generalize fuzzy conditions in the same way: a fuzzy
condition C is expressed (C, C) or $[C, C]$. These two types of conditions only differ
on point 3 which gives different importance to the two poles.

Since w (resp. c) plays the same role as e (resp. f), it is interesting to study the
behavior of (c, w) and $[w, c]$. First of all, we notice that in the boolean case, both
formalisms have the same meaning, as shown in the following example.

Example 1. Let x, y and z be three elements attached respectively to the following
pair of grades: $(1, 1)$, $(1, 0)$ and $(0, 0)$ with respect to conditions c and w (the pair
$(0,1)$ is discarded because it does not satisfy the coherence property; the pair $(0,0)$ is
kept to show that such a—null— satisfaction leads to the least position when using
the lexicographical order). The lexicographical order delivers the same order in both
situations (c, w) and $[w, c]$: $x > y > z$. •

However, when c and w are fuzzy conditions, the two formalisms do not express
the same semantics because the lexicographical order does not deliver the same order
(cf. Example 2).

Example 2. Table 1 shows that tuples t_1, t_2, t_3 and t_4 are not sorted according to the
same order, depending on whether the formalism (c, w) or $[w, c]$ is used. •

The basic difference between these two formalisms is the fact that the formalism
"c, and if possible w" gives more importance to the non rejected elements (which
means that satisfaction with respect to the condition of acceptance c is privileged),

whereas the formalism "*e*, or else *f*" gives more importance to the optimal elements (i.e. the satisfaction with respect to the optimal condition *e* is privileged).

In this context, a fuzzy bipolar condition of the form (c, w) can be defined as a pair of fuzzy conditions, which define a set of optimal values (w) and a set of acceptable values (c) under the consideration that the non-rejection is more important than the optimality; and a fuzzy bipolar condition of the form $[e, f]$ is defined as a pair of fuzzy conditions, which define a set of optimal values (e) and a set of acceptable values (f) under the consideration that the optimality is more important than the non-rejection.

Fuzzy bipolar conditions of types "and if possible" and "or else" consists of two parts: the former is the most important concept (the concept attached to *c* for the fuzzy bipolar condition (c, w) and the concept attached to *e* for the fuzzy bipolar condition $[e, f]$), the latter is the least important concept of the bipolar condition. This means that it is possible to use the lexicographical order to compare and to handle couples of scores of these two types of fuzzy bipolar conditions. In so doing, the first choice is made on the condition which corresponds to the most important concept in the fuzzy bipolar condition, and the second choice is made on the condition which corresponds to the least important concept of the same fuzzy bipolar condition. In other words, it becomes possible to express fuzzy bipolar conditions of both forms "and if possible" and "or else" together in the single bipolar query.

In other words, the satisfaction with respect to a fuzzy bipolar condition can be rewritten using the same syntax (<a,b> to denote either [a,b] or (a,b)):

- <a,b> with a < b represents the satisfaction with respect to a fuzzy bipolar condition of type "or else",
- <a,b> with a = b represents the satisfaction with respect to a fuzzy condition,
- <a,b> with a > b represents the satisfaction with respect to a fuzzy bipolar condition of type "and if possible".

When querying a relation R with a fuzzy bipolar condition, each tuple t from R is then attached with a pair of grades denoted $< a, b >$ that expresses the degrees of its satisfaction to the wish w and to the constraint c, or to the constraint c and to the wish w (and a so-called fuzzy bipolar relation is obtained). A tuple t is then denoted $< \mu_a(t), \mu_b(t) > /t$ and any tuple t such that $< \mu_a(t), \mu_b(t) >=< 0, 0 >$ does not appear in the fuzzy bipolar relation. Fuzzy bipolar relations are generalizations of fuzzy and-if-possible-bipolar relations and fuzzy or-else-bipolar relations.

Couples <a,b> can be ranked using the lexicographical order to state that tuple t_1 is preferred to tuple t_2 if and only if:

$$(\mu_a(t_1) > \mu_a(t_2)) \text{or}((\mu_a(t_1) = \mu_a(t_2)) \wedge (\mu_b(t_1) > \mu_b(t_2))),$$

which is noted $< \mu_a(t_1), \mu_b(t_1) >\succ< \mu_a(t_2), \mu_b(t_2) >$.

We note $(< \mu_a(t_1), \mu_b(t_1) >\succeq< \mu_a(t_2), \mu_b(t_2) >)$ when $(< \mu_a(t_1), \mu_b(t_1) >\succ< \mu_a(t_2), \mu_b(t_2) >)$ or $(< \mu_a(t_1), \mu_b(t_1) >=< \mu_a(t_2), \mu_b(t_2) >)$.

Table 2 Example of extension of the relation R

#Journey	From	To	Duration	Departure	Mode	...
10	Paris	Brest	1h30	9h30 am	Plane	...
11	Paris	Brest	2h30	6h30 am	Plane	...
12	Paris	Brest	3h50	8h am	Car	...
13	Paris	Brest	3h10	9h15 am	Train	...
14	Paris	Brest	3h15	8h15 am	Train	...

Based on the lexicographical order, the conjunction (resp. disjunction) of bipolar conditions and the intersection (resp. union) of bipolar relations can be defined by the *lmin* (resp. *lmax*) operator [3, 9]. They are respectively defined as follows (where (μ, η) and (μ', η') are two pairs of satisfaction degrees with respect to fuzzy bipolar conditions):

$$lmin(<\mu, \eta>, <\mu', \eta'>) = <\mu, \eta>$$
$$if \, \mu < \mu' \vee (\mu = \mu' \wedge \eta < \eta'),$$
$$= <\mu', \eta'> \text{ otherwise.}$$

$$lmax(<\mu, \eta>, <\mu', \eta'>) = <\mu, \eta>$$
$$if \, \mu > \mu' \vee (\mu = \mu' \wedge \eta > \eta'),$$
$$= <\mu', \eta'> \text{ otherwise.}$$

The *lmin* (resp. *lmax*) operator is commutative, associative, idempotent and monotonic. The pair of grades $< 1, 1 >$ is the neutral (resp. absorbing) element of the operator *lmin* (resp. *lmax*) and the pair $< 0, 0 >$ is the absorbing (resp. neutral) element of the operator *lmin* (resp. *lmax*).

Example 3. Let R be a relational table about *journeys* from Paris to Brest (see Table 2), derived from a transport information system. A user can express a bipolar query as in the following query:

"Find journeys from Paris to Brest which are *(fast, and if possible very fast)* or journeys which have *(an early departure, or else a morning departure)*".

To evaluate this query, we define the following fuzzy predicates and their membership functions respectively:

- *fast*: is a fuzzy predicate defined by the following membership function:
  $if \, d \in [0, 2], \mu_{Fast}(d) = 1; \, if \, d \in [2, 5], \mu_{Fast}(d) = \frac{-d}{3} + \frac{5}{3};$ otherwise $\mu_{Fast}(d) = 0,$

  where d is the journey time expressed in hours. It defines a fuzzy relation of fast journeys, where:

$$\mu_{Fast\,Journey}(t) = \mu_{Fast}(t.duration).$$

We define a fuzzy predicate *very fast* by:

$$\mu_{Very\_Fast}(d) = (\mu_{Fast}(d))^2.$$

- *early*: is a fuzzy predicate defined by the following membership function:
  $if\ d \in [5h\,am, 7h\,am],\ \mu_{Early}(d) = 1;\ if\ d \in [7h\,am, 10h\,am],\ \mu_{Early}(d) = \frac{-d}{3} + \frac{10}{3};$ otherwise, $\mu_{Early}(d) = 0,$

where d is the journey departure. It defines a fuzzy relation of early departure journeys, where:

$$\mu_{Early\,Journey}(t) = \mu_{Early}(t.departure).$$

- *morning*: is a fuzzy predicate defined by the following membership function:
  $if\ d \in [5h\,am, 9h\,am],\ \mu_{Morning}(d) = 1;\ if\ d \in [9h\,am, 12h],\ \mu_{Morning}(d) = \frac{-d}{3} + 4;$ otherwise, $\mu_{Morning}(d) = 0,$

where d is the journey departure. It defines a fuzzy relation of morning departure journeys, where:

$$\mu_{Morning\,Journey}(t) = \mu_{Morning}(t.departure).$$

The above fuzzy predicates allow us to define from the relation R the following bipolar relations $Journey_{(fast, VeryFast)}$ and $Journey_{[Early, Morning]}$ (see Table 3).

	Fuzzy bipolar relation $Journey_{(Fast,VeryFast)}$	
#Journey	$\mu_{Fast}(\#Journey)$	$\mu_{VeryFast}(\#Journey)$
10	1	1
11	0.8	0.64
12	0.4	0.16
13	0.6	0.36
14	0.58	0.33
	Fuzzy bipolar relation $Journey_{[Early,Morning]}$	
#Journey	$\mu_{Early}(\#Journey)$	$\mu_{Morning}(\#Journey)$
10	0.33	0.83
11	1	1
12	0.66	1
13	0.25	0.91
14	0.58	1

Table 3 The obtained fuzzy bipolar relations

The set of answers to the query is obtained by using the *lmax* operator for each tuple of both fuzzy bipolar relations as follows:

- #journey = 10: $lmax((1, 1), [0.33, 0.83]) = (1, 1)$,
- #journey = 11: $lmax((0.8, 0.64), [1, 1]) = [1, 1]$,
- #journey = 12: $lmax((0.4, 0.16), [0.66, 1]) = [0.66, 1]$,
- #journey = 13: $lmax((0.6, 0.36), [0.25, 0.9]) = (0.6, 0.36)$,
- #journey = 14: $lmax((0.58, 0.33), [0.58, 1]) = [0.58, 1]$,

We notice that journeys are retrieved depending on the degree of satisfaction to the most important fuzzy condition: *fast* for the fuzzy bipolar condition (*Fast*, *VeryFast*) and *early* for the fuzzy bipolar condition [*Early*, *Morning*]. Indeed, in the case of tuples #10 and #11, the maximal couple of degrees is attached to answers because they fully satisfy the requirement of the query. In the case of tuples #12 and #13, the returned couple of degrees corresponds to the couple in which the satisfaction with respect to the most important fuzzy condition is the highest. In the case of tuple #14, the satisfaction is the same with regard to the most important fuzzy condition in both fuzzy bipolar conditions; therefore, the least important condition is used to determine which couple of degrees to attach to the resulting tuple. •

3.2 The Negation Operator

To be consistent, any negation operator of a bipolar condition $< a, b >$ must verify the following properties [2]:

Property 1 (Order reversing). The negation operator must deliver a reverse order:

$$< a, b >\succ< c, d > \Leftrightarrow \neg < c, d >\succ \neg < a, b > .$$

Property 2 (Consistency). The negation of a bipolar condition must also be a bipolar condition.

Property 3 (Involutivity). The negation operator must be an involutive operator:

$$\neg(\neg < a, b >) =< a, b > .$$

In the next subsection, we show that the negation of the fuzzy bipolar condition (c, w) is the fuzzy bipolar condition $[\neg c, \neg w]$ and, reciprocally, the negation of the fuzzy bipolar condition $[w, c]$ is the fuzzy bipolar condition $(\neg w, \neg c)$. Then, a negation operator is proposed for the general case.

3.2.1 Negation of (c, w)

We recall that a bipolar condition (c, w) is defined by the following properties:

Table 4 Example of tuples with their couple of grades for the fuzzy bipolar condition $(Young, Young \wedge WellPaid)$

#Employee	$\mu_{Young}(\#Employee)$	$\mu_{Young}(\#Employee) \wedge \mu_{WellPaid}(\#Employee)$
10	1	1
11	0.75	0.56
12	0.6	0.36
13	0.5	0.25

1. $\neg c$ corresponds to the rejection (c denotes acceptable elements),
2. w corresponds to the optimal values,
3. The acceptability condition c is more important than the optimality condition w.

In addition, the set of optimal values is included in the set of acceptable values ($w \subseteq c$).

If we express these properties in the context of a negation $\neg(c, w)$, we obtain:

1. $\neg c$ corresponds to the optimal values (c contains rejected and acceptable elements),
2. w corresponds to the rejected values ($\neg w$ denotes acceptable elements),
3. The optimality condition $\neg c$ is more important than the acceptability condition $\neg w$.

In addition, since $w \subseteq c$, we have $\neg c \subseteq \neg w$.

Therefore, we obtain a fuzzy bipolar condition which fits within the definition of the "or else" fuzzy bipolar conditions and we denote it by $[\neg c, \neg w]$. That means:

$$\neg(c, w) = [\neg c, \neg w].$$

It is important to notice that in (c, w), the importance is put on c and the concept attached to c is still the most important for $\neg(c, w)$. As example, when considering "*young and if possible young and well-paid* employees", the importance is put on the *age* (young). In its negation, the *age* is still the most important aspect to consider. **Example 4.** Let $(Young, Young \wedge WellPaid)$ be a fuzzy bipolar condition of type "and if possible" which defines *young* and if possible *young and well-paid* employees. Let Table 4 be a set of returned tuples from a relational table. The negation of the fuzzy bipolar condition $(Young, Young \wedge WellPaid)$ is defined as *[notYoung,not(Young $\wedge$ WellPaid)]*, which corresponds to "employees which are *not young*, or else are not (*young* and *well-paid*)". We notice that concepts used to define the fuzzy bipolar condition $(Young, Young \wedge WellPaid)$ are the same ones used to express its negation *[notYoung,not(Young $\wedge$ WellPaid)]*. More precisely, the concept *age* is used to define the most important fuzzy condition in both fuzzy bipolar condition $(Young, Young \wedge WellPaid)$ and $[notYoung, not(Young \wedge WellPaid)]$.

Table 5 Obtained tuples with their couple of degrees for the fuzzy bipolar condition $[not\,Young, not\,(Young \wedge Well\,Paid)]$

#Employee	$\mu_{not\,Young}(\#Employee)$	$\mu_{not\,(Young\ and\ WellPaid)}(\#Employee)$
13	0.5	0.75
12	0.4	0.64
11	0.25	0.44

Table 5 shows obtained couples of degrees for the fuzzy bipolar condition $[not\,Young, not\,(Young \wedge Well\,Paid)]$. The tuple #10 is completely discarded and the order reversing can be checked. •

3.2.2 Negation of $[w, c]$

We recall that a fuzzy bipolar condition $[w, c]$ is defined by the following properties:

- $\neg c$ corresponds to the rejection (c denotes the acceptable elements),
- w corresponds to the optimal values,
- The optimality condition w is more important than the acceptability.

In addition, the set of optimal values is included in the set of acceptable values ($w \subseteq c$).

If we express these properties in the context of a negation, we obtain:

- $\neg c$ corresponds to the optimal values,
- w corresponds to the rejected values ($\neg w$ denotes the acceptable elements),
- The acceptability condition $\neg w$ is the more important condition.

In addition, since $w \subseteq c$, we have $\neg c \subseteq \neg w$. This property states that, in the context of negation of a fuzzy bipolar condition of the form $[w, c]$, the set of optimal values is included in the set of acceptable values.

Therefore, we obtain a fuzzy bipolar condition, which fits within the definition of the "and if possible" fuzzy bipolar conditions and we denote it by $(\neg w, \neg c)$. That means:

$$\neg[w, c] = (\neg w, \neg c).$$

As for bipolar conditions of type "and if possible", we notice that in $[w, c]$, the importance is put on w and the concept attached to w is still the most important for $\neg[w, c]$.

3.2.3 The Negation of the General Form

From the previous two subsections, we get:

$$\neg < a, b >=< \neg a, \neg b > .$$

This form of negation is involutive and reverses the lexicographical order.

Proof. *Reversing of the lexicographical order.*

Let x_1, x_2, y_1 and y_2 be four values from $[0, 1]$. We have to prove that :
$$< x_1, y_1 >\succ< x_2, y_2 >\Leftrightarrow< 1 - x_2, 1 - y_2 >\succ< 1 - x_1, 1 - y_1 > .$$
$$< x_1, y_1 >\succ< x_2, y_2 >$$
$$\Leftrightarrow (x_1 > x_2) \text{ or } ((x_1 = x_2) \text{ and } (y_1 > y_2))$$
$$\Leftrightarrow (1 - x_1 < 1 - x_2) \text{ or } ((x_1 = x_2) \text{ and } (1 - y_1 < 1 - y_2))$$
$$\Leftrightarrow < 1 - x_2, 1 - y_2 >\succ< 1 - x_1, 1 - y_1 >$$
$$\Leftrightarrow \neg < x_2, y_2 >\succ \neg < x_1, y_1 >$$

Endproof.

Proof. *Involutivity.*
$$\neg(\neg < a, b >) = \neg < \neg a, \neg b >=< \neg\neg a, \neg\neg b >=< a, b >.$$

Endproof.

3.3 The Satisfaction of De Morgan's Laws

In the context of fuzzy bipolar conditions, the two De Morgan's laws holds (where $< a, b >$ and $< a', b' >$ are two satisfactions with respect to fuzzy bipolar conditions):

$$lexmax(\neg < a, b >, \neg < a', b' >) = \neg lexmin(< a, b >, < a', b' >),$$
$$lexmin(\neg < a, b >, \neg < a', b' >) = \neg lexmax(< a, b >, < a', b' >).$$

Proof. It is assumed that $< a, b >\succ< a', b' >$ (the proof is obvious in the case of the equality).

If $< a, b >\succ< a', b' >$ and since the negation reverses the order we get the values for the left parts of the equalities:

$$lexmax(\neg < a, b >, \neg < a', b' >) = \neg < a', b' >,$$
$$lexmin(\neg < a, b >, \neg < a', b' >) = \neg < a, b > .$$

And it is obvious that same results are obtained for the right parts:

$$\neg lexmin(< a, b >, < a', b' >) = \neg < a', b' >,$$
$$\neg lexmax(< a, b >, < a', b' >) = \neg < a, b > .$$

Endproof.

4 The Extension of Algebraic Operators

The extended algebraic operators to bipolarity apply on regular relations or on relations issued from a querying with bipolar conditions. As a consequence, the concept of fuzzy bipolar relation is introduced in Sect. 4.1. As pointed out, they are a generalization of regular and fuzzy relations. Then, the extended algebraic operators (the complement, the intersection, the union, the Cartesian product, the selection, the joint operator, the difference and the projection) are introduced in Sect. 4.2.

4.1 Fuzzy Bipolar Relations and Bipolar Queries

A bipolar query is a query expressing at least one fuzzy bipolar condition. As a consequence, the satisfaction with respect to a fuzzy bipolar query is a pair of degrees.

We recall that a fuzzy bipolar relation R is defined as a relation where each tuple t is attached with a pair of degrees denoted $< \mu_{R1}(t), \mu_{R2}(t) >$ reflecting its satisfaction with respect to a bipolar query. As a consequence, for a given tuple t of R:

- $< \mu_{R1}(t), \mu_{R2}(t) >$ with $\mu_{R1}(t) < \mu_{R2}(t)$ represents a satisfaction of type "or else",
- $< \mu_{R1}(t), \mu_{R2}(t) >$ with $\mu_{R1}(t) = \mu_{R2}(t)$ represents a satisfaction with respect to a fuzzy condition,
- $< \mu_{R1}(t), \mu_{R2}(t) >$ with $\mu_{R1}(t) > \mu_{R2}(t)$ represents a satisfaction of type "and if possible".

A fuzzy bipolar relation is denoted $R = \{< \mu_{R1}(t), \mu_{R2}(t) > /t\}$. Obviously, regular relations (initial relations) can be expressed with couple $< 1, 1 >$ to state a maximum satisfaction.

4.2 Extended Algebraic Operators

The complement. The complement of a fuzzy bipolar relation $R = \{< \mu_{R1}(t), \mu_{R2}(t) > /t\}$ is a fuzzy bipolar relation $\neg R$ computed using the negation operator. We get:

$$\neg R = \{< 1 - \mu_{R1}(t), 1 - \mu_{R2}(t) > /t\}.$$

It is important to notice that tuples t such that $< \mu_{R1}(t), \mu_{R2}(t) >=< 0, 0 >$ should be considered to apply this definition.

The intersection of fuzzy bipolar relations. The intersection of two fuzzy bipolar relations $R = \{< \mu_{R1}(t), \mu_{R2}(t) > /t\}$ and $S = \{< \mu_{S1}(t'), \mu_{S2}(t') > /t'\}$ is a fuzzy bipolar relation computed with the *lmin* operator:

$$R \cap S = \{lmin(< \mu_{R1}(t''), \mu_{R2}(t'') >, < \mu_{S1}(t''), \mu_{S2}(t'') >)/t''$$
$$|< \mu_{R1}(t''), \mu_{R2}(t'') > /t'' \in R \wedge < \mu_{S1}(t''), \mu_{S2}(t'') > /t'' \in S\}.$$

The union. The union of two fuzzy bipolar relations $R = \{< \mu_{R1}(t), \mu_{R2}(t) > /t\}$ and $S = \{< \mu_{S1}(t'), \mu_{S2}(t') > /t'\}$ is a fuzzy bipolar relation computed with the *lmax* operator:

$$R \cup S = \{lmax(< \mu_{R1}(t''), \mu_{R2}(t'') >, < \mu_{S1}(t''), \mu_{S2}(t'') >)/t''$$
$$|< \mu_{R1}(t''), \mu_{R2}(t'') > /t'' \in R \vee < \mu_{S1}(t''), \mu_{S2}(t'') > /t'' \in S\}.$$

The Cartesian product. The Cartesian product of two fuzzy bipolar relations $R = \{< \mu_{R1}(t), \mu_{R2}(t) > /t\}$ and $S = \{< \mu_{S1}(t'), \mu_{S2}(t') > /t'\}$ is a fuzzy bipolar relation computed with the *lmin* operator:

$$R \times S = \{lmin(< \mu_{R1}(t), \mu_{R2}(t) >, < \mu_{S1}(t'), \mu_{S2}(t') >)/t \oplus t'$$
$$|< \mu_{R1}(t), \mu_{R2}(t) > /t \in R \wedge < \mu_{S1}(t'), \mu_{S2}(t') > /t' \in S\}$$

where $\oplus$ denotes the concatenation of tuples.

The selection. The selection of a fuzzy bipolar relation $R = \{< \mu_{R1}(t), \mu_{R2}(t) > /t\}$ by a fuzzy bipolar predicate $\phi =< \mu, \mu' >$ is a fuzzy bipolar relation defined by:

$$R : \phi = \{lmin(< \mu_{R1}(t), \mu_{R2}(t) >, < \mu(t), \mu'(t) >)/t$$
$$|< \mu_{R1}(t), \mu_{R2}(t) > /t \in R\}.$$

The join operator. The θ-join of two fuzzy bipolar relations $R = \{< \mu_{R1}(t), \mu_{R2}(t) > /t\}$ and $S = \{< \mu_{S1}(t'), \mu_{S2}(t') > /t'\}$ is defined as a selection on a Cartesian product.

The comparator θ applies on attribute A of R and B of S. It can be regular ($\in \{<, >, \leq, \geq, =, \neq\}$) or fuzzy (*around, much greater than, ...*). In both cases, the satisfaction with respect to θ can be represented in a bipolar way ($< \theta(R.A, S.B), \theta(R.A, S.B) >$). The θ-join of R and S is defined by:

$$\bowtie (R, S, \theta, A, B) = \{lmin(< \mu_{R1}(t), \mu_{R2}(t) >, < \mu_{S1}(t'), \mu_{S2}(t') >,$$
$$< \theta(R.A, S.B), \theta(R.A, S.B) >)/t \oplus t'$$
$$|< \mu_{R1}(t), \mu_{R2}(t) > /t \in R \wedge < \mu_{S1}(t'), \mu_{S2}(t') > /t' \in S\},$$

where $\oplus$ denotes the concatenation of tuples and the *lmin* operator is extended to three arguments ($lmin(a, b, c) = lmin(a, lmin(b, c))$).

The difference. The difference of two fuzzy bipolar relations $R = \{< \mu_{R1}(t), \mu_{R2}(t) > /t\}$ and $S = \{< \mu_{S1}(t), \mu_{S2}(t) > /t\}$ is a fuzzy bipolar relation which can be computed with the previously extended operators:

$$R - S = R \cap \neg S.$$

We get:

$$R - S = \{lmin(< \mu_{R1}(t''), \mu_{R2}(t'') >, < 1 - \mu_{S1}(t''), 1 - \mu_{S2}(t'') >)/t''$$
$$|< \mu_{R1}(t''), \mu_{R2}(t'') > /t'' \in R \wedge < \mu_{S1}(t''), \mu_{S2}(t'') > /t'' \in S\}.$$

The projection. The projection of a fuzzy bipolar relation $R = \{< \mu_{R1}(t), \mu_{R2}(t) > /t\}$ on the set of attributes ATT from R is defined by:

$$R[ATT] = \{< \mu_{R[ATT]1}(att), \mu_{R[ATT]2}(att) > /att$$
$$|< \mu_{R[ATT]1}(att), \mu_{R[ATT]2}(att) >$$
$$= lmax_{t \in R, t[ATT] = att} < \mu_{R1}(t), \mu_{R2}(t) >\}.$$

For a given att, we keep the highest value $< \mu_{R1}(t), \mu_{R2}(t) >$ among tuples t from R such that $t[ATT] = att$.

5 Linguistic Quantifiers Extended to Fuzzy Bipolar Conditions

Linguistic quantifiers [16] describe an intermediate attitude between the universal quantifier $\forall$ and the existential quantifier $\exists$ and they correspond to linguistic expressions as *almost all, around 4, few* etc. They provide an interesting trade-off between the conjunction and the disjunction and they are used to build complex fuzzy conditions called quantified statements. As a consequence, the extension of linguistic quantifiers to bipolar conditions provides an aggregation of such conditions sets between the *lexmin* and the *lexmax* operators. Section 5.1 is a recall about linguistic quantifiers and quantified statements while Sect. 5.2 introduces our propositions to apply linguistic quantifiers to fuzzy bipolar conditions.

5.1 Linguistic Quantifiers

A linguistic quantifier [16] can be relative (it refers to a proportion, as in *around the half*) or absolute (it refers to a number, as in *about 2*). A relative (resp. absolute) linguistic quantifier Q is defined by a fuzzy set with a membership function μ_Q from [0, 1] to [0, 1] (resp. from the set of real numbers to [0, 1]). The value $\mu_Q(x)$

Fig. 1 The linguistic quantifier *most of*

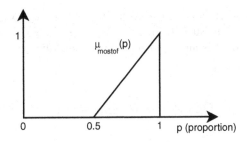

expresses the extent to which proportion x (resp. the cardinality x) is in agreement
with the quantifier. As a consequence, linguistic quantifiers can be considered as
fuzzy conditions which are defined on cardinalities or proportions.

Example 5. The relative linguistic quantifier *most of* is given by Fig. 1. According to
this linguistic quantifier, a proportion less than 50 % cannot be considered in agreement with *most of* (since $\mu_{mostof}(p)$ is 0 for $p \leq 0.5$). In other words, when less than
50% of the elements from a given referential X satisfy a property A, the statement
"*most of* elements in X satisfy A" is entirely false. For a proportion between 50 and
100%, the closer to 100 % the proportion, the more it is in agreement with *most of*.
As example, when 75 % of a given referential X satisfy a property A, the statement
"*most of* the elements in X satisfy A" is true at degree 0.5 (since $\mu_{mostof}(0.75) =
0.5$). •

It is also possible to distinguish increasing quantifiers from decreasing ones. The
representation of an increasing linguistic quantifier Q satisfies: (1) $\mu_Q(0) = 0$, (2)
$\exists k$ such that $\mu_Q(k) = 1$, (3) $\forall a, b$ if $a > b$ then $\mu_Q(a) \geq \mu_Q(b)$. Figure 1 provides
an example of an increasing relative quantifier.

A decreasing linguistic quantifier Q is defined by: (1) $\mu_Q(0) = 1$, (2) $\exists k$ such that
$\mu_Q(k) = 0$, (3) $\forall a, b$ if $a > b$ then $\mu_Q(a) \leq \mu_Q(b)$.

Two types of quantified statements can be distinguished, quantified statements of
type "Q X are A" and of type "Q B X are A" where Q is a linguistic quantifier, A
and B are two gradual predicates defined by fuzzy sets. In this article, we restrict
ourselves to the first type since it is the most useful.

A quantified statement of the type "Q X are A" means that, among the elements
of set X, a quantity Q satisfies the fuzzy predicate A. Such a statement can be more
or less true and many approaches can be used to compute the degree of truth of such
a quantified statement (i.e. to interpret the quantified statement). Examples of interpretation are provided by the use of OWA operators [14] or Yager's decomposition
based approach [13] (among others). The one we retain is the decomposition-based
approach because it relies only on operators already extended to bipolarity. This
approach considers an increasing quantifier Q and the truth value for "Q X are A"
is given by the following expression (where Q is an absolute quantifier):

$$max_{i=1,...,n}min(\mu_A(x_{\sigma(i)}), \mu_Q(i))$$

where n is the cardinality of set $X = \{x_1, x_2, \ldots, x_n\}$ and σ a permutation of $\{1, 2, \ldots, n\}$ such that $\mu_A(x_{\sigma(i)}) \geq \mu_A(x_{\sigma(i+1)})$, for i in $\{1, 2, \ldots, n-1\}$.
If Q is an increasing relative quantifier, it is necessary to replace each $\mu_Q(i)$ by $\mu_Q(\frac{i}{n})$.

5.2 Quantified Statements and Bipolarity

We consider three types of quantified statements involving bipolarity (bipolar quantified statements).

In the type 1, the linguistic quantifier is fuzzy while the predicate is bipolar. Such statements are denoted "Q X are $< a, b >$" where Q is a quantifier, X a set of elements and $< a, b >$ a fuzzy bipolar condition (either of type "and if possible" or of type "or else"). The statement "Q X are $< a, b >$" means that, among the elements of set X, a quantity Q satisfy the bipolar condition $< a, b >$. An example is provided by the following condition "*at least half* X are (*young* and if possible *young* and *well paid*)" where X is a set of employees. This statement means that among the employees in X, a quantity of *at least half* are (*young* and if possible *young* and *well paid*). This example uses an "and if possible" condition but a "or else" bipolar condition can also be considered.

The second type (type 2) concerns statements where the linguistic quantifier is bipolar while the predicate is a fuzzy predicate as in "(*at least half* and if possible *most of*) X are *well-paid*", X being a set made of employees. These statements can be expressed "$< Q_1, Q_2 >$ X are A" and the linguistic quantifier is bipolar since it is a bipolar condition defined on a cardinality or a proportion. When the linguistic quantifier is a and-if-possible bipolar condition, we should have $(Q_2 \subseteq Q_1)$ as in "*at least half* and if possible *most of*". When it is an or-else bipolar condition, we should have $(Q_1 \subseteq Q_2)$ as in "*most of* or else *at least half*".

The last type (type 3) concerns statements where both the linguistic quantifier and the condition are bipolar. It is written "$< Q_1, Q_2 >$ X are $< a, b >$" and an example is provided by "(*at least half* and if possible *most of*) X are (*young* and if possible *young* and *well paid*)".

Section 5.2.1 (resp. 5.2.2) is devoted to the evaluation of bipolar quantified propositions of the three types according to a decomposition-based approach (resp. a distribution-based approach). The decomposition-based approach is similar to Yager's decomposition based approach to interpret quantified statements, while the distribution-based approach leads to transform the quantified statement into a bipolar condition.

5.2.1 Fuzzy Bipolar Quantified Statements and the Decomposition-Based Approach

Let "Q X are $< a, b >$" be a bipolar quantified proposition such that Q is an increasing quantifier. Since Q is increasing, we propose an interpretation similar to the decomposition based approach of Yager [13]. More precisely, the interpretation of this proposition delivers a pair of grades of satisfaction (where Q is an absolute quantifier):

$$lmax_{i=1,...,n}lmin(< \mu_a(x_{\sigma(i)}), \mu_b(x_{\sigma(i)}) >, < \mu_Q(i), \mu_Q(i) >)$$

where n is the cardinality of set $X = \{x_1, x_2, \ldots, x_n\}$ and σ a permutation of $\{1, 2, \ldots, n\}$ such that $< \mu_a(x_{\sigma(i)}), \mu_b(x_{\sigma(i)}) > \geq < \mu_a(x_{\sigma(i+1)}), \mu_b(x_{\sigma(i+1)}) >$, for i in $\{1, 2, \ldots, n - 1\}$. A proof of the validity of this expression is given hereafter. If Q is an increasing relative quantifier, it is necessary to replace each $\mu_Q(i)$ by $\mu_Q(\frac{i}{n})$.

Proof. *Validity of this interpretation.*

Since Q is an increasing quantifier, the evaluation of the quantified proposition can be based on the evaluation of the best subset E of X which satisfies *at most* the following two conditions:

1. each element in E satisfies the bipolar condition $< a, b >$,
2. the cardinality of E satisfies the linguistic quantifier Q.

Let $E \subseteq X$, the satisfaction with respect to the first condition is:

$$lmin_{x \in E} < \mu_a(x), \mu_b(x) >,$$

and the satisfaction with respect to the second condition is:

$$\mu_Q(|E|).$$

This latter is rewritten $< \mu_Q(|E|), \mu_Q(|E|) >$ in the context of bipolarity. The satisfaction of the set E to both conditions is then:

$$lmin(lmin_{x \in E} < \mu_a(x), \mu_b(x) >, \mu_Q(|E|)).$$

The evaluation of the bipolar quantified proposition is the best satisfaction with respect to the different sets E:

$$\delta = lmax_{E \subseteq X}lmin(lmin_{x \in E} < \mu_a(x), \mu_b(x) >, \mu_Q(|E|)).$$

It is possible to simplify this expression. First it is necessary to distinguish the different cardinalities $i \in \{1, 2, \ldots, n\}$ and to denote E_i the different subsets of X having the same cardinality i. We get:

$$\delta = lmax_{i=1,2,...,n}(lmax_{E_i \subseteq X}lmin(lmin_{x \in E_i} < \mu_a(x), \mu_b(x) >, \mu_Q(i))).$$

Obviously, for a given i, the maximal value $lmin_{x \in E_i} < \mu_a(x), \mu_b(x) >$ among sets E_i is obtained for the set $\{x_{\sigma(1)}, x_{\sigma(2)}, \ldots, x_{\sigma(i)}\}$ when denoting $X = \{x_1, x_2, \ldots, x_n\}$

and σ a permutation of $\{1, 2, \ldots, n\}$ such that $< \mu_a(x_{\sigma(i)}), \mu_b(x_{\sigma(i)}) > \succeq$ $< \mu_a(x_{\sigma(i+1)}), \mu_b(x_{\sigma(i+1)}) >$, for i in $\{1, 2, \ldots, n-1\}$. As a consequence, we obtain the final result:

$$lmax_{i=1,\ldots,n}lmin(< \mu_a(x_{\sigma(i)}), \mu_b(x_{\sigma(i)}) >, < \mu_Q(i), \mu_Q(i) >)$$

where $X = \{x_1, x_2, \ldots, x_n\}$ and σ a permutation of $\{1, 2, \ldots, n\}$ such that $< \mu_a(x_{\sigma(i)}), \mu_b(x_{\sigma(i)}) > \succeq < \mu_a(x_{\sigma(i+1)}), \mu_b(x_{\sigma(i+1)}) >$, for i in $\{1, 2, \ldots, n-1\}$.
Endproof.

Example 6. We consider the linguistic quantifier *most of* of Fig. 1 and the following bipolar query:
find journeys where *most of* their steps are situated in a "*cultural* and if possible *big*" city with a "*comfortable* or else a *cheap*" hotel.
Each journey is an answer to the query with a satisfaction expressed by the truth value of the following fuzzy bipolar quantified statement:

$$\text{"most of } X \text{ are } < a, b >\text{"},$$

where X is the set made of steps of the journey, $< a, b >$ being the conjunction of the two bipolar conditions "situated in a *cultural* and if possible *big*" city and with a "*comfortable* or else a *cheap*" hotel.
We consider a journey with 5 steps: $X=\{x_1, x_2, x_3, x_4, x_5\}$ such that:

$$< \mu_a(x_1), \mu_b(x_1) >=< 1, 1 >, < \mu_a(x_2), \mu_b(x_2) >$$
$$=< 1, 0.9 >, < \mu_a(x_3), \mu_b(x_3) >=< 0.9, 1 >, < \mu_a(x_4), \mu_b(x_4) >$$
$$=< 0.9, 0.6 >, < \mu_a(x_5), \mu_b(x_5) >=< 0.7, 0.6 > .$$

Satisfactions with respect to $< a, b >$ are already ordered (to simplify the presentation) and the satisfaction of the bipolar quantified statement "*most of* X are $< a, b >$" is:

$$lmax$$
$$lmin(< \mu_a(x_1), \mu_b(x_1) >, < \mu_{mostof}(1/5), \mu_{mostof}(1/5) >)$$
$$lmin(< \mu_a(x_2), \mu_b(x_2) >, < \mu_{mostof}(2/5), \mu_{mostof}(2/5) >)$$
$$lmin(< \mu_a(x_3), \mu_b(x_3) >, < \mu_{mostof}(3/5), \mu_{mostof}(3/5) >)$$
$$lmin(< \mu_a(x_4), \mu_b(x_4) >, < \mu_{mostof}(4/5), \mu_{mostof}(4/5) >)$$
$$lmin(< \mu_a(x_5), \mu_b(x_5) >, < \mu_{mostof}(1), \mu_{mostof}(1) >).$$

We get:

$lmax$

$lmin(< 1, 1 >, < 0, 0 >), lmin(< 1, 0.9 >, < 0, 0 >),$

$lmin(< 0.9, 1 >, < 0.2, 0.2 >), lmin(< 0.9, 0.6 >, < 0.6, 0.6 >),$

$lmin(< 0.7, 0.6 >, < 1, 1 >)$

$=< 0.7, 0.6 >.$

This journey satisfies the query at a rather high level ($< 0.7, 0.6 >$). It can be checked that 4 (out of 5) steps highly satisfy the constraint $< a, b >$ and the result is close to $\mu_{mostof}(4/5) = 0.6$ ($< 0.6, 0.6 >$). •

A similar approach can be used to interpret fuzzy bipolar quantified statement of the type "$< Q_1, Q_2 > X$ are A" (where Q_1 and Q_2 are both increasing). The truth value of such a statement is then:

$$lmax_{i=1,...,n}lmin(< \mu_A(x_{\sigma(i)}), \mu_A(x_{\sigma(i)}) >, < \mu_{Q_1}(i), \mu_{Q_2}(i) >)$$

where n is the cardinality of set $X = \{x_1, x_2, \ldots, x_n\}$ and σ a permutation of $\{1, 2, \ldots, n\}$ such that $< \mu_a(x_{\sigma(i)}), \mu_b(x_{\sigma(i)}) > \geq < \mu_a(x_{\sigma(i+1)}), \mu_b(x_{\sigma(i+1)}) >$, for i in $\{1, 2, \ldots, n - 1\}$.

If Q_1 and Q_2 are increasing relative quantifiers, it is necessary to replace each $\mu_{Q_1}(i)$ by $\mu_{Q_1}(\frac{i}{n})$ and each $\mu_{Q_2}(i)$ by $\mu_{Q_2}(\frac{i}{n})$. Obviously, the proof of the validity of this interpretation is similar to the previous one.

Example 7. We consider the linguistic quantifier *most of* of Fig. 1 and the increasing quantifier *at least half* such that $\mu_{atleasthalf}(1/4) = 0.5$ and $\mu_{atleasthalf}(2/4) = \mu_{atleasthalf}(3/4) = \mu_{atleasthalf}(1/4) = 1$. They can be used to express a bipolar query such that:

find journeys where "*at least half* and if possible *most of*" their steps have a *comfortable* hotel.

Each journey is an answer to the query with a satisfaction expressed by the truth value of the following fuzzy bipolar quantified statement:

"*at least half* and if possible *most of* X are A",

where X is the set made of the steps of the journey and A the fuzzy predicate "with a *comfortable* hotel". We consider a journey with four steps: X=$\{x_1, x_2, x_3, x_4\}$ such that:

$$\mu_A(x_1) = 1, \quad \mu_A(x_2) = 0.8, \quad \mu_A(x_3) = 0.3, \quad \mu_A(x_4) = 0.1.$$

Satisfactions with respect to A are already ordered (to simplify the presentation) and the satisfaction of the bipolar quantified statement "*at least half* and if possible *most of* X are A" is:

$lmax$

$lmin(< \mu_A(x_1), \mu_A(x_1) >, < \mu_{atleasthalf}(1/4), \mu_{mostof}(1/4) >)$

$lmin(< \mu_A(x_2), \mu_A(x_2) >, < \mu_{atleasthalf}(2/4), \mu_{mostof}(2/4) >)$

$$lmin(< \mu_A(x_3), \mu_A(x_3) >, < \mu_{atleasthalf}(3/4), \mu_{mostof}(3/4) >)$$
$$lmin(< \mu_A(x_4), \mu_A(x_4) >, < \mu_{atleasthalf}(1), \mu_{mostof}(1) >).$$

We get:

$$lmax$$
$$lmin(< 1, 1 >, < 0.5, 0 >), lmin(< 0.8, 0.8 >, < 1, 0 >),$$
$$lmin(< 0.3, 0.3 >, < 1, 0.5 >), lmin(< 0.1, 0.1 >, < 1, 1 >)$$
$$=< 0.8, 0.8 > .$$

This journey satisfies the query at a rather high level ($< 0.8, 0.8 >$). It can be checked that the statement "*at least half* X are *A*" is intuitively rather true while "*most of* X are *A*" is intuitively rather false. As a consequence, we obtain a result ($< 0.8, 0.8 >$) for the statement "*at least half* and if possible *most of* X are *A*" which is rather close to $< 1, 0 >$ (when the lexicographical order is considered).•

A similar approach can be used to interpret a fuzzy bipolar quantified statement of type "$< Q_1, Q_2 >$ X are $< a, b >$" (where Q_1 and Q_2 are both increasing). The truth value of such a statement is then:

$$lmax_{i=1,...,n} lmin(< \mu_a(x_{\sigma(i)}), \mu_b(x_{\sigma(i)}) >, < \mu_{Q_1}(i), \mu_{Q_2}(i) >)$$

where n is the cardinality of set $X = \{x_1, x_2, \ldots, x_n\}$ and σ a permutation of $\{1, 2, \ldots, n\}$ such that $< \mu_a(x_{\sigma(i)}), \mu_b(x_{\sigma(i)}) > \succeq < \mu_a(x_{\sigma(i+1)}), \mu_b(x_{\sigma(i+1)}) >$, for i in $\{1, 2, \ldots, n - 1\}$.
If Q_1 and Q_2 are increasing relative quantifiers, it is necessary to replace each $\mu_{Q_1}(i)$ by $\mu_{Q_1}(\frac{i}{n})$ and each $\mu_{Q_2}(i)$ by $\mu_{Q_2}(\frac{i}{n})$. Obviously, the proof of the validity of this interpretation is similar to the previous one.

5.2.2 Fuzzy Bipolar Quantified Statements and the Distribution-Based Approach

Let "Q X are $< a, b >$" be a bipolar quantified proposition such that Q is an increasing quantifier. The idea of the distribution-based approach is to evaluate the bipolar quantified statement "Q X are $< a, b >$" as the condition $<$"Q X are a","Q X are b"$>$. As example, the bipolar condition "*at least half* X are (*young* and if possible *young* and *well paid*)" (where X is a set of employees) is evaluated by "(*at least half* X are *young*) and if possible (*at least half* X are *young* and *well paid*)".

More precisely, "Q X are $< a, b >$" is evaluated by the bipolar satisfaction:

$$< \delta, \gamma >,$$

where δ and γ are respectively the evaluation of "Q X are a" and "Q X are b" (both computed by the same approach, and we propose either an OWA operator [14] or Yager's decomposition-based approach [13]).

It is easy to see that the obtained result is coherent (in the sense of the inclusion of the wished values in the constrained values) since the linguistic quantifier Q is increasing and since an OWA operator or Yager's decomposition-based approach is used to compute δ and γ.

Similarly, the linguistic statement "$< Q_1, Q_2 >$ X are A" (where Q_1 and Q_2 are both increasing) is evaluated by the bipolar satisfaction:

$$< \delta', \gamma' >,$$

where δ' and γ' are respectively the evaluation of "Q_1 X are A" and "Q_2 X are A" (both computed by the same approach, and we propose either an OWA operator [14] or Yager's decomposition-based approach [13]). The bipolar linguistic quantifier $< Q_1, Q_2 >$ being coherent, it is easy to see that the obtained result is also coherent (in the sense of the inclusion of the wished values in the constrained values). This is due to the monotonicity property of the OWA operator and Yager's decomposition-based approach used to compute δ' and γ'.

The distribution-based approach to evaluate $< Q_1, Q_2 >$ X are $< a, b >$" leads to evaluate four quantified statements: (1) "Q_1 X are a", (2) "Q_1 X are b", (3) "Q_2 X are a", (4) "Q_2 X are b". How to get a bipolar evaluation $< \delta'', \gamma'' >$ from these four quantified statements is out of the scope of this chapter and is left for future work.

6 Conclusion

This chapter has proposed a generalization for fuzzy bipolar conditions of type "if possible" and "or else" based on a hierarchical interpretation of these two different types of fuzzy bipolar conditions. This generalization is an extension of fuzzy conditions and a conjunction (*lexmin*) and a disjunction (*lexmax*) of such generalized conditions have been proposed (these two operators are extensions of the triangular norm min and the triangular co-norm max). All the algebraic operators have been extended to this generalization and is has been shown that the definition of the complement (negation) satisfies the De Morgan's law. In addition three types of quantified statements involving bipolarity have been distinguished and two different types of evaluation have been proposed. As a consequence, we have at our disposal several aggregations of bipolar conditions which aim at delivering a result sets between the *lexmin* and the *lexmax* operator. Their comparison is planned as future work.

It is then possible to define a query language expressing preferences defined by bipolar conditions. This new approach to flexible querying can be considered as an extension of the approach based on fuzzy conditions. It presents the advantage to offer a more powerful flexible querying, still in the context of an algebraic language.

However, a particular case appears when the fuzzy bipolar condition is made of a fuzzy condition and a Boolean condition included in the core of this fuzzy condition (as in "*cheap* price and if possible a price <100" or "price <100 or else a *cheap* price" assuming that the interval $[0, 100[$ is included in the core of the fuzzy set "*cheap* price"). One may wonder if such a fuzzy bipolar condition is different from the fuzzy condition (is "*cheap* price and if possible a price <100" really different from "*cheap* price" ?). However, we think that these two conditions are clearly different since the fuzzy condition describes the best elements and its negation the rejected elements while it is not the case for the fuzzy bipolar condition. As an example, when considering the fuzzy condition "*cheap* price", the perfect elements are described by this fuzzy set, while when considering "*cheap* price and if possible a price <100" (or "price <100 or else a *cheap* price") the best elements are the prices under 100. As a consequence, the use of "*cheap* price" or "*cheap* price and if possible a price <100" (respectively "price <100 or else a *cheap* price") in a query does not lead to the same order of the answers. To illustrate this behaviour, we can consider the query: retrieve the trips having a "*cheap* price" and a "departure *in the morning* and if possible an *early* departure". The trip t_1 (respectively t_2) satisfies "*cheap* price" at degree 0.7 (respectively 0.8) while its satisfaction with respect to "a departure *in the morning* if possible an *early* departure" is $< 0.9, 0.5 >$ (respectively $< 0.2, 0.1 >$). As a consequence, trip t_1 has a satisfaction of $lmin(< 0.7, 0.7 >, < 0.9, 0.5 >) =< 0.7, 0.7 >$ and is preferred to trip t_2 whose satisfaction is $lmin(< 0.8, 0.8 >, < 0.2, 0.1 >) =< 0.2, 0.1 >$. When replacing "*cheap* price" by "price <100 or else a *cheap* price" in the previous query the order is inverted since the satisfaction of t_1 is $lmin(< 0, 0.7 >, < 0.9, 0.5 >) =< 0, 0.7 >$ while that of t_2 is $lmin(< 0, 0.8 >, < 0.2, 0.1 >) =< 0, 0.8 >$.

As future work, we aim to define an extension of the SQLf language to generalized fuzzy bipolar conditions (it has been already made for "and if possible" conditions [12]). Other approaches for the definition of fuzzy bipolar quantified propositions can also be studied.

References

1. Bosc, P., Pivert, O.: SQLf: A relational database langage for fuzzy querying. IEEE Trans. Fuzzy Syst. **3**(1), 1–17 (Feb 1995)
2. Bosc, P., Pivert, O.: A propos de la négation de conditions bipolaires floues. On the negation of fuzzy bipolar conditions. In: Rencontre francophone sur la logique floue et ses applications, pp. 21–28 (2010)
3. Bosc, P., Pivert, O., Liétard, L., Mokhtari, A.: Extending relational algebra to handle bipolarity. In: 25th ACM Symposium on Applied Computing, SAC'10, pp. 1717–1721 (2010)
4. de Tré, G., Zadrozny, S., Matthé, T., Kacprzyk, J., Bronselaer, A.: Dealing with positive and negative query criteria in fuzzy database quering bipolar satisfaction degrees. LNAI FQAS **5822**, 593–604 (2009)
5. Dubois, D., Prade, H.: Bipolarité dans un processus d'interrogation flexible. In: Actes des Rencontres Francophones sur la Logique Floue et ses Applications (LFA'02), pp. 127–134 (2002)

6. Dubois, D., Prade, H.: Bipolarity in flexible querying. LNAI **2522**, 174–182 (2002)
7. Dubois, D., Prade, H.: Handling bipolar queries in fuzzy information processing. In: Galindo, J. (ed.) Handbook of Research on Fuzzy Information Processing in Databases, pp. 97–114. Information Science Reference, Hershey (2008)
8. Dubois, D., Prade, H.: An introduction to bipolar representations of information and preference. Int. J. Intell. Syst. **23**, 866–877 (2008)
9. Liétard, L., Rocacher, D.: On the definition of extended norms and co-norms to aggregate fuzzy bipolar conditions. In: IFSA/EUSFLAT, pp. 513–518 (2009)
10. Liétard, L., Rocacher, D., Bosc, P.: On the extension of SQL to fuzzy bipolar conditions. In: The 28th North American Information Processing Society Annual Conference (NAFIPS'09) (2009)
11. Liétard, L., Tamani, N., Rocacher, N.: Fuzzy bipolar conditions of type "or else". In: The 20th IEEE International Conference on Fuzzy Systems (FUZZZ-IEEE'11), pp. 2546–2551 (2011)
12. Tamani, N., Liétard, L., Rocacher, D.: Bipolar SQLf: A flexible querying language for relational databases. In: The 9th International Conference on Flexible Query Answering Systems (FQAS'11), LNAI, vol. 7022, Springer, pp. 472–484 (2011)
13. Yager, R.R.: Quantifiers in the formulation of multiple objective decision functions. Inf. Sci. **31**, 107–139 (1983)
14. Yager, R.R.: On ordered weighted averaging aggregation operators in multicriteria decision-making. IEEE Trans. Syst. Man Cybern. **18**, 183–190 (1988)
15. Zadeh, L.: Fuzzy sets. Inf. Control **8**(3), 338–353 (1965)
16. Zadeh, L.: A computational approach to fuzzy quantifiers in natural languages. Comput. Math. Appl. **9**, 149–184 (1983)
17. Zadrozny, S.: Bipolar queries revisited. LNAI MDAI **3558**, 387–398 (2005)
18. Zadrozny, S., Kacprzyk, J.: Bipolar queries and queries with preferences (invited paper). In: DEXA'06: 17th International Conference on Database and Expert Systems Applications (2006)

Chapter 4
Bipolarity in Database Querying: Various Aspects and Interpretations

Sławomir Zadrożny and Janusz Kacprzyk

Abstract A crucial problem in database querying is how to devise a query to best reflect the very intentions and preferences of the user. A new line of research in this area aims at taking into account the polarity of preferences what should considerably enhance the functionality and usefulness of flexible database querying systems. Bipolar queries constitute an important concept in this area. They are meant here, in general, as queries involving negative and positive information. In a special, promising interpretation they can be viewed in terms of necessary and possible conditions. The purpose of this paper is to critically analyze, recast in a unified perspective and clarify with respect to conceptual, algorithmic and implementation related aspects of various ways to deal with bipolarity. This should open new perspectives for research and commercial applications of bipolar and related queries which should provide more comprehensive, enhanced and more human consistent querying capabilities.

1 Introduction

A crucial problem in database querying is how to formulate, and then represent and process a query to best reflect the very intentions and preferences of the user. Traditionally, databases are meant to store highly structured information and to support

To Patrick, Professor Patrick Bosc, a friend and peer, who has been for a long time stimulating and amplifying our interest in flexible database querying.

S. Zadrożny (✉) · J. Kacprzyk
Systems Research Institute, Polish Academy of Sciences, Warsaw, Poland
e-mail: Slawomir.Zadrozny@ibspan.waw.pl

J. Kacprzyk
e-mail: Janusz.Kacprzyk@ibspan.waw.pl

S. Zadrożny
Kazimierz Pulaski University of Technologies and Humanities in Radom, Radom, Poland

O. Pivert and S. Zadrożny (eds.), *Flexible Approaches in Data, Information and Knowledge Management*, Studies in Computational Intelligence 497, DOI: 10.1007/978-3-319-00954-4_4, © Springer International Publishing Switzerland 2014

equally highly structured and precise query languages. However, nowadays databases find their application in many various settings and are expected to provide information to growing population of end users without a relevant IT/ICT training. Moreover, a growing complexity of application domains requires some more sophisticated forms of queries to make it possible to reflect the real users' intentions, preferences and interests, the semantics of which is not obvious and straightforward. Advanced graphical user interfaces (GUIs) alleviate to some extent the problem of man-machine communication, but, alone, do not help much with respect to the semantic representation of the users' information needs. Thus, some more conceptually advanced and sophisticated information access methods are needed to make a full use of the potential brought by the vast amount of information gathered in modern databases. First of all, a query is usually first conceived in natural language—the primary, maybe the only fully natural way of articulation, communication and "information processing" for the human being. Then, it has to be translated to a form required by a given database management system (DBMS). This translation is often lossy and thus those users are in need of another access path. This need is addressed by the traditional research on query answering systems which ultimately aims at providing the users with a fully natural language based query interface. An important issue is the very modeling of linguistic terms which may be used in queries and here many interesting approaches have been developed by the fuzzy logic community.

An important novel line of research concerning advanced querying of databases addresses the issue of the bipolar nature of conditions describing data sought by the user. Namely, the user looking for data usually can specify some disqualifying (negative information) and some desired (positive information) features of data. Classical query formalisms do not allow to express such requirements. The problem becomes particularly complex when these features are specified in an imprecise way. There is quite a rich literature dealing with this problem in the framework of fuzzy logic in general, or for some specific applications, notably decision making. A few other chapters in this volume also belong to this direction.

Various existing approaches to the representation and processing of such bipolar queries are based on different assumptions, paradigms and formal tools. As always in such a case, there is an acute need for some deeper analysis of views and perspectives within which various authors deal with the problem so that crucial differences and similarities be discovered. Such a comprehensive study of many various possible interpretations of bipolarity related to database queries seems to be missing. This is the purpose of this chapter in which we first provide a quick review of known approaches and interpretations and then propose our own contribution to the understanding of this phenomenon with special emphasis on the discussion of various scales of bipolarity which play a particular role for our purpose.

2 Background

2.1 Basic Concepts

The starting point for our considerations is the seminal Zadeh's concept of a *fuzzy set* [44] which may be conveniently identified with its *membership function*. Namely, a fuzzy set A in a universe U will be in what follows usually identified with its membership function:

$$\mu_A : U \to [0, 1]$$

such that $\mu_A(x)$ denotes degree to which element $x \in U$ is a member of the fuzzy set A.

We will skip most of the basic concepts related to fuzzy sets as they are clearly superfluous in this volume. We will just remind briefly a few concepts which will be important for the further discussion. We refer an interested reader to a vast literature, notably to the recent paper by Dubois and Prade [26] who provide a perspective on the notions of fuzziness, uncertainty and bipolarity which are all very important in the context of data modeling and database querying.

The *support* and the *core* of a fuzzy set A in universe U, denoted $Support(A)$ and $Core(A)$ respectively, are "classical" (crisp) sets defined as follows:

$$x \in Support(A) \Leftrightarrow \mu_A(x) > 0 \tag{1}$$

$$x \in Core(A) \Leftrightarrow \mu_A(x) = 1 \tag{2}$$

The concept of the *twofold fuzzy set* is an extension of the concept of the regular fuzzy set [21]. A twofold fuzzy set A over a given universe of discourse U is defined by two membership functions $\pi_A : U \to [0, 1]$ and $\eta_A : U \to [0, 1]$ such that:

$$\eta_A(x) > 0 \Rightarrow \pi_A(x) = 1 \quad \forall x \in U \tag{3}$$

Intuition behind the condition (3) is such that $\pi_A(x)$ may denote the degree to which it is possible that x belongs to A, while $\eta_A(x)$ may denote the degree to which it is necessary that x belongs to A, and then (3) is a natural consequence of the essence of possibility theory. This condition may also be expressed in the following way. Assuming that both membership functions specify regular fuzzy sets, it is required that the support of the fuzzy set defined by η_A must be contained in the core of the fuzzy set defined by π_A.

Another extension of the concept of the regular fuzzy set is *Atanassov's intuitionistic fuzzy set* (AIFS, for short) [1]; for a debate about the appropriateness of the term "intuitionistic", cf. Dubois et al. [20]. An AIFS A over a given universe of discourse U is defined by two functions: a membership function $\mu_A : U \to [0, 1]$ and a non-membership function $\nu_A : U \to [0, 1]$ such that:

$$\mu_A(x) + \nu_A(x) \leq 1 \quad \forall x \in U \tag{4}$$

Thus in this approach the membership and the non-membership of an element to an AIFS may be determined to some extent independently. The consistency condition (4) finds an interesting interpretation in the context of the bipolar queries against the database, discussed in this paper, cf. [23].

2.2 Classical and Flexible Queries: A Brief Overview

We adopt the basic terminology of the *relational data model*. In particular, we will mainly refer to a single *relation* (or, more precisely, relational variable), comprising a set of tuples $T = \{t_i\}$, characterized by a set of attributes $At = \{X, Y, \ldots\}$.

In this chapter we focus on the conditions in a query which specify which data is sought. Looking from the perspective of the SQL language, we are concerned with the WHERE clause of a simple SELECT-FROM-WHERE query. Some aspects discussed here may be further extended to the case of more complex SQL queries, e.g., involving the join operator though this goes beyond the scope of the current paper. Our study concerns the following scenario. A classical ("crisp") relational database is considered, against which queries of the SELECT-FROM-WHERE type are addressed. However, these queries may contain some non-standard conditions in their WHERE clause. First, some imprecision (fuzziness) may be present, exemplified by a query "Find all middle-aged employees", which is meant to express in a better and more direct way the user requirements than, e.g., a query "Find all employees whose age falls in the interval between 35 and 45". We follow here the line of research of Bosc and Pivert [3, 4, 6–8, 10, 12–14, 16, 17] and Kacprzyk et al. [30–35, 46, 47, 51] in which it is assumed that such imprecise conditions are construed using some linguistic terms as, e.g., "middle-aged" in the previous example, which are modeled using fuzzy logic. Second, we assume that the condition may be composed of some positive and negative components, what is meant to reflect bipolarity of user preferences. A study of the latter feature is the main topic of this chapter.

The simplest form of bipolarity may be ascribed to any classical "crisp" query against a database. Let us consider a simple query involving just one atomic condition concerning one attribute. Let this condition be:

$$\text{price} \leq \text{USD 500K,} \tag{5}$$

as in Example 1 (cf. p. 7). Then, the values in the domain of the attribute price which are above 500 K are rejected (i.e., treated as "negative"), while the values lower or equal than that are accepted ("positive"). This is however a very specific type of bipolarity which is of a lesser interest, at least due to the two following reasons:

1. the rejection/acceptance is binary
2. there is no notion of "neutrality", and, what is closely related, an element of the domain under consideration which is not "negative" is necessarily "positive" and vice versa.

The classical fuzzy approach to querying obviously alleviates the first limitation but not the second: still, if an element x of the domain is indicated as "positive" to a degree $\mu(x)$, then it is automatically treated as "negative" to the degree $1 - \mu(x)$. This type of bipolarity may be referred to, after Dubois and Prade [24, 26], a *symmetric univariate bipolarity*. There is still another feature of the bipolarity which is not properly addressed by the classical, either crisp or fuzzy, approaches to the querying. Namely, there should be available for the user some specific aggregation operators, which take into account the bipolarity of preferences while computing the overall matching degree. In order to be in accord with psychological observations, such operators should make it possible to treat negative and positive preferences in a different way [29].

3   Unipolar and Bipolar Fuzzy Conditions

3.1   Classical Fuzzy Approach to the Modeling of Query Conditions and Bipolarity

A fuzzy logic based perspective has been adopted to model conditions of a query against a database since the early days; cf., e.g., Tahani [40]. The flexibility of modeling provided by the concept of a fuzzy set finds an immediate application for the purposes of query conditions specification. Here we will only very briefly summarize the advantages of the classical fuzzy logic based approach so as to clearly show later the difference that taking into account of bipolarity makes.

Let us consider a database of real-estate properties offer for sale by a real estate agency. Let the particular houses be characterized by some attributes exemplified by: `price`, `location`, and `size` (in square meters). Let us further assume that a customer of the agency is looking for a house of a *low* price (usually a customer will require a few conditions that should be met by the house he or she is looking for, but here for the sake of presentation clarity we will focus on one attribute). Using, e.g., a flexible querying interface provided by Kacprzyk and Zadrożny's FQUERY for Access [32, 33], he or she can form a query using the *linguistic term* "low" directly to express the constraint on the price. The set of acceptable prices will be modeled by a fuzzy set A in a universe U, characterized by its membership function:

$$\mu_A : U \to [0, 1] \tag{6}$$

In our example, the term "low" will be modeled by such a fuzzy set A (in the universe U which is identified then with the domain of given attribute; here: `price`) that $\mu_A(x)$ denotes to which degree a given price x is low, and this degree will be treated as the matching degree of a house with the given price against the query under consideration. Thus the user is released from artificially distinguishing the prices which (fully) are low from those which are not. This way of modeling is clearly more human consistent, i.e., more in line with human perception of such linguistic terms as "low", "moderate", "high", etc. A direct consequence is then the possibility to order the tuples of a database (real estate properties, in the case of our example) according to their matching degrees of the query condition.

We will denote a classical fuzzy query concerning attribute X and using a linguistic term modeled by a fuzzy set A as:

$$X \text{ is } A \tag{7}$$

Referring to our previous example, X in (7) denotes the attribute `price`, while fuzzy set A represents the linguistic term "low".

From the point of view of this chapter the interpretation of the membership degrees related to (6) is the most interesting. It is worthwhile to note that in the classical fuzzy approach a unipolar scale is tacitly associated with (6). Namely, $\mu_A(x)$ denotes the degree to which a given attribute value is compatible with the meaning of a given linguistic term and, in consequence, the degree to which this value satisfies query condition. There is no explicit distinction between "negative" ("rejected", "bad") and "positive" ("accepted", "good") values. It may be argued that such a distinction is usually made by the user in the framework of the classical fuzzy approach but it is of a slightly different nature than the one considered here. Namely, usually the query languages and interfaces proposed (c.f., e.g., [14, 33]) presume the use of a threshold in a query which indicates that only tuples matching the query to a degree higher than this threshold are shown to the user. Thus, in a sense, the mechanism related to such a threshold makes the distinction between "negative" and "positive" tuples which may be further interpreted as the distinction concerning the attribute values—if the query condition is atomic and refers to just one attribute (there are also approaches in which such a threshold may be associated with each atomic condition separately [39]). However, this is a binary distinction and is made externally with respect to the query condition, and thus of a lesser interest to us here.

Following the arguments mentioned in the introduction we assume that very often the user preferences are inherently bipolar. This bipolarity may manifest itself *at the level of each attribute domain* or *at the level of the comprehensive evaluation* [29] of the whole tuple. In the former case, the user may see particular elements of the domain as "negative", "positive" or "neutral", to a degree. This classification should, of course, influence the matching degree of a tuple having a particular element of the domain as the value of the attribute under consideration. In the latter case the user is expected to express some conditions, involving possibly many attributes, which when satisfied by a tuple (to a degree) make it "negative" or "positive" (to a degree). Thus, effectively, in the latter case some combinations of multiple attributes values are seen as "negative"/"positive". The former case may be seen as a special case of the latter, when both the "negative" and "positive" conditions concern the same attribute and thus demarcate the "negative" and "positive" values of a given attribute. However, the distinguishing of the former case is worthwhile as it is somehow less intuitive but still practically useful and there do exist some formal means which provide for its elegant formal representation which will be discussed later on. Moreover, the distinguishing of this case makes it easier to study various approaches known in the literature. Let us illustrate that on two examples.

Example 1 Let us consider a customer of our real-estate property agency. He or she may have the following view on the domain of the `price` attribute:

(a) the price above USD 500 K is definitely negative,
(b) the price below USD 300 K is definitely positive,
(c) the remaining prices are neither negative nor positive, i.e., are neutral.

In Example 1 the bipolarity is defined in the crisp way on the level of the attribute domain. Now let us consider more complex preferences of the user.

Example 2 Let us consider another customer of our real-estate property agency. He or she finds:

(a) the properties more expensive than USD 500 K and, at the same time, of the size less than 100 sq. m. as definitely negative,
(b) the properties located in Waterfront as definitely positive,
(c) the remaining properties as neither negative nor positive, i.e., as neutral.

In Example 2 the bipolarity concerns a combination of attribute values, or, equivalently the whole tuples (here: the real-estate properties) possessing these combinations of values. In this example (and also in Example 1, which is however possibly less obvious) the point (b) requires some discussion concerning the compatibility of condition given there with the condition given in the point (a)—such a discussion will be provided later on while presenting alternative ways to formally represent the bipolarity in query conditions.

Both Examples 1 and 2 are crisp, however may be easily "fuzzified" in practical scenarios by using such (subjectively defined) linguistic terms as "very expensive", "rather cheap", "small" instead of the numbers expressing the price and size of the property. What is worth noting is that the user preferences expressed both in the crisp and fuzzy versions of these examples cannot be properly expressed using the classical crisp or fuzzy approaches to database querying. On the other hand, one may argue that some special cases of such preferences are representable using the classical approaches. Namely, while considering a modification of the first example, one can claim that the condition "price <USD 500 K" properly represents such bipolar preferences if the values of the price higher than USD 500 K are "negative" and the values below or equal to the same threshold are "positive". More generally, in the specific case where the subset of "positive" values in the domain of given attribute is the complement of the subset of "negative" values, it may be seen as corresponding

to the bipolarity defined by using the symmetric bipolar univariate scale (called a symmetric univariate bipolarity by Dubois and Prade [25]).

Example 1 clearly explains the terminology often used in the literature (cf., e.g., [23, 48]) when referring to the negative and positive parts of the scale: the former is used as the scale for the required conditions and the latter as the scale for the desired conditions.

Thus, the classical fuzzy approach makes it possible for the user to clearly specify in the query a distinction between the "negative" data he or she rejects (to a degree) and the "positive" data he or she accepts (to a degree). It is worth to emphasize this concept of the negative/positive traits of data the user has in mind, which is, of course, relative to a given query, and which is accompanied with some affect. This distinction between the negative and positive traits of data should then be properly taken into account during the computation of the overall matching degree using appropriate aggregation operators—we will discuss this in more detail later on. Without this distinction one can argue that already in the traditional, "crisp" approaches a query defines the set of rejected and accepted data.

3.2 Bipolarity: Which Scale to Use

3.2.1 Bipolarity in the Query Condition via a Univariate Bipolar Scale

In this case it is assumed that for the data under consideration, being either an attribute domain element or the whole tuple, the user may evaluate its "negative" and "positive" sides and he or she is in a position to combine these evaluations and expresses an overall evaluation on one univariate bipolar scale. It may be instructive to consider two cases, depending on the level at which this bipolarity is expressed.

Univariate bipolarity at the level of an attribute domain element

Here we assume that the user has a bipolar evaluation of each element of a domain dom_X of a given attribute X. For convenience, we assume that such an evaluation is of the form (cf., e.g., [29] for a discussion):

$$\xi_X : dom_X \rightarrow [-1, 1] \tag{8}$$

and for $x \in dom_X$ the value $\xi_X(x) > 0$ denotes x's degree of "positiveness", $\xi_X(x) < 0$ denotes its degree of "negativeness" and $\xi_X(x) = 0$ means that x is neutral from the point of view of the user, concerning given query.

In such a case, the user preferences may be properly modeled using a twofold fuzzy set (3) in dom_X. Thus in (7) fuzzy set A will be now replaced by a twofold fuzzy set. This twofold fuzzy set will be interpreted as follows:

- the membership function π_A is used to represent the negative evaluations of the elements of dom_X; its values equal the evaluation for these elements plus 1:

$$\pi_A(x) = \min(1 + \xi_X(x), 1) \tag{9}$$

More precisely, the values of $\pi_A(x)$ form a reversed negative scale: value 1 denotes no negative evaluation, value 0 the strongest negative evaluation, while intermediate values represent some degrees of negative evaluation—the closer to 0 they are the stronger negative evaluation it is.

- the membership function η_A is used to represent the positive evaluations of the elements of dom_X:

$$\eta_A(x) = \max(\xi_X(x), 0) \tag{10}$$

i.e., the value 0 of $\eta_A(x)$ denotes no positive evaluation, the value 1 the strongest positive evaluation, while intermediate values represent some degrees of positive evaluation—the closer to 1 they are the stronger positive evaluation it is.

The idea is illustrated in Fig. 1. Thanks to the very property of the twofold fuzzy set (3) there is a one-to-one mapping between the degrees of evaluation ξ_X given by (8) and the membership functions of the corresponding twofold fuzzy set. The mapping from $\xi_X(x)$ to a pair $(\pi_A(x), \eta_A(x))$ is given by Eqs. (9)–(10). The reverse transformation is given by the following formula:

$$(\pi_A(x), \eta_A(x)) \rightarrow \xi_X(x) \tag{11}$$

$$\xi_X(x) = \begin{cases} \eta_A(x) & \text{for } \pi_A(x) = 1 \\ \pi_A(x) - 1 & \text{otherwise} \end{cases} \tag{12}$$

In this scenario the user is assumed to express his or her bipolar preferences with respect to an attribute X using a univariate bipolar scale. In order to do so one can choose two linguistic terms from the dictionary (or define them; cf., e.g., details of the user interface of the FQUERY for Access system [32, 33]) which are represented by fuzzy sets forming together a twofold fuzzy sets, i.e., whose membership functions satisfy condition (3). An illustration is shown in Example 3.

Example 3 Let us consider a customer who does not like small houses and would be most satisfied with a house of the size around 350 sq. m. Then he or she may express his or her preferences defining or choosing from the dictionary two linguistic terms "small" and "around 350 sq. m." and form a twofold fuzzy set A with the following membership functions $(\pi_A(x), \eta_A(x))$:

$$\pi_A(x) = \mu_{\text{"not small"}}(x)$$
$$\eta_A(x) = \mu_{\text{"around 350 sq. m."}}(x)$$

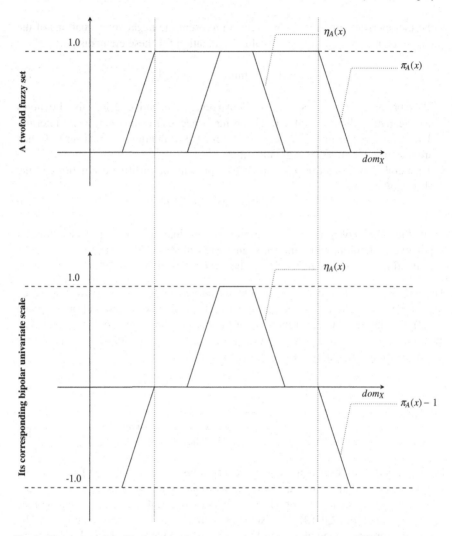

Fig. 1 Illustration of the bipolar univariate scale representation using the twofold fuzzy set

assuming that the support of the fuzzy set representing "around 350 sq. m." is a subset of the core of the complement of the fuzzy set representing the linguistic term "small", i.e., the fuzzy set representing the linguistic term "not small".

Note, that in this case it may be fairly easily checked that the membership functions $\pi_A(x)$ and $\eta_A(x)$ really form a twofold fuzzy set, i.e., satisfy the condition (3). Thus, it is reasonable to assume that the user, properly supported by the user interface, picks up an appropriate pair of fuzzy sets.

Univariate bipolarity at the level of a tuple (at the comprehensive evaluation level)

This is a more general case than the previous one as now it is assumed that the user has a comprehensive evaluation of the whole tuple expressed using a univariate bipolar scale. Thus, as previously, we assume an evaluation ξ_T ranging over the interval $[-1, 1]$ but this time its domain is the set of tuples T:

$$\xi_T : T \rightarrow [-1, 1] \tag{13}$$

Again, a formal representation of this evaluation is obtained using a twofold fuzzy set denoted by a pair of membership functions $(\pi_T(t), \eta_T(t))$. Now we will assume that the user defines two conditions denoted $C(t)$ and $P(t)$, respectively, that will in turn define these membership functions, i.e.,

$$\pi_T(t) = C(t) \tag{14}$$
$$\eta_T(t) = P(t) \tag{15}$$

Here, and in what follows, we will denote by C and P both fuzzy predicates identified by the respective conditions and the fuzzy sets of tuples satisfying, to a degree, these predicates. Moreover, by $C(t)$ and $P(t)$ we will denote the membership function values of the particular tuples $t \in T$ to these fuzzy sets. Let us illustrate that with a "fuzzified" version of Example 2, which is given below as Example 4.

In this case the link between (13), and (14) and (15) is analogous as in the case of univariate bipolarity at the level of an attribute domain elements as discussed earlier.

Example 4 Let us consider a customer of our real-estate property agency. He or she finds:

(a) very expensive and, at the same time, small properties as definitely negative,
(b) properties located in eastern districts of the city as definitely positive.

In Example 4 the fuzzy predicates C and P are defined as "not (*very expensive and small*)", and "located in *eastern* districts", respectively (we assume that "eastern districts" is a gradual notion, well represented by a fuzzy predicate). Note that in the case of a comprehensive evaluation it is rather unreasonable to expect that the respective membership functions satisfy condition (3), or—to put that more precisely—that the user may be somehow aware if they do or do not. For example, there may exist a property located in the "totally eastern" (i.e., to the degree 1) district but very expensive and small. Thus, in fact, (15) has to be modified so as to force the satisfaction of (3). The simplest way to do that is to use the following variant of (15):

$$\eta_T(t) = \begin{cases} P(t) & \text{for } C(t) = 1 \\ 0 & \text{otherwise} \end{cases} \tag{16}$$

3.2.2 Bipolarity in the Query Condition via a Bivariate Bipolar Scale

In this approach the user is assumed to define separately positive and negative traits of the data sought. This may be done again, as in the previous case, at the level of an attribute domain element or at the level of a tuple. However, here still another distinction should be made regarding the semantics of these separate positive and negative evaluations. Namely, we will distinguish two cases in which:

1. the negative evaluation is treated as related to the violation (to a degree) of a constraint and the positive evaluation is treated as related to the satisfaction of a desire, i.e., of a somehow supplementary condition; thus the positive evaluation plays here a subsidiary role—the elements violating the constraint are thus treated as rejected (to a degree); in what follows we will refer to this case as the "required/desired semantics",
2. both evaluations are treated "equally".

Thus, in the first case a specific semantics of "positive" and "negative" evaluations is assumed. This implies a need for some consistency conditions which express the fact that something may be desired at most to a degree to which it satisfies the constraints, i.e., to a degree to which it is not rejected. We will discuss that issue in Sect. 4 in a more detailed way. It should be noted that this semantics is adopted in most of the works related to bipolar queries; cf., e.g. [23, 48]. It is definitely very intuitive and of a high practical value. However, second case, in which the treatment of bipolarity is more general in the sense of just reflecting the existence of a positive and negative condition without any specific interpretation of their relations and interplay, deserves more attention and research.

Bivariate bipolarity at the level of an attribute domain element—the general case

Here it is assumed that the user has a bipolar evaluation of each element of a given attribute X domain dom_X and can separately evaluate its positive and negative traits (his "liking" and "disliking" of an element). For convenience, we assume that such an evaluation is expressed via two functions and such that

$$\xi_X^+ : dom_X \rightarrow [0, 1] \tag{17}$$

$$\xi_X^- : dom_X \rightarrow [0, 1] \tag{18}$$

where ξ_X^+ and ξ_X^- denote how "good" and "bad", respectively, the element x is. Let us illustrate this case with the following example.

Example 5 Let us consider a customer, who especially cares for the location of a house which is given in the database by the name of the district of a city. For each district he can list some "pros" and "cons" (possibly, of varying importance/strength). For example, district D is well communicated with the rest of the city but is known for its relatively high crime level. The user is able to separately aggregate the lists of the arguments and to come up with a separate positive and negative evaluation of each location.

It should be noted that such a bipolar bivariate evaluation makes sense basically only in the case when there is a set of criteria that may be related to the elements of the domain in question but are not directly represented in a database. If the locations of the houses were represented in the database of Example 5 at a more detailed level (including the communication convenience, crime level etc.), then the preferences of the user would probably be better expressed using the bipolar univariate or even just unipolar scale, with respect to the domains of attributes comprising this more detailed representation.

Bivariate bipolarity at the level of an attribute domain element—the required/desired semantics

Here it is assumed that the user has a bipolar evaluation of each element of a given attribute X domain dom_X in the sense that he or she can distinguish a (fuzzy) set R of rejected elements and the (fuzzy) set P of really desired (preferred) elements. For convenience, such an evaluation may be expressed by two (membership) functions:

$$\xi_X^R : dom_X \rightarrow [0, 1] \tag{19}$$

$$\xi_X^P : dom_X \rightarrow [0, 1] \tag{20}$$

and for $x \in dom_X$ the value $\xi_X^R(x)$ denotes degree to which x is rejected, while $\xi_X^P(x)$ denotes degree to which it is desired. It should be noticed that if the sets R and P are complements of each other then only one of them have to be specified, i.e., it refers to the case of the classical fuzzy logic based querying: ξ_X^P may be identified with (6). Thus, this case is interesting only if $C = \bar{R} \neq P$ and such a bipolar query may be represented equivalently by the pairs of (fuzzy) sets (R, P) or (C, P).

Due to the postulated required/desired semantics it is rational to impose a consistency condition $P \subseteq C$, which states that an element has first to be non-rejected before it can be desired (preferred). Thus, the user preferences may be here properly modeled using an AIFS (4) in dom_X. Thus in (7) fuzzy set A will be now replaced by an AIFS which will be interpreted as follows:

- the membership function μ_A is used to define the degree to which a particular element is desired:

$$\mu_A(x) = \xi_X^P(x)$$

- the non-membership function ν_A is used to define the degree to which a particular element is rejected:

$$\nu_A(x) = \xi_X^R(x)$$

The consistency condition $P \subseteq C$, which may be expressed equivalently as $P \subseteq \bar{R}$ or $\xi_X^P(x) \le 1 - \xi_X^R(x)$ coincides with the condition (4) characteristic for the AIFS's.

Thus, in this scenario the user is assumed to express his or her bipolar preferences with respect to an attribute X using a bivariate bipolar scale. In order to do that one can choose two linguistic terms from the dictionary, representing sets R and P, respectively, the membership functions of which have to satisfy condition (4). An illustration is shown in Example 6 which is a modified version of Example 3.

Example 6 Let us consider a customer who does not like small houses and would be most satisfied with a house of the size around 350 sq. m. Then, he or she may expresses his or her preferences by defining or choosing from the dictionary two linguistic terms "small" and "around 350 sq. m." and by forming the following AIFS A $(\mu_A(x), \nu_A(x))$:

$$\nu_A(x) = \mu_{\text{"small"}}(x)$$
$$\mu_A(x) = \mu_{\text{"around 350 sq. m."}}(x)$$

assuming that for all values representing the size the sum of its membership degrees to these two fuzzy sets is not larger than 1.0.

Note, that in this case the preferences of the user may be expected to be consistent and, if the two above mentioned fuzzy sets adequately represent his or her subjective understanding of the linguistic terms "small" and "around 350 sq. m", then the query formed in such a way will represent the user's preferences in a fair way.

4 Semantics of the Bipolar Bivariate Conditions: An Aggregation Perspective

In the previous sections we were mainly concerned with the identification of various forms of bipolarity in queries and their representations. Here we discuss an interpretation of bipolar queries in terms of an ordering of tuples they imply. In particular, we focus on a specific semantics of bipolar conditions, referred to as "required/desired" semantics in the previous section. Here we will discuss this semantics in a more detailed way, in a specific perspective.

4.1 A General View

The most general interpretation of the bivariate bipolarity in queries is the one mentioned in Sect. 3.2.2—with the positive and negative conditions treated as equally important and independent. Thus, we have two conditions and each tuple is evaluated against them yielding a pair of satisfaction (matching) degrees. The natural question is then how to order data in an answer to such a query.

Basically, while doing that we should take into account the very nature of both matching degrees, i.e., the fact that they correspond to the positive and negative conditions. The situation here may be compared to that of decision making under risk. Namely, in the latter context a decision maker who is risk-averse may not accept actions leading with some non-zero probability to a loss. On the other hand, a risk-prone decision maker may ignore risk of an even serious loss as long as there are prospects for a high gain. Similar considerations apply in the case of bipolar queries. Some users may be more concerned about negative aspects and will reject a piece of data with a non-zero matching degree of the negative condition. Some other users may be more oriented towards the satisfaction of the positive conditions and may be ready to accept the fact that a given piece of data satisfies to some extent the negative conditions.

The conclusion from the above considerations is such that the bipolar query meant in such a general sense should be evaluated in a database in a way strongly dependent on the specific attitude of the user. In the extreme cases, the above-mentioned analogs of "risk-averse" and "risk-prone" attitudes would be represented by lexicographic orders. In the former case the lexicographic ordering would be first non-decreasing with respect to the negative condition matching degree and then non-increasing with respect to the positive condition matching degree. The less extreme attitudes of the users may be represented by various aggregation operators producing a scalar overall matching degree of a bipolar query.

An approach to a comprehensive treatment of such generally meant bipolar queries has been proposed by De Tré and Matthé [38], and further developed in [19, 37]. In this approach a pair of matching degrees of the positive and negative conditions is referred to as a *bipolar satisfaction degree* (BSD). The respective matching degrees are denoted as s and d, and called the *satisfaction degree* and the *dissatisfaction degree*, respectively. The ranking of data retrieved against a bipolar query in this approach may be obtained in various ways. One of the options is based on the difference $s - d$ of the two matching degrees. In this case a "risk-neutral" attitude of the user is modeled: he or she does not favor neither the positive nor the negative evaluation.

The BSDs are assumed to be assigned at the attribute level and then are aggregated so that an overall BSD for the whole query is obtained. In [19, 37] it is proposed how such an aggregation should be carried out in case of the standard logical connectives. See also a paper by Matthé et al. in this volume which reports on further developments in this research direction.

4.2 The Required/Desired Semantics Once Again

The semantics in question supports the following interpretation of the *positive* and *negative* conditions in the bipolar query: the data items sought have to satisfy the complement of the latter conditions unconditionally while the former conditions is of somehow secondary importance. For example, a house the user is looking for may have to be cheap and then among cheap houses those which are closer to a railway station are preferred. The negative condition is here "not being cheap" while the positive condition is "being close to the railway station". Usually, the complement of the negative condition will be specified in such a query (denoted C), which may therefore be interpreted as a *required* condition. On the other hand, the positive condition is expressed directly and may be referred to as a *desired* condition (denoted C). It is worth noting that this interpretation is close to the mode of *aggregation* of a hierarchy of conditions proposed in 1987 in a seminal work of Lacroix and Lavency [36], of course without any reference to the notion of bipolarity at that time. This type of aggregation may be seen as based on the "and possibly" operator: to satisfy the required conditions *and if possible also* the desired conditions. We develop this idea further in the following sections.

Whatever the interpretation of the positive and negative conditions is adopted, the main practical problem is how to order the tuples based on their satisfaction degrees of these conditions. In case of the "required/desired" semantics we denote a pair of these conditions as (C, P). The problem mentioned may be solved in many ways.

The simplest approach is to use the matching degree with respect to the desired condition just to order the data items which satisfy the required condition. This idea leads to the use of the lexicographic order which is promoted by many authors, notably Dubois and Prade; cf., e.g., [23]. This interpretation is in fact predominant in the literature dealing with bipolar queries. The early works of Bosc and Pivert [9, 11] which aim at introducing a fuzzified version of the operator for aggregating the conditions in the spirit of Lacroix and Lavency also belong to this category. In those papers, as well as in the sophisticated possibility theory based interpretations by Dubois and Prade [24, 25] focus is on a proper treatment of *multiple* required and preferred conditions, basically assuming the lexicographic order as the way of combining the required (negative) and desired (positive) conditions, cf. also Bosc et al. [18]. However, if a fine (detailed) scale for the satisfaction of the required condition is adopted then a smallest possible dominance of one tuple over another with respect to the satisfaction of the required condition makes it "better" even if the other tuple is much better with respect to the desired condition. The solution proposed is to use a coarser scale of required condition satisfaction degrees but still it is a rather artificial solution.

Another approach consists in employing an aggregation operator which combines the degrees of matching (satisfaction) of conditions C and P and yields an overall matching degree which is then used to order tuples in the usual way. In particular the operator introduced by Lacroix and Lavency [36] may be used. Then, the whole query may be interpreted as expressing the following condition:

$$C \text{ and possibly } P \tag{21}$$

In the literature such aggregation operators have been studied by many authors under different names, and sometimes in slightly different contexts. However, in the framework of database querying Lacroix and Lavency proposed first such an approach. Zadrożny [45] proposed a direct "fuzzification" of the approach by Lacroix and Lavency, Zadrożny and Kacprzyk [49, 52] studied some properties of that solution.

4.3 The "and possibly" Operator Based Aggregation

The essence of the "and possibly" operator consists in taking into account the whole dataset while combining the matching degrees related to the required and desired conditions. Namely, if there is a piece of data which satisfies both conditions, then and only then it is actually *possible* to satisfy both of them and each piece of data has to meet both of them. Thus, the (C, P) query reduces to the usual conjunction $C \wedge P$. On the other hand, if there is no such a piece of data, then it is *not possible* to satisfy both conditions and the desired one can be disregarded. Thus, the (C, P) query reduces to C. These are however two extreme cases and actually it may be the case that the two conditions may be simultaneously satisfied to some degree. Then, the matching degree of the (C, P) query against a piece of data lies somewhere between its matching degrees of $C \wedge P$ and C. This may be formally written for the crisp case as [36]:

$$C(t) \underline{\text{ and possibly }} P(t) \equiv C(t) \wedge \exists s (C(s) \wedge P(s)) \Rightarrow P(t) \tag{22}$$

and for the fuzzy case as [45, 48]:

$$C(t) \underline{\text{ and possibly }} P(t) \equiv \min \left(C(t), \max(1 - \max_{s \in T} \min(C(s), P(s)), P(t)) \right) \tag{23}$$

where T denotes the whole dataset being queried.

The formula (23) is derived from (22) using the classic fuzzy interpretation of the logical connectives via the maximum and minimum operators. In Zadrożny and Kacprzyk [49, 50, 52] we analyze the properties of the counterparts of (23) obtained by using a broader class of operators modeling the logical connectives.

The "and possibly" aggregation operator that is implicit in the Lacroix and Lavency's proposal [36] has been later proposed independently by Dubois and Prade [22] in the context of default reasoning and by Yager [42, 43] in the context of multi-criteria decision making for the case of so-called *possibilistically qualified criteria*. Yager [43] intuitively characterizes a possibilistically quantified criterion as such which should be satisfied unless it interferes with the satisfaction of other criteria. This is in fact the essence of the aggregation operator "and possibly" as we understand it here. The concept of this operator was also used by Bordogna and Pasi [2] in the context of textual information retrieval.

Recently, the modeling of the aggregations operators in the spirit of the "and possibly" operator is gaining a broad interest. Usually, they lack the dependence on the whole data set what is a distinguishing characteristic feature of the operator based on the Lacroix and Lavency approach. However, they may have some importance for the implementation of bipolar queries and some of them are proposed to this aim. Dujmović [27] already in 1979 defined an aggregation operator combining two arguments in such a way that one of them controls the influence of the other ones on the result of their combination. Bosc and Pivert [15] also consider similar operators. Tudorie [41] introduced the "among" operator which is similar to the "and possibly" operator and is used to form queries such as "find data satisfying a condition P among those satisfying a condition C". The evaluation of a query with the "among" operator is expressed in terms of the rescaling of fuzzy predicates used to specify condition P.

5 Concluding Remarks

The idea of taking into account bipolarity of user preferences expressed in the form of database queries is gaining a growing popularity. However, there are still some basic questions open. This paper is an attempt to describe the very essence of bipolarity in the considered context, in a slightly more general way by concentrating on the presentation of various possible views and perspectives, and then attempting to find a unifying view. We also briefly review relevant literature to support our line of reasoning and views, and to show a line of logical developments which have occurred in the research efforts related to bipolar queries. In particular, we distinguish various possible approaches depending on the following aspects:

1. the type of a bipolar scale used to express preferences,
2. the existence (and type of) or lack of consistency constraints imposed on the positive and negative preferences, and
3. the level of data at which these evaluations are given.

We hope that this provides a better perspective on the research on bipolar queries. In particular, it shows that the approaches currently predominant in the literature cover only a part of the spectrum of possible interpretations.

The concepts and relations developed have been illustrated by numerous partial examples. However, due to space limitation, it has been impossible to present an in-depth analysis of one of many applications of the method proposed, notably in the area of querying real estate databases. Basically, due to the very essence of this domain and a relevance of interaction with the human customer, the presentation to be meaningful would have required a detailed coverage of many aspects exemplified by dictionaries of terms, analyses of preferences, multicriteria choice processes, etc.

Acknowledgments This contribution is partially supported by the Foundation for Polish Science under International PhD Projects in Intelligent Computing. Project financed from The European Union within the Innovative Economy Operational Programme (2007–2013) and European Regional Development Fund.

References

1. Atanassov, K.: Intuitionistic fuzzy sets. Fuzzy Sets Syst. **20**, 87–96 (1986)
2. Bordogna, G., Pasi, G.: Linguistic aggregation operators of selection criteria in fuzzy information retrieval. Int. J. Intel. Syst. **10**(2), 233–248 (1995)
3. Bosc, P., Galibourg, M.: Flexible selection among objects: A framework based on fuzzy sets. In: Proceedings of the SIGIR Conference, Grenoble, France (1988)
4. Bosc, P., Galibourg, M., Hamon, G.: Fuzzy querying with SQL: extensions and implementation aspects. Fuzzy Sets Syst. **28**, 333–349 (1988)
5. Bosc, P., Kacprzyk, J. (eds.): Fuzziness in Database Management Systems. Physica, Heidelberg (1995)
6. Bosc, P., Lietard, L.: Quantified statements and some interpretations for the OWA operators. In: Yager, R., Kacprzyk, J. (eds.) The Ordered Weighted Averaging Operators: Theory and Applications, pp. 241–257. Kluwer, Boston (1997)
7. Bosc, P., Lietard, L., Pivert, O.: Quantified statements and database fuzzy querying. In: Bosc, Kacprzyk [5], pp. 275–308
8. Bosc, P., Lietard, L., Prade, H.: An ordinal approach to the processing of fuzzy queries with flexible quantifiers. In: Hunter, A., Parsons, S. (eds.) Applications of Uncertainty Formalisms, LNAI, vol. 1455, pp. 58–75. Springer, Berlin (1998)
9. Bosc, P., Pivert, O.: Discriminated answers and databases: fuzzy sets as a unifying expression means. In: Proceedings of the IEEE International Conference on Fuzzy Systems (FUZZ-IEEE), pp. 745–752, San Diego, USA (1992)
10. Bosc, P., Pivert, O.: Fuzzy querying in conventional databases. In: Zadeh, L., Kacprzyk, J.(eds.) Fuzzy Logic for the Management of Uncertainty, pp. 645–671. Wiley, New York (1992)
11. Bosc, P., Pivert, O.: An approach for a hierarchical aggregation of fuzzy predicates. In: Proceedings of the Second IEEE International Conference on Fuzzy Systems (FUZZ-IEEE'93), pp. 1231–1236, San Francisco, USA (1993)
12. Bosc, P., Pivert, O.: Flexible queries, discriminated answers and fuzzy sets. In: Proceedings of the Fifth IFSA World Congress, pp. 525–528. Seoul, Korea (1993).
13. Bosc, P., Pivert, O.: On the evaluation of simple fuzzy relational queries: principles and measures. In: Lowen, R., Roubens, M. (eds.) Fuzzy Logic: State of the Art, pp. 355–364. Kluwer Academic Publishers, Boston (1993)
14. Bosc, P., Pivert, O.: SQLf: a relational database language for fuzzy querying. IEEE Trans. Fuzzy Syst. **3**(1), 1–17 (1995)
15. Bosc, P., Pivert, O.: On three fuzzy connectives for flexible data retrieval and their axiomatization. In: Chu, W.C., Wong, W.E., Palakal, M.J., Hung, C.C. (eds.) SAC, pp. 1114–1118. ACM (2011)
16. Bosc, P., Pivert, O., Farquhar, K.: Integrating fuzzy queries into an existing database management system: an example. Int. J. Intel. Syst. **9**, 475–492 (1994)
17. Bosc, P., Pivert, O., Lietard, L.: Aggregate operators in database flexible querying. In: Proceedings of the IEEE International Conference on Fuzzy Systems (FUZZ-IEEE 2001), pp. 1231–1234, Melbourne, Australia (2001)

18. Bosc, P., Pivert, O., Mokhtari, A., Lietard, L.: Extending relational algebra to handle bipolarity. In: Shin, S.Y., Ossowski, S., Schumacher, M., Palakal, M.J., Hung, C.C. (eds.) SAC, pp. 1718–1722. ACM (2010)

19. De Tré, G., Zadrożny, S., Matthe, T., Kacprzyk, J., Bronselaer, A.: Dealing with positive and negative query criteria in fuzzy database querying. Lect. Notes Comput. Sci. **5822**, 593–604 (2009)

20. Dubois, D., Gottwald, S., Hájek, P., Kacprzyk, J., Prade, H.: Terminological difficulties in fuzzy set theory—the case of "intuitionistic fuzzy sets". Fuzzy Sets Syst. **156**(3), 485–491 (2005)

21. Dubois, D., Prade, H.: Twofold fuzzy sets and rough sets—some issues in knowledge representation. Fuzzy Sets Syst. **23**, 3–18 (1987)

22. Dubois, D., Prade, H.: Default reasoning and possibility theory. Artif. Intel. **35**(2), 243–257 (1988)

23. Dubois, D., Prade, H.: Bipolarity in flexible querying. In: Andreasen, T., Motro, A., Christiansen, H., Larsen, H.L. (eds.) FQAS 2002, LNAI, vol. 2522, pp. 174–182. Springer, Berlin (2002)

24. Dubois, D., Prade, H.: Handling bipolar queries in fuzzy information processing. In: Galindo [28], pp. 97–114

25. Dubois, D., Prade, H.: An overview of the asymmetric bipolar representation of positive and negative information in possibility theory. Fuzzy Sets Syst. **160**(10), 1355–1366 (2009)

26. Dubois, D., Prade, H.: Gradualness, uncertainty and bipolarity: Making sense of fuzzy sets. Fuzzy Sets Syst. **192**, 3–24 (2012)

27. Dujmović, J.: Partial absorption function. J. teh Univ. Belgrade EE Dept. **659**, 156–163 (1979)

28. Galindo, J. (ed.): Handbook of Research on Fuzzy Information Processing in Databases. Information Science Reference, New York (2008)

29. Grabisch, M., Greco, S., Pirlot, M.: Bipolar and bivariate models in multicriteria decision analysis: descriptive and constructive approaches. Int. J. Intel. Syst. **23**, 930–969 (2008)

30. Kacprzyk, J.: Fuzzy logic in DBMSs and querying. In: Proceedings of Second New Zealand International Two-Stream Conference on Artificial Neural Networks and Expert Systems, Dunedin, New Zealand, pp. 106–109. IEEE Computer Society Press, Los Alamitos (1995)

31. Kacprzyk, J., Zadrożny, S.: Fuzzy querying for Microsoft Access. In: Proceedings of the Third IEEE Conference on Fuzzy Systems (FUZZ-IEEE'94), vol. 1, pp. 167–171, Orlando, USA (1994)

32. Kacprzyk, J., Zadrożny, S.: FQUERY for access: fuzzy querying for a windows-based DBMS. In: Bosc and Kacprzyk [5], pp. 415–433

33. Kacprzyk, J., Zadrożny, S.: Computing with words in intelligent database querying: standalone and internet-based applications. Inf. Sci. **134**(1–4), 71–109 (2001)

34. Kacprzyk, J., Zadrożny, S.: SQLf and FQUERY for access. In: Proceedings of the Conference IFSA/NAFIPS 2001, pp. 2464–2469, Vancouver, Canada (2001)

35. Kacprzyk, J., Zadrożny, S., Ziółkowski, A.: FQUERY III+: a "human consistent" database querying system based on fuzzy logic with linguistic quantifiers. Inf. Syst. **14**(6), 443–453 (1989)

36. Lacroix, M., Lavency, P.: Preferences: putting more knowledge into queries. In: Proceedings of the 13 International Conference on Very Large Databases, pp. 217–225, Brighton, UK (1987)

37. Matthé, T., De Tré, G., Zadrożny, S., Kacprzyk, J., Bronselaer, A.: Bipolar database querying using bipolar satisfaction degrees. Int. J. Intel. Syst. **26**(10), 890–910 (2011)

38. Matthé, T., Tré, G.D.: Bipolar query satisfaction using satisfaction and dissatisfaction degrees: bipolar satisfaction degrees. In: Shin, S.Y., Ossowski, S. (eds.) SAC, pp. 1699–1703. ACM (2009)

39. Medina, J.M., Pons, O., Miranda, M.A.V.: Gefred: a generalized model of fuzzy relational databases. Inf. Sci. **76**(1–2), 87–109 (1994)

40. Tahani, V.: A conceptual framework for fuzzy query processing: a step toward very intelligent database systems. Inf. Proces. Manage. **13**(5), 289–303 (1977)

41. Tudorie, C.: Qualifying objects in classical relational database querying. In: Galindo [28], pp. 218–245

42. Yager, R.: Higher structures in multi-criteria decision making. Int. J. Man Mach. Stud. **36**, 553–570 (1992)
43. Yager, R.: Fuzzy logic in the formulation of decision functions from linguistic specifications. Kybernetes **25**(4), 119–130 (1996)
44. Zadeh, L.: Fuzzy sets. Inf. Control **8**(3), 338–353 (1965)
45. Zadrożny, S.: Bipolar queries revisited. In: Torra, V., Narukawa, Y., Miyamoto, S. (eds.) Modelling Decisions for Artificial Intelligence (MDAI 2005), LNAI, vol. 3558, pp. 387–398. Springer, Berlin (2005)
46. Zadrożny, S., Kacprzyk, J.: Fuzzy querying using the 'query-by-example' option in a windows-based DBMS. In: Proceedings of Third European Congress on Intelligent Techniques and Soft Computing EUFIT'95, vol. 2, pp. 733–736, Aachen, Germany (1995)
47. Zadrożny, S., Kacprzyk, J.: Multi-valued fields and values in fuzzy querying via FQUERY for access. In: Proceedings of the Fifth International Conference on Fuzzy Systems (FUZZ-IEEE'96), vol. 2, pp. 1351–1357, New Orleans, USA (1996)
48. Zadrożny, S., Kacprzyk, J.: Bipolar queries and queries with preferences. In: Proceeding of the 17th International Conference on Database and Expert Systems Applications (DEXA'06), pp. 415–419. IEEE Computer Society, Krakow, Poland (2006)
49. Zadrożny, S., Kacprzyk, J.: Bipolar queries using various interpretations of logical connectives. In: Melin, P., Castillo, O., Aguilar, L., Kacprzyk, J., Pedrycz, W. (eds.) Foundations of Fuzzy Logic and Soft Computing, Lecture Notes in Computer Science, pp. 181–190. Springer (2007)
50. Zadrożny, S., Kacprzyk, J.: Bipolar queries: an approach and its various interpretations. In: Carvalho, J.P., Dubois, D., Kaymak, U., da Costa Sousa, J.M. (eds.) IFSA/EUSFLAT Conference, pp. 1288–1293 (2009)
51. Zadrożny, S., Kacprzyk, J.: Issues in the practical use of the OWA operators in fuzzy querying. J. Intel. Inf. Syst. **33**(3), 307–325 (2009)
52. Zadrozny, S., Kacprzyk, J.: Bipolar queries: an aggregation operator focused perspective. Fuzzy Sets Syst. **196**, 69–81 (2012)

Part II
Ontology-based Data Access

Chapter 5
On the Top-k Retrieval Problem
for Ontology-Based Access to Databases

Umberto Straccia

Abstract The chapter is a succinct summary on the problem of evaluating ranked top-k queries in the context of ontology-based access over relational databases. An ontology layer is used to define the relevant abstract concepts and relations of the application domain, while facts with associated score are stored into a relational database. Queries are conjunctive queries with ranking aggregates and scoring functions. The results of a query may be ranked according to the score and the problem is to find efficiently the top-k ranked query answers.

1 Introduction

Managing uncertainty and fuzziness is starting to play an important role in Semantic Web (SW) research, and has been recognised by a large number of research efforts in this direction (see, e.g., [59, 62] for a concise overview).

We recall for the inexpert reader that there has been a long-lasting misunderstanding in the literature of artificial intelligence and uncertainty modelling, regarding the role of probability/possibility theory and vague/fuzzy theory. A clarifying chapter is [21]. Specifically, under *uncertainty theory* fall all those approaches in which statements rather than being either true or false, are true or false to some *probability* or *possibility* (for example, "it will rain tomorrow"). That is, a statement is true or false in any world/interpretation, but we are "uncertain" about which world to consider as the right one, and thus we speak about e.g. a probability distribution or a possibility distribution over the worlds. On the other hand, under *fuzzy theory* fall all those approaches in which statements (for example, "the hotel is cheap") are true to some *degree*, which is taken from a truth space (usually [0, 1]). That is, an interpretation maps a statement to a truth degree, since we are unable to establish whether a

U. Straccia (✉)
ISTI-CNR, Via G. Moruzzi 1, 56124 Pisa, PI, Italy
e-mail: straccia@isti.cnr.it

O. Pivert and S. Zadrożny (eds.), *Flexible Approaches in Data, Information and Knowledge Management*, Studies in Computational Intelligence 497, DOI: 10.1007/978-3-319-00954-4_5, © Springer International Publishing Switzerland 2014

statement is entirely true or false due to the involvement of vague concepts, such as "cheap" (we cannot always say whether a hotel is cheap or not). Here, we will focus on fuzzy logic only.

In the SW, the standard Semantic Web Languages (SWLs) such as *triple languages* RDF & RDFS [9], *conceptual languages* or *frame-based languages* of the OWL 2 family [43] and *rule languages* such as RIF [50] are playing a dominant role. It also emerges that often in SW contexts, data are typically very large and dominate the intentional level of the ontologies. Hence, in that case one could still accept reasoning, specifically query answering, that is exponential on the intentional part, but it is mandatory that reasoning and query answering is polynomial in the data size, i.e. in *data complexity* [67].

In this chapter, we will briefly discuss a relatively novel issue for SWLs with a huge data repository, namely the problem of *evaluating ranked top-k queries*. Usually, an answer to a query is a set of tuples that satisfy a query. Each tuple does or does not satisfy the predicates in the query. However, very often the information need of a user involves so-called *fuzzy predicates* [32]. For instance, a user may need: "Find *cheap* hotels *near* to the conference location". Here, *cheap* and *near* are scoring predicates. Unlike the classical case, tuples satisfy now these predicates to a degree. In the former case the degree depends, e.g., on the price, while in the latter case it depends e.g. on the distance between the hotel location and the conference location. Therefore, a major problem we have to face with in such cases is that now an answer is a set of tuples *ranked* according to their *degree*. This poses a non-negligible challenge in case we have to deal with a huge amount of data records. Indeed, virtually every tuple may satisfy a query with a non-zero degree and, thus, has to be ranked. Computing all these degrees, ranking them and then selecting the top-k ones is not feasible in practice for large size databases [32].

While there are many works addressing the top-k problem for vague queries in databases (cf. [10, 14, 22, 23, 30, 31, 34, 35, 39]), little is known for the corresponding problem in knowledge representation and reasoning and specifically for SWLs. For instance, [68] considers non-recursive logic programs in which the score combination function is a function of the score of the atoms in the body. The work [55] considers non-recursive logic programs as well, though the score combination function is more expressive and may consider so-called expensive fuzzy relations (the score may depend on the value of an attribute, see [14]). However, a score combination function is allowed in the query rule only. We point out that in the case of non-recursive rules, we may rely on a query rewriting mechanism, which, given an initial query, rewrites it, using rules and/or axioms of the KB, into a set of new queries until no new query rule can be derived (this phase may require exponential time relative to the size of the KB, but is polynomial in the size of the facts). The obtained queries may then be submitted directly to a top-k retrieval database engine. The answers to each query are then merged using the disjunctive threshold algorithm (DTA) given in [55]. The works [54, 56, 58, 63] (see also [61]) address the top-k retrieval problem for the description logic *DL-Lite/DLR-Lite* [7, 12], though recursion is allowed among the axioms. Again, the score combination function may consider expensive fuzzy relations. However, a score combination function is allowed in the query only. The work [60] shows an application of top-k retrieval to the case of multimedia information retrieval by relying on a fuzzy variant of *DLR-Lite*. [57] addresses the top-k retrieval for general (recursive) LPs and is closest to this work. [37] slightly extends [57] as it allows also DLR-Lite axioms to occur and tries to rely as much as possible on current top-k database technology. However, these two works exhibits incorrect algorithms, which have been corrected in [64]. In this latter work, it is additionally shown that we can smoothly extend the top-k problem to the top-k-n problem. This latter problem has been shown to be fundamental in electronic Matchmaking [47, 48]. Moreno et al. [45] uses a threshold mechanism in the query for reducing the amount of computation needed to answer a propositional query in a tabulation-based procedure for propositional multi-adjoint logic programs and, thus, does not address the top-k retrieval problem. Chortaras et al. and Damasio et al. [15, 16, 18, 19] propose query answering procedures, which are based on unification to compute answers, but do not address the top-k retrieval problem. It is unclear yet whether unification-based query driven query answering procedures can be combined with a threshold mechanism in such a way to compute top-k answers.

In this chapter, we will provide the basic notions and salient references to be known concerning the top-k retrieval problem for Semantic Web languages.[1]

In the following, we will proceed as follows. We overview briefly SWLs and relate them to their logical counterpart. Then, we briefly sketch a general framework for ontology-based access to databases, the related top-k retrieval problem and algorithmic solutions to this problem.

2 Semantic Web Languages: Overview

Semantic Web Languages (SWL) are standard languages used to provide a formal description of concepts, terms, and relationships within a given knowledge domain. There are essentially three family of languages: namely, *triple languages* RDF & RDFS [9] (*Resource Description Framework*), *conceptual languages* of the OWL 2 family (*Ontology Web Language*) [43] and *rule languages* of the RIF family (*Rule Interchange Format*) [50]. While their syntactic specification is based on XML [69], their semantics is based on logical formalisms, which will be the focus here: briefly,

- RDFS is a logic having intensional semantics and the logical counterpart is ρdf [41].
- OWL 2 is a family of languages that relate to *Description Logics* (DLs) [7].
- RIF relates to the *Logic Programming* (LP) paradigm [36].
- Both OWL 2 and RIF have an extensional semantics.

---

[1] By purpose we will neglect the details of these works (including description of algorithms and implementations).

2.1 RDF and RDFS

The basic ingredients of *RDF* are *triples*, which are of the form

$$(s, p, o) \,,$$

such as *(umberto, likes, tomato)*, stating that *subject s* has *property p* with *value o*. In *RDF Schema* (RDFS), which is an extension of RDF, additionally some special keywords (subclass, subproperty, property domain and range and instance of specifications) may be used as properties to further improve the expressivity of the language. For instance we may also express that the class of 'tomatoes is a subclass of the class of vegetables', *(tomato, sc, vegetables)*, while Zurich is an instance of the class of cities, *(zurich, type, city)*.

From a computational point of view, one computes the so-called *closure* (denoted $cl(\mathcal{K})$) of a set of triples $\mathcal{K}$. That is, one infers all possible triples using inference rules [40, 41, 49], such as

$$\frac{(A, \mathsf{sc}, B), (X, \mathsf{type}, A)}{(X, \mathsf{type}, B)}$$

if A subclass of B and X instance of A then infer that X is instance of B,

and then store all inferred triples into a relational database to be used then for querying. Note that making all implicit knowledge explicit is viable due to the low complexity of the closure computation, which is $\mathcal{O}(|\mathcal{K}|^2)$ in the worst case.

2.2 OWL Family

The Web Ontology Language *OWL* [42] and its successor *OWL 2* [17, 43] are "object oriented" languages for defining and instantiating ontologies. An OWL ontology may include descriptions of classes, properties and their instances, such as

```
class Person partial Human
        restriction (hasName someValuesFrom String)
        restriction (hasBirthPlace someValuesFrom Geoplace)
```

The class Person is a subclass of class Human and has two attributes: hasName having a string as value, and hasBirthPlace whose value is an instance of the class Geoplace.

Given such an ontology, the OWL formal semantics specifies how to derive its logical consequences. For example, if an individual Peter is an instance of the class Student, and Student is a subclass of Person, then one can derive that Peter is also an instance of Person in a similar way as it happens for RDFS. However, let us note that OWL is more expressive than RDFS, as the decision problems for OWL are in higher complexity classes [46] than for RDFS.

OWL 2 [17, 43] is an update of OWL 1 adding several new features, including an increased expressive power. OWL 2 also defines several *OWL 2 profiles*, i.e. OWL 2 language subsets that may better meet certain computational complexity requirements or may be easier to implement. The choice of which profile to use in practice will depend on the structure of the ontologies and the reasoning tasks at hand. The OWL 2 profiles are:

OWL 2 EL.   It is particularly useful in applications employing ontologies that contain very large numbers of properties and/or classes (basic reasoning problems can be performed in time that is polynomial with respect to the size of the ontology [4]). The EL acronym reflects the profile's basis in the $\mathcal{EL}$ family of description logics [4].

OWL 2 QL.   It is aimed at applications that use very large volumes of instance data, and where query answering is the most important reasoning task. In OWL 2 QL, conjunctive query answering can be implemented using conventional relational database systems. Using a suitable reasoning technique, sound and complete conjunctive query answering can be performed in LOGSPACE with respect to the size of the data (assertions) [3, 13]. The QL acronym reflects the fact that query answering in this profile can be implemented by rewriting queries into a standard relational Query Language such as SQL [66].

OWL 2 RL.   It is aimed at applications that require scalable reasoning without sacrificing too much expressive power. OWL 2 RL reasoning systems can be implemented using rule-based reasoning engines as a mapping to *Logic Programming* [36], specifically *Datalog* [66], exists. The RL acronym reflects the fact that reasoning in this profile can be implemented using a standard rule language [24]. The computational complexity is the same as for Datalog [20] (polynomial in the size of the data, EXPTIME w.r.t. the size of the knowledge base).

2.3 RIF Family

The *Rule Interchange Format* (RIF) aims at becoming a standard for exchanging rules, such as

> Forall?Buyer ?Item ?Seller
>   buy(?Buyer ?Item ?Seller): −sell(?Seller ?Item ?Buyer)

Someone buys an item from a seller if the seller sells that item to the buyer

among rule systems, in particular among Web rule engines. RIF is in fact a family of languages, called *dialects*, among which the most significant are:

RIF-BLD. The *Basic Logic Dialect* is the main logic-based dialect. Technically, this dialect corresponds to Horn logic with various syntactic and semantic extensions. The main syntactic extensions include the frame syntax and predicates with named arguments. The main semantic extensions include datatypes and externally defined predicates.

RIF-PRD. The *Production Rule Dialect* aims at capturing the main aspects of various production rule systems. Production rules, as they are currently practiced in mainstream systems like Jess[2] or JRules,[3] are defined using ad hoc computational mechanisms, which are not based on a logic. For this reason, RIF-PRD is not part of the suite of logical RIF dialects and stands apart from them. However, significant effort has been extended to ensure as much sharing with the other dialects as possible. This sharing was the main reason for the development of the RIF Core dialect;

RIF-Core. The *Core Dialect* is a subset of both RIF-BLD and RIF-PRD, thus enabling limited rule exchange between logic rule dialects and production rules. RIF-Core corresponds to Horn logic without function symbols (i.e., Datalog) with a number of extensions to support features such as objects and frames as in F-logic [33].

RIF-FLD. The *Framework for Logic Dialects* is not a dialect in its own right, but rather a general logical extensibility framework. It was introduced in order to drastically lower the amount of effort needed to define and verify new logic dialects that extend the capabilities of RIF-BLD.

3 Ontology-Based Databases

We illustrate here a simple, though general enough framework to present the top-k retrieval problem for the various SWLs sketched in the previous section.

To start with, the scoring space is the set (n positive integer)

$$L_n = \left\{ 0, \frac{1}{n}, \ldots, \frac{n-2}{n-1}, 1 \right\}.$$

Then a *Knowledge Base* (KB) $\mathcal{K} = \langle \mathcal{F}, \mathcal{O}, \mathcal{M} \rangle$ consists of a *facts component* $\mathcal{F}$, an *Ontology component* $\mathcal{O}$ and an *mapping component* $\mathcal{M}$, which are defined below. Informally, the facts component is the relational database in which the extensional data is stored, the ontology component is the intentional level and describes the important relations about the world we are modelling and the mapping component defines the bridge between the low-level database schema vocabulary and the abstract ontology vocabulary. While the facts and mapping component is syntactically equal for all SWLs, the ontology component is language dependent. The same applies for

---

[2] http://www.jessrules.com/

[3] http://www.ilog.com/products/jrules/

the query language. It is also assumed that relations occurring in $\mathscr{F}$ do not occur in $\mathscr{O}$ (so, we do not allow that database relation names occur in $\mathscr{O}$).

3.1 The Facts Layer

$\mathscr{F}$ is a finite set of expressions of the form

$$R(c_1, \ldots, c_n) : s \ ,$$

where R is an n-ary relation, every c_i is a constant, and s is a degree in L_n.

For each relation R, we represent the facts $R(c_1, \ldots, c_n) : s$ in $\mathscr{F}$ by means of a relational $n + 1$-ary table T_R, containing the records $\langle c_1, \ldots, c_n, s \rangle$. We assume that there cannot be two records $\langle c_1, \ldots, c_n, s_1 \rangle$ and $\langle c_1, \ldots, c_n, s_2 \rangle$ in T_R with $s_1 \neq s_2$ (if there are, then we remove the one with the lower degree). Each table is sorted in descending order with respect to the degrees. For ease, we may omit the degree component and in such case the value 1 is assumed.

3.2 The Mapping Layer

The mapping component is a set of "mapping statements" that allow to connect classes and properties to physical relational tables. Essentially, this component is used as a wrapper to the underlying database and, thus, prevents that relational table names occur in the ontology. Formally, a *mapping statement* (see [63]) is of the form

$$R \mapsto (c_1, \ldots, c_n) : c_{score}.sql \ ,$$

where sql is a SQL statement returning n-ary tuples $\langle c_1, \ldots, c_n \rangle$ $(n = 1, 2)$ with score determined by the c_{score} column. The tuples have to be ranked in decreasing order of score and, as for the fact component, we assume that there cannot be two records $\langle \mathbf{c}, s_1 \rangle$ and $\langle \mathbf{c}, s_2 \rangle$ in the result set of sql with $s_1 \neq s_2$ (if there are, then we remove the one with the lower score). The score c_{score} may be omitted and in that case the score 1 is assumed for the tuples. We assume that R occurs in $\mathscr{O}$, while all of the relational tables occurring in the SQL statement occur in $\mathscr{F}$.

3.3 An RDFS Ontology Layer

The ontology component is used to define the relevant abstract concepts and relations of the application domain by means of axioms and defines the so-called *intentional level*. To what concerns us here, we will consider the essential features of RDFS only and follow [25, 40, 41] by considering the "core" part of RDFS, called ρdf [41] (read rho-df, the ρ from restricted RDF). Specifically, $\mathcal{O}$ is a finite set of *axioms* having the form

$$(s, p, o) \, ,$$

where p is a property that may belong to ρdf$^-$:

$$\rho\text{df}^- = \rho\text{df} \setminus \{\text{type}\} = \{\text{sp, sc, dom, range}\} \, ,$$

The keywords in ρdf$^-$ may be used in triples as properties. Informally,

- (p, sp, q) means that property p is a *sub-property* of property q;
- (c, sc, d) means that class c is a *subclass* of class d;
- (p, dom, c) means that the *domain* of property p is c; and
- (p, range, c) means that the *range* of property p is c.

Note that we don't allow the use of ρdf triples of the form (a, type, b) with meaning "a is of *type* b" in the intentional level.

3.4 An OWL 2 Ontology Layer

The OWL 2 ontology layer $\mathcal{O}$ consists of a finite set of OWL 2 class and property axioms (see, [43]) of the form

$$C \sqsubseteq D$$
$$R \sqsubseteq P$$

where C, D are OWL 2 classes (*concepts*) and R and P are OWL 2 properties . The intuition of an axiom of the form $C \sqsubseteq D$ (resp. $R \sqsubseteq P$) is that any instance of class C (resp. role R) is an instance of class D (resp. P) as well. For computational reasons (to have a query answering procedure that is worst case polynomial), we will consider here specifically the case of the logical counterpart of the three main OWL 2 profiles, namely OWL 2 EL, OWL 2 QL and OWL 2 RL only.

3.4.1 The Case of OWL 2 EL

The importance of the $\mathcal{EL}$ DL family [4, 6, 26] is due to the fact that it is the logical counterpart of the OWL 2 EL profile [44], i.e. OWL 2 EL constructs can be mapped into the DL $\mathcal{EL}^{++}$(d). We recall that it enables polynomial time algorithms for all the standard reasoning tasks[4] and, thus, it is particularly suitable for applications where very large ontologies are needed.

For illustrative purposes, we consider here the following sublanguage of $\mathcal{EL}^{++}$(d) together with its FOL reading [4–6]:

Concept expressions	FOL-reading
A	$A(x)$
$C \sqcap D$	$C(x) \wedge D(x)$
$\exists R.C$	$\exists y.R(x, y) \wedge C(x)$
Axioms	**FOL-reading**
$C \sqsubseteq D$	$\forall x.C(x) \Rightarrow D(x)$
$R_1 \circ R_2 \sqsubseteq R$	$\forall x \forall y \forall z.R_1(x, z) \wedge R_2(z, y) \Rightarrow R(x, y)$
$\mathsf{dom}(R) \sqsubseteq C$	$\forall x.R(x, y) \Rightarrow C(x)$
$\mathsf{ran}(R) \sqsubseteq C$	$\forall y.R(x, y) \Rightarrow C(y)$
$\mathsf{ref}(R)$	$\forall x.R(x, x)$.

3.4.2 The Case of OWL 2 QL

The importance of the DL-Lite DL family [3, 11–13] is due to the fact that it is the logical counterpart of the OWL 2 QL profile [44], i.e.OWL 2 QL constructs can be mapped into the DL DL-Lite$_R$(d), which, we recall, was designed so that sound and complete query answering is in LOGSPACE (more precisely, in AC^0) with respect to the size of the data, while providing many of the main features necessary to express conceptual models such as UML class diagrams and ER diagrams.

We next recap succinctly the DL-Lite DL family [13]. We start with the language DL-Lite$_{core}$ that is the core language for the whole family. Concepts and roles are formed according to the following syntax (A is an atomic concept, P is an atomic role and P^- is its inverse):[5]

$$B \longrightarrow A \mid \exists R$$
$$C \longrightarrow B \mid \neg B$$
$$R \longrightarrow P \mid P^-$$
$$E \longrightarrow R \mid \neg R .$$

B denotes a *basic concept*, that is, a concept that can be either an atomic concept or a concept of the form $\exists R$, where R denotes a *basic role*, that is, a role that is either an atomic role or the inverse of an atomic role. C denotes a concept, which can be a basic concept or its negation, whereas E denotes a role, which can be a basic role or its negation. Sometimes we write $\neg C$ (resp., $\neg E$) with the intended meaning that

---

[4] That is, the ontology satisfiability problem, the subsumption problem and the instance checking problem.

[5] The FOL-reading of concept $\exists R$ is: set of x such that $\exists y.R(x, y)$.

$\neg C = \neg A$ if $C = A$ (resp., $\neg E = \neg R$ if $E = R$), and $\neg C = A$, if $C = \neg A$ (resp., $\neg E = R$, if $E = \neg R$).[6]

Inclusion axioms are of the form

$$B \sqsubseteq C$$

We might include $B_1 \sqcup B_2$ (FOL-reading: set of x such that $B_1(x) \vee B_2(x)$) in the constructs for the left-hand side of inclusion axioms and $C_1 \sqcap C_2$ in the constructs for the right-hand side. In this way, however, we would not extend the expressive capabilities of the language, since these constructs can be simulated by considering that $B_1 \sqcup B_2 \sqsubseteq C$ is equivalent to the pair of assertions $B_1 \sqsubseteq C$ and $B_2 \sqsubseteq C$, and that $B \sqsubseteq C_1 \sqcap C_2$ is equivalent to $B \sqsubseteq C_1$ and $B \sqsubseteq C_2$. Similarly, we might add $\bot$ to the constructs for the left-hand side and $\top$ to those for the right-hand side.

Eventually, DL-Lite$_R$ is now obtained by extending DL-Lite$_{core}$ with the ability of specifying inclusion axioms between roles of the form

$$R \sqsubseteq E.$$

where R and E are defined as above.

3.4.3 The Case of OWL 2 RL

The importance of the *Horn*-DL family [24, 65] is due to the fact that it is the logical counterpart of the OWL 2 RL profile [44], i.e.OWL 2 RL constructs can be mapped into Horn-DL. This is achieved by defining a syntactic subset of OWL 2, which is amenable to implementation using rule-based technologies. Essentially, the restrictions are designed so as to avoid the need to infer the existence of individuals not explicitly present in the knowledge base, and to avoid the need for nondeterministic reasoning. Here we report a subset of the DL specification of OWL 2 RL. Specifically, concepts are formed according to the following syntax (the FOL-reading of concept $\forall R.C$ is: set of x such that $\forall y.R(x, y) \Rightarrow C(y)$):

$$B \longrightarrow A \mid B_1 \sqcap B_2 \mid B_1 \sqcup B_2 \mid \exists R.B$$
$$C \longrightarrow A \mid C_1 \sqcap C_2 \mid \neg B \mid \forall R.C \mid$$
$$R \longrightarrow P \mid P^-$$

Inclusion axioms have the form

---

[6] Of course, for any interpretation $\mathcal{I}$, $(\neg R)^{\mathcal{I}} = \Delta^{\mathcal{I}} \times \Delta^{\mathcal{I}} \setminus R^{\mathcal{I}}$.

$$B \qquad\qquad \sqsubseteq C$$
$$R_1 \qquad\qquad \sqsubseteq R_2$$
$$R_1 \qquad\qquad = R_2$$
$$\mathsf{dom}(R) \quad\; \sqsubseteq C$$
$$\mathsf{ran}(R) \quad\;\; \sqsubseteq C$$
$$R \circ R \sqsubseteq R.$$

3.5 A RIF Ontology Layer

A RIF ontology layer $\mathcal{O}$ consists of a finite set of RIF rules. For illustrative purposes, we consider here RIF-Core rules only and recall that RIF-Core corresponds to Horn logic without function symbols (i.e., Datalog [66]) with a number of extensions to support features such as objects and frames as in F-logic [33]. To what concerns us, a rule is of the form

$$p(\mathbf{x}) \leftarrow \exists \mathbf{y}.\varphi(\mathbf{x}, \mathbf{y}),$$

where $\varphi(\mathbf{x}, \mathbf{y})$ is a conjunction[7] of n-ary predicates $p_i(\mathbf{z}_i)$ and $\mathbf{z}_i$ is a vector of distinguished or non-distinguished variables. Specifically, we say that $p(\mathbf{x})$ is the *head* and $\exists \mathbf{y}.\varphi(\mathbf{x}, \mathbf{y})$ is the *body* of the rule, $\mathbf{x}$ is a vector of variables occurring in the body, called the *distinguished variables*, $\mathbf{y}$ are so-called *non-distinguished variables* and are distinct from the variables in $\mathbf{x}$, each variable occurring in p_i is either a distinguished or a non-distinguished variable. If clear from the context, we may omit the existential quantification $\exists \mathbf{y}$. The intended meaning of a rule such as (3.5) is that the head $p(\mathbf{x})$ is true whenever the body $\exists \mathbf{y}.\varphi(\mathbf{x}, \mathbf{y})$ is true.

4 Top-k Queries

Having defined how extensional data (a database) and intentional data (an ontology) may be represented, it remains to define how we may query the data. To this end, we define the notion of conjunctive query, which is at the heart of the standard Semantic Web query language SPARQL [52, 53]. Strictly speaking, SPARQL is a query language for data that is stored natively as RDFS or viewed as RDF via middleware. From a logical point of view, its logical counterpart are the well-known notions of *conjunctive/disjunctive* queries. As such, we may see SPARQL essentially as a query language for databases and, indeed, has much in common with SQL.

While SPARQL has originally been proposed to query RDFS graphs only, in the meanwhile, by relying on the representation of OWL and RIF in RDFS, SPARQL is being used to query OWL 2 and RIF ontologies as well, via the definition of the so-called *entailment regimes*. In fact, what correct answers to a SPARQL query

---

[7] We use the symbol ',' to denote conjunction in the rule body.

are depends on the used entailment regime [51] and the vocabulary from which the resulting answers can be taken.

To what concerns our presentation here, we will consider the essential logical counterpart of SPARQL: namely, conjunctive queries. Specifically, a *simple query* is of the rule-like form

$$q(\mathbf{x}) \leftarrow \exists \mathbf{y}.\varphi(\mathbf{x}, \mathbf{y})$$

where $\varphi(\mathbf{x}, \mathbf{y})$ is a conjunction of atoms whose notion is SWL dependent. Specifically, For RDFS, an atom is a triple in which variables and constants may occur; for OWL 2, an atom is n-ary FOL atom ($n = 1, 2$) in which variables and constants may occur; while for RIF, an atom is an n-ary FOL atom in which variables and constants may occur.

We additionally allow built-in atoms involving build-in predicates (properties) having a *fixed interpretation*. For instance, an RDFS query is

$$q(x_1, x_2) \leftarrow (x, works For, google), (x, has Salary, s), (s, <, 23000)$$

and is asking for Google employees earning less than 23000. Here $<$ is a build-in predicate.

For convenience, we write "functional built-in predicates"[8] as *assignments* of the form $x := f(\mathbf{z})$.

As next, we extend the query language by allowing so-called aggregates to occur in a query. Essentially, aggregates may be like the usual SQL aggregate functions such as SUM, AVG, MAX, MIN.

For instance, suppose we are looking for employees that work for some company. We would like to know the average salary of their employment. Such a query may be expressed as

$$q(x, avgS) \leftarrow (x, works For, y), (x, has Salary, s),$$
$$\text{GroupedBy}(x),$$
$$avgS := \text{AVG}[s] .$$

Essentially, we group by the employee, consider for each employee the salaries, and compute the average salary value for each group. That is, if $g = \{\langle t, t_1 \rangle, \ldots, \langle t, t_n \rangle\}$ is a group of tuples with the same value t for employee x, and value t_i for s, then the value of $avgL$ for the group g is $(\sum_i t_i)/n$.

Formally, let @ be an aggregate function with

$$@ \in \{\text{SUM, AVG, MAX, MIN, COUNT}\}$$

then a query with aggregates is of the form

---

[8] A predicate $p(\mathbf{x}, y)$ is functional if for any $\mathbf{t}$ there is *unique* t' for which $p(\mathbf{t}, t')$ is true.

$$q(\mathbf{x}, \alpha) \leftarrow \exists \mathbf{y}.\varphi(\mathbf{x}, \mathbf{y}),$$
$$\text{GroupedBy}(\mathbf{w}), \tag{1}$$
$$\alpha := @[f(\mathbf{z})]$$

where $\mathbf{w}$ are variables in $\mathbf{x}$ or $\mathbf{y}$, each variable in $\mathbf{x}$ occurs in $\mathbf{w}$ and any variable in $\mathbf{z}$ occurs in $\mathbf{y}$.

Eventually, we further allow to order answers according to some ordering functions. For instance, assume that additionally we would like to order the employee according to the average salary of employment. Then such a query will be expressed as

$$q(x, avgS) \leftarrow (x, worksFor, y), (x, hasSalary, s),$$
$$\text{GroupedBy}(x),$$
$$avgS := \text{AVG}[s],$$
$$\text{OrderBy}(avgS).$$

Formally, a query with ordering is of the form

$$q(\mathbf{x}, z) \leftarrow \exists \mathbf{y}.\varphi(\mathbf{x}, \mathbf{y}), \text{OrderBy}(z)$$

or, in case grouping is allowed as well, it is of the form

$$q(\mathbf{x}, z, \alpha) \leftarrow \exists \mathbf{y}.\varphi(\mathbf{x}, \mathbf{y}),$$
$$\text{GroupedBy}(\mathbf{w}),$$
$$\alpha := @[f(\mathbf{z})], \tag{2}$$
$$\text{OrderBy}(z).$$

Eventually, we define a *top-k query* as a query limiting the result to the top-k scoring answers, i.e.

$$q(\mathbf{x}, z, \alpha) \leftarrow \exists \mathbf{y}.\varphi(\mathbf{x}, \mathbf{y}),$$
$$\text{GroupedBy}(\mathbf{w}),$$
$$\alpha := @[f(\mathbf{z})], \tag{3}$$
$$\text{OrderBy}(z),$$
$$\text{Limit}(k)$$

We refer the reader to e.g. [2] for an exact formal definition in case of RDFS, to [63] for the case of OWL 2 and to [64] for the case of RIF.

5 Top-k Query Answering Methods

Having now illustrated what a top-k query is, it remains to illustrate the basic methods in computing the top-k answers efficiently. The methods depend on the chosen SWL. In the following, let $\mathscr{K} = \langle \mathscr{F}, \mathscr{O}, \mathscr{M} \rangle$ be a KB.

5.1 The RDFS Case

So far, we have here essentially two options. The first option is as follows.

1. Convert all facts in $\mathcal{F}$ into a set $T_{\mathcal{F}}$ of triples, by using the mapping layer $\mathcal{M}$. Consequently, $\mathcal{K} = \langle \mathcal{F}, \mathcal{O}, \mathcal{M} \rangle$ can be mapped into a pure RDFS triple set

$$T_{\mathcal{K}} = \mathcal{O} \cup T_{\mathcal{F}},$$

called *RDFS graph*, which has the same order of size.
2. Given $T_{\mathcal{K}}$, we then compute its closure $cl(T_{\mathcal{K}})$, whose size is bounded by $\mathcal{O}(|T_{\mathcal{K}}|^2)$ and store it into a relational database. Note that there are several ways to store the closure in a database (see [1, 28]). Essentially, either we may store all the triples in table with four columns *subject, predicate, object, score*, or we use a table for each predicate, where each table has columns *subject, object, score*. The latter approach seems to be better for query answering purposes. Top-k query answering for RDFS reduces then to top-k query answering over relational databases for which efficient solutions exists already.

Another option consists in using top-k retrieval technologies for rule-based KBs and is defined as follows.

1. Compute the closure $cl(\mathcal{O})$ of $\mathcal{O}$.
2. Map all the triples in $cl(\mathcal{O})$ into Datalog rules using e.g. the mapping rules below (see also e.g. [27–29])

$$(p, \mathsf{sp}, q) \mapsto q(x) \leftarrow p(x)$$
$$(c, \mathsf{sc}, d) \mapsto d(x) \leftarrow c(x)$$
$$(p, \mathsf{dom}, c) \mapsto c(x) \leftarrow p(x, y)$$
$$(p, \mathsf{range}, c) \mapsto c(y) \leftarrow p(x, y).$$

obtaining the rule layer $\mathcal{O}_{\mathcal{K}}$. Let $\mathcal{K}' = \langle \mathcal{F}, \mathcal{O}_{\mathcal{K}}, \mathcal{M} \rangle$ be the resulting rule-based KB.
3. Apply a top-k procedure for rule-based KBs to $\mathcal{K}'$ (see, e.g. [64] and Sect. 5.3).

5.2 The OWL 2 Profile Cases

5.2.1 OWL EL

An option consists in relying on the query reformulation method proposed in [38] to answer conjunctive queries by making use of standard relational database management systems (RDBMSs), and has some commonalities to the OWL QL case (next section). The central idea is to incorporate the consequences of $\mathcal{O}$ into the facts component $\mathcal{F}$.

To capture this formally, the notion of combined first-order (FO) rewritability has been introduced. A DL enjoys combined FO rewritability if it is possible to effectively rewrite

1. $\mathcal{F}$ and $\mathcal{O}$ into an FO structure (independently of the query q); and
2. q and (possibly) $\mathcal{O}$ into a FO query q' (independently of $\mathcal{F}$) such that query answers are preserved, i.e., the answers to q' over the FO structure is the same as the answers to q over $\mathcal{F}$ and $\mathcal{O}$.

The connection to RDBMSs then relies on the well-known equivalence between FO structures and relational databases, and FO queries and SQL queries. The notion of combined FO rewritability generalises the notion of FO reducibility, where $\mathcal{O}$ is incorporated into the query q rather than into $\mathcal{F}$, as it happens for the OWL QL case, while the facts component $\mathcal{F}$ itself is used as a relational instance without any modification [13].

Hence, a top-k query answering procedure for OWL EL may consists of

1. rewriting, once for all, $\mathcal{F}$ and $\mathcal{O}$, using $\mathcal{M}$, into an relational database $DB_{\mathcal{K}}$;
2. rewriting a top-k query q using (possibly) $\mathcal{O}$ and $\mathcal{M}$ into a top-k SQL query to be submitted to the RDBMS containing $DB_{\mathcal{K}}$.

5.2.2 OWL QL

The OWL QL case proceeds similarly as for the OWL EL case, as DL-Lite [3, 13] enjoys the FO reducibility property. Specifically, a top-k query answering procedure for OWL QL may consists of (see [63]) rewriting a top-k query q using (possibly) $\mathcal{O}$ and $\mathcal{M}$ into a top-k SQL query to be submitted to the RDBMS containing $DB_{\mathcal{K}}$.

5.2.3 OWL RL

Concerning OWL RL, we employ the close connection between Horn-DL and Datalog. Specifically, an option to compute the top-k answers consist in using top-k retrieval technologies for rule-based KBs. To this end, we now define a recursive mapping function σ which takes a set of inclusion axioms and maps them into the following expressions:[9]

---

[9] For the sake of ease of presentation, we are not going to present the whole mapping for Horn-DL, but for a significant subset only that is sufficient to illustrate the main idea behind this translation.

$$\sigma(R_1 \sqsubseteq R_2) \mapsto \sigma_{role}(R_2, x, y) \leftarrow \sigma_{role}(R_1, x, y)$$
$$\sigma_{role}(R, x, y) \mapsto R(x, y)$$
$$\sigma_r(R^-, x, y) \mapsto R(y, x)$$

$$\sigma(B \sqsubseteq C) \mapsto \sigma_h(C, x) \leftarrow \sigma_b(B)$$
$$\sigma_h(A, x) \mapsto A(x)$$
$$\sigma_h(C_1 \sqcap C_2, x) \mapsto \sigma_h(C_1, x) \wedge \sigma_h(C_2, x)$$
$$\sigma_h(\forall R.C, x) \mapsto \sigma_h(C, x) \leftarrow \sigma_{role}(R, x, y)$$
$$\sigma_b(A, x) \mapsto A(x)$$
$$\sigma_b(C_1 \sqcap C_2, x) \mapsto \sigma_b(C_1, x) \wedge \sigma_b(C_2, x)$$
$$\sigma_b(C_1 \sqcup C_2, x) \mapsto \sigma_b(C_1, x) \vee \sigma_b(C_2, x)$$
$$\sigma_b(\exists R.C, x) \mapsto \sigma_{role}(R, x, y) \wedge \sigma_b(C, y)$$

where y is a new variable.

We then transform the above generated expressions into rules by applying recursively the following mapping:

$$\sigma_r((H \wedge H') \leftarrow B) \mapsto \sigma_r(H \leftarrow B), \sigma_r(H' \leftarrow B)$$
$$\sigma_r((H \leftarrow H') \leftarrow B) \mapsto \sigma_r(H \leftarrow (B \wedge H'))$$
$$\sigma_r(H \leftarrow (B_1 \vee B_2)) \mapsto \sigma_r(H \leftarrow B_1), \sigma_r(H \leftarrow B_2)$$

Eventually, if none of the above three rules can be applied then

$$\sigma_r(H \leftarrow B) \mapsto H \leftarrow B .$$

Therefore, a top-k retrieval method of OWL RL is as follows:

1. Map $\mathcal{O}$ into Datalog rules using e.g. the mapping rules above obtaining the rule layer $\mathcal{O}_{\mathcal{H}}$. Let $\mathcal{K}' = \langle \mathcal{F}, \mathcal{O}_{\mathcal{H}}, \mathcal{M} \rangle$ be the resulting rule-based KB.
2. Apply a top-k procedure for rule-based KBs to $\mathcal{K}'$ (see, e.g. [64] and Sect. 5.3).

5.3 The RIF case

Concerning the RIF case, by exploiting the relationship to Datalog (specifically of RIF-Core), we may opt for a solution inherited from the logic programming context. We have already reported in Sect. 1 about various proposal developed so far. We recap here the main principle behind the so far most general approach [64].

The basic reasoning idea stems from the database literature (see, e.g. [35]) and consists in retrieving iteratively query answers and simultaneously computing a threshold

δ. The threshold δ has the fundamental property that any newly retrieved tuple will have a score less or equal than δ. As a consequence, as soon as we have retrieved k tuples greater or equal to δ, we may stop. Note that a distinguishing feature of this query answering procedure is that it does not determine all answers, but collects, during the computation, answers incrementally together and stops as soon as it has gathered k answers greater or equal than a computed threshold δ. The finiteness of the truth space guarantees the termination of this process, which otherwise may not terminate.

6 Conclusions

In this work, we briefly discussed about a relatively novel issue for SWLs with a huge data repository, namely the problem of evaluating ranked top-k queries. We have illustrated how this problem may be currently approached within the context of RDFS (the triple language), OWL 2 Profiles (frame-based languages) and RIF (rule language).

While for relational databases a non negligible amount of solutions have been proposed so far, in the context of knowledge representation and reasoning, the development is still in its infancy, both from an algorithmic and implementation point of view. So far, for the languages RDFS, OWL QL, OWL EL and RIF ad hoc solution have been worked out, while for OWL RL a reduction to RIF is required. Note that for RDFS and OWL QL, a reduction to RIF exists as well and, thus, top-k techniques for this latter can be applied.

We believe that with the growth in size of data repositories accessible via SWLs, the top-k retrieval problem will emerge as a significant problem, as much as the top-k retrieval problem is for current Information Retrieval Systems [8].

References

1. Abadi, D.J., Marcus, A., Madden, S., Hollenbach, K.: Sw-store: a vertically partitioned dbms for semantic web data management. VLDB J. **18**(2), 385–406 (2009)
2. Zimmermann, A.P.A., Lopes, N., Straccia, U.: A general framework for representing, reasoning and querying with annotated semantic web data. J. Web Semant. **11**, 72–95 (March 2012)
3. Artale, A., Calvanese, D., Kontchakov, R., Zakharyaschev, M.: The DL-Lite family and relations. J. Artif. Intell. Res. **36**, 1–69 (2009)
4. Baader, F., Brandt, S., Lutz, C.: Pushing the $\mathcal{EL}$ envelope. In: Proceedings of the 19th International Joint Conference on Artificial Intelligence IJCAI-05, pp. 364–369, Edinburgh, UK. Morgan-Kaufmann Publishers, San Francisco (2005)
5. Baader, F.: Terminological cycles in a description logic with existential restrictions. In: Proceedings of the 18th International Joint Conference on Artificial intelligence, pp. 325–330. Morgan Kaufmann, San Francisco (2003)

6. Baader, F., Brandt S., Lutz, C.: Pushing the $\mathcal{EL}$ envelope further. In: Clark, K., Patel-Schneider, P.F. (eds.) Proceedings of the OWLED 2008 DC Workshop on OWL: Experiences and Directions (2008)

7. Baader, F., Calvanese, D., McGuinness, D., Nardi, D., Patel-Schneider, P.F. (eds.). The Description Logic Handbook: Theory, Implementation, and Applications. Cambridge University Press, Cambridge (2003)

8. Baeza-Yates, R.A., Ribeiro-Neto, B.: Modern Information Retrieval. Addison-Wesley Longman, Boston (1999)

9. Brickley, D., Guha, R.V.: RDF vocabulary description language 1.0: RDF schema. In: W3C Recommendation, W3C (2004). http://www.w3.org/TR/rdf-schema/

10. Bruno, N., Chaudhuri, S., Gravano, L.: Top-k selection queries over relational databases: mapping strategies and performance evaluation. ACM Trans. Database Syst. **27**(2), 153–187 (2002)

11. Calvanese, D., De Giacomo, G., Lembo, D., Lenzerini, M., Rosati, R.: DL-Lite: tractable description logics for ontologies. In: Proceedings of the 20th National Conference on Artificial Intelligence (AAAI 2005) (2005)

12. Calvanese, D., De Giacomo, G., Lembo, D., Lenzerini, M., Rosati, R.: Data complexity of query answering in description logics. In: Proceedings of the 10th International Conference on Principles of Knowledge Representation and Reasoning (KR-06), pp. 260–270 (2006)

13. Calvanese, D., Giacomo, G., Lembo, D., Lenzerini, M., Rosati, R.: Tractable reasoning and efficient query answering in description logics: the dl-lite family. J. Autom. Reasoning **39**(3), 385–429 (2007)

14. Chang, K.C.-C., won Hwang, S.: Minimal probing: supporting expensive predicates for top-k queries. In: Proceedings of the SIGMOD Conference, pp. 346–357 (2002)

15. Chortaras, A., Stamou, G.B., Stafylopatis, A.: Integrated query answering with weighted fuzzy rules. In: Proceedings of the 9th European Conference on Symbolic and Quantitative Approaches to Reasoning with Uncertainty (ECSQARU-07), vol. 4724 in Lecture Notes in Computer Science, pp. 767–778. Springer (2007)

16. Chortaras, A., Stamou, G.B., Stafylopatis, A.: Top-down computation of the semantics of weighted fuzzy logic programs. In: Proceedings of the 1st International Conference on Web Reasoning and Rule Systems (RR-07), pp. 364–366 (2007)

17. Cuenca-Grau, B., Horrocks, I., Motik, B., Parsia, B., Patel-Schneider, P.F., Sattler, U.: OWL 2: the next step for OWL. J. Web Semant. **6**(4), 309–322 (2008)

18. Damásio, C.V., Medina, M., Ojeda-Aciego, J.: A tabulation procedure for first-order residuated logic programs. In: Proceedings of the IEEE World Congress on Computational Intelligence (Sect. Fuzzy Systems) (WCCI-06), pp. 9576–9583 (2006)

19. Damásio, C.V., Medina, M., Ojeda-Aciego, J.: A tabulation procedure for first-order residuated logic programs. In: Proceedings of the 11th International Conference on Information Processing and Managment of Uncertainty in Knowledge-Based Systems, (IPMU-06) (2006)

20. Dantsin, E., Eiter, T., Gottlob, G., Voronkov, A.: Complexity and expressive power of logic programming. ACM Comput. Surv. **33**(3), 374–425 (2001)

21. Dubois, D., Prade, H.: Possibility theory, probability theory and multiple-valued logics: a clarification. Ann. Math. Artif. Intell. **32**(1–4), 35–66 (2001)

22. Fagin, R.: Combining fuzzy information: an overview. SIGMOD Rec. **31**(2), 109–118 (2002)

23. Fagin, R., Lotem, A., Naor, M.: Optimal aggregation algorithms for middleware. In: Symposium on Principles of Database Systems (2001)

24. Grosof, B.N., Horrocks, I., Volz, R., Decker, S.: Description logic programs: combining logic programs with description logic. In: Proceedings of the 12th International Conference on World Wide Web, pp. 48–57. ACM Press (2003)

25. Gutierrez, C., Hurtado, C., Mendelzon, A.O.: Foundations of semantic web databases. In: Proceedings of the 23rd ACM SIGMOD-SIGACT-SIGART Symposium on Principles of Database Systems (PODS-04). ACM Press (2004)

26. Haase, C., Lutz, C.: Complexity of subsumption in the $\mathcal{EL}$ family of description logics: acyclic and cyclic tboxes. In: Ghallab, M., Spyropoulos, C.D., Fakotakis, N., Avouris, N. (eds.) Proceedings of the 18th European Conference on Artificial Intelligence (ECAI08), vol. 178 of Frontiers in Artificial Intelligence and Applications, pp. 25–29. IOS Press (2008)

27. Ianni, G., Krennwallner, T., Martello, A., Polleres, A.: Dynamic querying of mass-storage RDF data with rule-based entailment regimes. In: Proceedings of the 8th International Semantic Web Conference (ISWC-09), vol. 5823 in Lecture Notes in Computer Science, pp. 310–327. Springer (2009)
28. Ianni, G., Krennwallner, T., Martello, A., Polleres, A.: A rule system for querying persistent rdfs data. In: Proceedings of the 6th European Semantic Web Conference on Semantic Web: Research and Applications (ESWC-2009), pp. 857–862 (2009)
29. Ianni, G., Krennwallner, T., Martello, A., Polleres, A.: A rule system for querying persistent rdfs data. In: Aroyo, L., Traverso, P., Ciravegna, F., Cimiano, P., Heath, T., Hyvönen, E., Mizoguchi, R., Oren, E., Sabou, M., Simperl, E.P.B. (eds.) ESWC, vol. 5554 of Lecture Notes in Computer Science, pp. 857–862. Springer (2009)
30. Ilyas, I.F., Aref, W.G., Elmagarmid, A.K.: Supporting top-k join queries in relational databases. In: Proceedings of the 29th International Conference on Very Large Data, Bases (VLDB-03), pp. 754–765 (2003)
31. Ilyas, I.F., Aref, W.G., Elmagarmid, A.K., Elmongui, H.G., Shah, R., Vitter, J.S.: Adaptive rank-aware query optimization in relational databases. ACM Trans. Database Syst. **31**(4), 1257–1304 (2006)
32. Ilyas, I.F., Beskales, G., Soliman, M.A.: A survey of top-k query processing techniques in relational database systems. ACM Comput. Surv. **40**(4), 1–58 (2008)
33. Kifer, M., Lausen, G.: Logical foundations of object-oriented and frame-based languages. J. ACM **42**(4), 741–843 (1995)
34. Li, C., Chang, K.C.-C., Ilyas, I.F.: Supporting ad-hoc ranking aggregates. In: Proceedings of the 2006 ACM SIGMOD International Conference on Management of Data (SIGMOD-06), pp. 61–72. ACM Press, New York (2006)
35. Li, C., Chang, K.C.-C., Ilyas, I.F., Song, S.: RankSQL: query algebra and optimization for relational top-k queries. In: Proceedings of the 2005 ACM SIGMOD International Conference on Management of Data (SIGMOD-05), pp. 131–142. ACM Press, New York (2005)
36. Lloyd, J.W.: Foundations of Logic Programming. Springer, Heidelberg (1987)
37. Lukasiewicz, T., Straccia, U.: Top-k retrieval in description logic programs under vagueness for the semantic web. In: Proceedings of the 1st International Conference on Scalable Uncertainty Management (SUM-07), vol. 4772 in Lecture Notes in Computer Science, pp. 16–30. Springer (2007)
38. Lutz, C., Toman, D., Wolter, F.: Conjunctive query answering in the description logic $\mathcal{EL}$ using a relational database system. In: Proceedings of the 21st International Joint Conference on Artificial Intelligence (IJCAI09). AAAI Press (2009)
39. Marian, A., Bruno, N., Gravano, L.: Evaluating top-k queries over web-accessible databases. ACM Trans. Database Syst. **29**(2), 319–362 (2004)
40. Marin, D.: A formalization of rdf. Technical report TR/DCC-2006-8, Deptartment of Computer Science, Universidad de Chile (2004). http://www.dcc.uchile.cl/cgutierr/ftp/draltan.pdf
41. Muñoz, S., Pérez, J., Gutiérrez, C.: Minimal deductive systems for rdf. In: Proceedings of the 4th European Semantic Web Conference (ESWC-07), vol. 4519 in Lecture Notes in Computer Science, pp. 53–67. Springer (2007)
42. OWL web ontology language overview. In: W3C (2004). http://www.w3.org/TR/owl-features/
43. OWL 2 web ontology language document overview. In: W3C (2009). http://www.w3.org/TR/2009/REC-owl2-overview-20091027/
44. OWL 2 web ontology language profiles. In: W3C (2009). http://www.w3.org/TR/2009/REC-owl2-profiles-20091027/
45. Moreno, G., Julian, P., Medina, J., Ojeda, M.: Efficient thresholded tabulation for fuzzy query answering. In: Foundations of Reasoning Under Uncertainty, vol. 249 in Studies in Fuzziness and Soft Computing, pp. 125–141. Springer (2010)
46. Papadimitriou, C.H.: Computational Complexity. Addison Wesley, Reading (1994)
47. Ragone, A., Straccia, U., Di Noia, T., Di Sciascio, E., Donini, F.M.: Vague knowledge bases for matchmaking in p2p e-marketplaces. In: Proceedings of the 4th European Semantic Web Conference (ESWC-07), vol. 4519 in Lecture Notes in Computer Science, pp. 414–428. Springer (2007)

48. Ragone, A., Straccia, U., Di Noia, T., Di Sciascio, E., Donini, F.M.: Fuzzy matchmaking in e-marketplaces of peer entities using Datalog. Fuzzy Sets Syst. **160**(2), 251–268 (2009)
49. RDF semantics. In: W3C (2004). http://www.w3.org/TR/rdf-mt/
50. Rule interchange format (RIF). In: W3C (2011). http://www.w3.org/2001/sw/wiki/RIF
51. SPARQL 1.1 entailment regimes. In: W3C (2011). http://www.w3.org/TR/2011/WD-sparql11-entailment-20110512/
52. SPARQL 1.1 query language. In: W3C (2012). http://www.w3.org/TR/sparql11-query/
53. SPARQL query language for RDF. In: W3C (2008). http://www.w3.org/TR/rdf-sparql-query/
54. Straccia, U.: Answering vague queries in fuzzy DL-Lite. In: Proceedings of the 11th International Conference on Information Processing and Managment of Uncertainty in Knowledge-Based Systems, (IPMU-06), pp. 2238–2245. E.D.K., Paris (2006)
55. Straccia, U.:. Towards top-k query answering in deductive databases. In: Proceedings of the 2006 IEEE International Conference on Systems, Man and Cybernetics (SMC-06), pp. 4873–4879. IEEE (2006)
56. Straccia, U.: Towards top-k query answering in description logics: the case of DL-Lite. In: Proceedings of the 10th European Conference on Logics in Artificial Intelligence (JELIA-06), vol. 4160 in Lecture Notes in Computer Science, pp. 439–451, Liverpool, UK. Springer (2006)
57. Straccia, U.: Towards vague query answering in logic programming for logic-based information retrieval. In: Proceedings of World Congress of the International Fuzzy Systems Association (IFSA-07), vol. 4529 in Lecture Notes in Computer Science, pp. 125–134, Cancun, Mexico. Springer (2007)
58. Straccia, U.: Fuzzy description logic programs. In: Marsala, C., Bouchon-Meunier, B., Yager, R.R., Rifqi, M. (eds.) Uncertainty and Intelligent Information Systems, Chap. 29, pp. 405–418. World Scientific, Singapore (2008)
59. Straccia, U.: Managing uncertainty and vagueness in description logics, logic programs and description logic programs. In: Proceedings of 4th International Summer School, Tutorial Lectures on Reasoning Web, vol. 5224 in Lecture Notes in Computer Science, pp. 54–103. Springer (2008)
60. Straccia, U.: An ontology mediated multimedia information retrieval system. In: Proceedings of the 40th International Symposium on Multiple-Valued Logic (ISMVL-10), pp. 319–324. IEEE Computer Society (2010)
61. Straccia, U.: Softfacts: a top-k retrieval engine for ontology mediated access to relational databases. In: Proceedings of the 2010 IEEE International Conference on Systems, Man and Cybernetics (SMC-10), pp. 4115–4122. IEEE Press (2010)
62. Straccia, U.: Fuzzy logic, annotation domains and semantic web languages. In: Proceedings of the 5th International Conference on Scalable Uncertainty Management (SUM-11), vol. 6929 in Lecture Notes in Computer Science, pp. 2–21. Springer (2011)
63. Straccia, U.: Top-k retrieval for ontology mediated access to relational databases. Inf. Sci. **198**, 1–23 (2012)
64. Straccia, U., Madrid, N.: A top-k query answering procedure for fuzzy logic programming. Fuzzy Sets Syst. **205**, 1–29 (2012)
65. ter Horst, H.J.: Completeness, decidability and complexity of entailment for rdf schema and a semantic extension involving the owl vocabulary. J. Web Semant. **3**(2–3), 79–115 (2005)
66. Ullman, J.D.: Principles of Database and Knowledge Base Systems, vols. 1, 2. Computer Science Press, Potomac (1989)
67. Vardi, M.: The complexity of relational query languages. In: Proceedings of the 14th ACM SIGACT Symposium on Theory of Computing (STOC-82), pp. 137–146 (1982)
68. Vojtás, P.: Fuzzy logic aggregation for semantic web search for the best (top-k) answer. In: Sanchez, E. (ed.) Fuzzy Logic and the Semantic Web, Capturing Intelligence, Chap. 17, pp. 341–359. Elsevier, Amsterdam (2006)
69. XML. In: W3C. http://www.w3.org/XML/

Chapter 6
Semantic Data Management Using Fuzzy Relational Databases

Jesús R. Campaña, Juan M. Medina and Maria A. Vila

Abstract This chapter presents a schema and a transformation algorithm to store OWL ontologies in Object Relational Databases. The database schema allows the storage of an ontology structure, while the transformation algorithm creates an appropriate schema to store its instances preserving all information. We allow the use of instance data of imprecise nature, mostly fuzzy numerical data. An OWL ontology is defined allowing numerical fuzzy datatypes as the range of properties. In order to manage all the information, instance data handling is delegated onto a Fuzzy ORDBMS, which is briefly described. We present here a complete description of the structures conforming the storage schema proposed, and the algorithms used to transform the OWL ontology to a database schema. We also discuss the role of ontologies as relational database design tools.

1 Introduction

The Semantic Web [13] provides a common framework to share and reuse data across applications, enterprises, and communities. The goal of the Semantic Web is to provide access to its contents to humans and machines alike, in an efficient and simple way.

In this new model of the Web, ontologies have proved to be of prime importance, due to their potential to describe the semantics of information and the capacity to solve heterogeneity problems. In recent years, several standard ontology languages for the web have emerged. Resource Description Framework (RDF) and Schema

J. R. Campaña (✉) · J. M. Medina · M. A. Vila
Department of Computer Science and Artificial Intelligence, University of Granada,
18071 Granada, Spain
e-mail: jesuscg@decsai.ugr.es

M. A. Vila
e-mail: vila@decsai.ugr.es

J. M. Medina
e-mail: medina@decsai.ugr.es

O. Pivert and S. Zadrożny (eds.), *Flexible Approaches in Data, Information*
and Knowledge Management, Studies in Computational Intelligence 497,
DOI: 10.1007/978-3-319-00954-4_6, © Springer International Publishing Switzerland 2014

(RDFS), the Web Ontology Language (OWL), and recently the new version OWL 2, are the leading languages of the Semantic Web.[1]

OWL is a language aimed at information processing on the web. It is designed to be interpreted by computers, not read by people. An OWL ontology is an RDF graph, which is a set of RDF triples. Each RDF triple contains a subject, a predicate (or property) that denotes a relationship, and an object. A triple represents a statement of a relationship between the subject and the object. An OWL ontology graph can be written using different syntaxes, the most usual is RDF/XML. OWL has three increasingly expressive sub-languages, OWL Lite, OWL DL and OWL Full.

The use of ontologies as knowledge representation formalisms and its role in the development of the Semantic Web, has motivated the creation of lots of ontologies containing huge quantities of instance data, e.g. LUBM,[2] DBPedia,[3] etc. All this instance data must be managed and stored. Current ontology reasoners are not capable of managing efficiently large amounts of instance data. In this context, it is a good idea to look for solutions to these problems in other well known systems.

Nowadays, Database Management Systems are the most widespread and efficient data storage solution. A vast amount of ontology based Semantic Web data is going to be stored in relational databases. Moreover, most of the data in the traditional Web is already stored in relational databases. If we want to perform a smooth transition from the current web to a Semantic Web model, we must adapt the current content and make new content suitable to be processed in a relational database.

Database Management Systems provide features that ease the implementation of basic reasoning tasks in the database. These features include the capacity for storing and sharing large amounts of data, optimizations for data processing (indexes, clusters, views, etc.), complex data handling features and extensibility for systems based on standard SQL:2003, importation/exportation of different data formats to ease interoperability, recursive queries as considered in SQL:2003 (useful for reasoning operations such as class subsumption) and complex query optimizers.

Relational databases are specifically designed to deal with huge amounts of data, and ontologies are good knowledge representation formalisms. We propose a two-tier model where the conceptual tier is defined by an ontology, while data management is delegated upon a RDBMS, which can manage huge volumes of data and optimize queries.

Our first goal is the use of ontologies expressed in OWL as design tools for database schemas. Database schema design is a non-trivial task that requires knowledge about the domain of the problem to model, and knowledge of the conceptual model to use. Design of domain ontologies is neither a trivial task, but it is compulsory in order to design Semantic Web systems. Once an ontology is defined in OWL, it is not necessary to create another conceptual model for database design i.e. Entity-Relationship (ER) model. The ontology contains all the knowledge required

---

[1] W3C Semantic Web: http://www.w3.org/standards/semanticweb/

[2] LUBM—Lehigh University Benchmark: http://swat.cse.lehigh.edu/projects/lubm/

[3] DBPedia: http://wiki.dbpedia.org/

to develop an appropriate schema. Several works study the role of domain ontologies as database design tools [26, 39].

In order to improve the access to instances while retaining the semantics of the ontology, the ontology and its instances must be stored in different structures. There are two main approaches to ontology storage in databases. One approach stores the complete ontology [23] whereas the other creates a database schema based on the ontology definition to store instances [2, 44]. Each approximation has its virtues and its flaws, as will be explained later. In order to mitigate the drawbacks, our proposal uses a combination of both approaches. Some basic reasoning capacities can be added to the system through the combined use of the original ontology and the instance data in the database schema.

Current state-of-the-art systems based on Description Logic (DL) are ready to serve as a core reasoning engine for Entity-Relationship model. DL-Lite is presented in [19]. This DL is specifically tailored to capture the main notions of conceptual data models such as the ER model, while keeping the worst-case complexity of sound and complete reasoning tractable. In [12] it is proved that current reasoners are EXPTIME-complete with respect to the ER model enhanced with ISA on entities and relationships. In [31] a comparison between OWL and databases is presented. To overcome the differences encountered an extended DL is defined.

Several works deal with the problem of translation between database and ontology. One trend is to focus on the expression of ER schemas as an ontology. In [25] a framework to transform ER diagrams into an OWL ontology is presented. The transformation is performed by means of rules. A similar but more restrictive set of rules is presented in [32] to transform well-formed ER diagrams to OWL Lite ontologies. These approaches focus on the adaption of already existent databases to the semantic web. What we propose on this chapter is the creation of a framework to handle and store ontologies. Storage schemas are created from the ontologies and not the other way around.

The transformation from OWL to a relational schema is performed by means of an algorithm that has as input a domain ontology and generates a relational database schema to store instances. The ontology itself (*TBox*) is stored in database catalog tables specifically designed for this purpose. Instances (*ABox*) are stored in the schema previously built. Mappings between ontology concepts and instance data are created in order to perform basic semantic concordance tasks.

Ontologies are generally used to represent precise concepts. However fuzzy and incomplete data is present on the Web, where the characteristics of some elements may be described using approximate values, ranges of values, upper and lower bounds, linguistic labels, etc. Due to this fact, fuzzy representations of ontologies have emerged to offer a more accurate view of certain concepts.

The use of fuzzy logic in ontologies and its use in conjunction with databases has been object of study. SoftFacts [38], is presented as an ontology mediated information retrieval system over relational databases. An ontology layer is used to define the relevant abstract concepts and relations of the application domain, while graded facts are stored into a relational database. Facts stored in the database contain a degree of truth (or score) in [0, 1]. This score represents to which extent the tuple is an instance of the relation. This approach is different to ours, as we propose to deal with fuzziness at instance level. In order to do so, a special kind of ontology that deals with fuzzy data as part of attributes definitions in classes is needed. We call these ontologies OWL Like ontologies because they are defined essentially in OWL, but the use of fuzzy data in datatype properties is not a standard. This lack of a standard has motivated other approaches, like [16] which proposes to represent fuzziness using OWL 2 annotation properties.

DLs are also applied in conjunction with fuzzy logic. In [46] a DL to represent and reason on fuzzy object-oriented database models is presented. The DL proposed makes it possible to improve reasoning on database models. This approach is different to ours, as we use an object-relational model.

Our system relies in a Fuzzy Object Relational Database Management System (FORDBMS). We take advantage of the possibilities the Object Relational paradigm offers, and the extension mechanisms that allow to create new datatypes and ways of dealing with them. We focus on the use of numerical fuzzy types (approximate values, data intervals, upper/lower bounds and trapezoidal distributions). Using the extension mechanisms, a Fuzzy Database is built over a classical object relational system, taking advantage of all features offered by the underlying system.

This chapter proposes a complete representation of OWL ontologies using fuzzy datatypes within a FORDBMS. The database schema is designed using the ontology, and the ontology itself and its instances are stored. Instance data can contain fuzzy datatypes as attributes. This chapter is organized as follows. Section 2 presents a discussion on the similarities of Knowledge Bases and Databases. Section 3 deals with the use of ontologies as Relational Database design tools. The core of the chapter is presented in Sect. 4, where the whole process of storage and management of ontologies using fuzzy datatypes is explained. Finally, conclusions are presented in Sect. 5.

2 Knowledge Bases and Databases

It is important to clarify the role that each of the components that we are going to use holds in the general schema of this proposal, and how we are going to use them.

In general, both knowledge bases (KB) and databases (DB) allow to define and represent a domain and store information about it. However, both deal with the information available in different ways. Databases are about extracting and displaying data, whereas knowledge bases are about learning and answering knowledge. According to these capabilities the services provided are completely different. Through reasoning the knowledge extracted can be used to change the status of the knowledge base because of the learning process involved. This is not possible with data, because databases present data but are not able to reason about the data. Knowledge bases change because new knowledge changes them, however database are static w.r.t. new incoming data. Databases provide data, while knowledge bases provide answers, recommendations and expert advice.

The distinction between ontology and knowledge base is clear. We can see an ontology as a particular case of a knowledge base where there is only information about the intensional part described by the domain. In our case, we will use the term knowledge base to refer to the intensional and extensional component of information, while the term ontology will be used to refer only to the intensional part of a knowledge base. In general, when we refer to knowledge bases we refer to Description Logic based knowledge bases.

There are many systems where databases coexist with DL. Databases handle huge volumes of instance data, while knowledge representation systems manage intensional information in memory. DL provides a formal context similar to that provided by semantic models such as the Entity-Relationship model [20]. A problem modeled in DL can be checked for consistency and correction using a reasoner.

DL knowledge bases contain two distinct components, intensional knowledge in the *TBox* and extensional knowledge in the *ABox*. The *TBox* contains structural information, general properties and relations between concepts e.g. *Wine, WineColor, Wine hasWineColor Color*. The *ABox* contains knowledge about instances of the domain e.g. *Wine(CabernetSauvignon), WineColor(Red), hasWineColor(Cabernet Sauvignon, Red)*, meaning that *CabernetSauvignon* is a *Wine, Red* is a *WineColor*, and *CabernetSauvignon* has *Red* color. This separation not only is done based on the different types of knowledge, it also takes into account temporal evolution criteria. Knowledge in the *TBox* does not change over time, while knowledge in the *ABox* has a temporal character and can be created, updated or deleted.

In a sense, databases also present a similar behaviour. The database schema encodes intensional knowledge about the domain of a problem, and instance data stored in tables is a sort of extensional definition of the schema. This way we can say that a database schema is like a *TBox* in a knowledge base, while data is the equivalent to the *ABox*. However, semantics in an *ABox* are not the same as those used in database instances. One important difference between DB and KB is that the former use *closed-world semantics* while the latter use *open-world semantics* [3]. Absence of information in databases is interpreted as negative information: if it is not in the database it does not exists. In a *ABox*, absent information is not interpreted as non existence, but just as lack of knowledge. The use of each of the previous semantics is also connected to the kind of application to develop.

Query answering in a database is not a logical reasoning, it is the proof of a finite model, that is, the evaluation of a formula in a fixed finite model. However, querying in an *ABox* is more complex, it requires a non trivial reasoning process where all models (sometimes an infinite number of models) must be taken into account.

Another important difference is the notion of the primary key concept. In the first version of OWL, two instances in the ontology with the same name could refer to different individuals. In OWL 2 a key axiom is introduced providing the idea of primary key to ontologies.

These differences force us to decide which of these behaviours to use. In this case, we think that a closed-world assumption for querying is the best solution, because in query answering scenarios response time is a key factor. This decision limits the reasoning tasks that can be performed, but it is essential if we want to perform query answering over instance data. In this scenario we can only do basic reasoning tasks as class subsumption.

There is an ongoing effort to bring together the best characteristics of the database and DL worlds. The complexity results obtained by the DL-Lite family of DLs [18] give support for efficient query answering over large amounts of instance data. In [31] OWL ontologies are considered as incomplete databases, while databases in practice are complete. To overcome these differences they propose a extended DL knowledge base, in which certain *TBox* axioms are designated as integrity constraints, which behave differently depending the type of reasoning performed (*ABox* or *TBox*).

3 Ontologies as a Tool for Relational Database Design

In order to fully develop the potential of ontologies and to enable efficient and flexible information gathering, persistent storage of ontologies and its retrieval is of vital importance. There are different approaches to store ontologies, use a specific purpose database system, exploit the modeling capabilities of object oriented database systems, or use relational database management systems.

This last approach is particularly useful due to the widespread use of RDBMSs as data repositories on the web. A huge amount of the data present in the web is already stored in databases. Additionally, ontology instance management is a complex task that is best performed by optimized data management software such as databases.

Semantic Web data relies heavily in ontology definitions. Actually, data is comprised of ontology instances. So, when a new application is developed and it is necessary to design a database schema for ontology instances storage, a domain ontology describing application data can be used as a conceptual model for the relational database schema design.

Relationships between existent database schemas and ontologies through mappings have been widely studied in literature [10, 14]. But there is no such profusion of works on using ontologies as conceptual models.

In [42] an automatic transformation from ontology to the conceptual ER data model is presented. The graph oriented approach used, converts an ontology expressed in OWL DL into an ER model. This approach delegates the DDL and DML sentences generation to a commercial tool. The problem with this solution is that, once the schema is designed, the ontology is useless given that there is no way of establishing a correspondence between the schema and the original ontology. This contingency automatically discards the method as a possible tool for designing Semantic Web systems.

An insightful and deeply overview of domain ontologies from a database perspective is realized in [26]. A new and more complete definition of domain ontology is proposed along with a taxonomy of domain ontologies. Also, the use of ontologies as conceptual models is discussed. A domain ontology can be used as a first level of database concept specification, which later can be refined to incorporate particular requirements and hence define a conceptual model.

Several works deal with the idea of domain ontologies as conceptual models. In [36] a design methodology for Semantic Web database based systems is presented. It is used to design new databases and to add semantics to existent systems. Among the different possibilities presented, we focus in that which uses an ontology as a starting point in the design of a database schema. A domain ontology is presented as an abstraction of the knowledge of the data source schemas. After a process of refinement is performed, the ontology is adapted to a local schema. The new ontology obtained will be the basis of the implementation of the database schema.

The work [39] studies the role of domain ontologies in database design thoroughly. In this case, the approach is slightly different, the ontology is used to advise the designer about the correctness and completeness of the conceptual model created by her/him using domain knowledge contained in the ontology.

Ontologies have also been used as formalizations of a fuzzy database schema structure [15, 28]. Instances from these ontologies represent schemas describing domain information in a database. These meta-models can be used to create the schema defined in different Fuzzy DBMSs according to the data representation features available.

In our proposal, we use OWL DL ontologies as a formalism to define conceptual models for database schema design. The design of a database schema is reduced to the appropriate design of a domain ontology that captures all requirements of data. The design of the ontology can be done by using CASE tools such as Protégé [1] and SWOOP [27]. Once the ontology is defined, trough a transformation process the ontology is expressed as a set of DDL and DML sentences implementing a relational database schema. An ontology expressed in OWL DL is enough to create the database schema, reducing the problem of database design to that of proper domain ontology definition.

Once the database schema is created, it is important to deal with data instances and their relation to the original ontology; in order to analyze this, next sections study different paradigms to ontology storage in database management systems.

4 OWL Ontology Storage in ORDBMS

This section deals with the implementation of ontology storage in ORDBMSs. Our proposal is based on two principles, efficient access to instance data and the preservation of the original ontology to offer basic reasoning capabilities. The proposal is oriented towards the efficient management of large volumes of instance data in a query oriented environment, where reasoning capabilities are not as important as efficient query response time.

We have selected a subset of the OWL DL constructs to design the storage procedure, which can be seen in Table 1. A complete description of the constructs used in OWL can be found in the document *OWL Web Ontology Language Overview*.[4]

---

[4] http://www.w3.org/TR/owl-features/

Table 1 OWL Constructs used in the proposal

`owl:Class`	A class defines a group of individuals grouped together because they share some properties
`rdfs:subClassOf`	This class axiom defines a subclass relation between two OWL classes
`ObjectProperty`	Property that relates two classes and thus their instances
`DatatypeProperty`	Property that defines relations between instances of classes and XML Schema datatypes
`rdfs:subPropertyOf`	States that a property is a subproperty of one or more other properties
`rdfs:domain`	Property that links a property to a class description. Values in the domain of a property are limited to the class specified
`rdfs:range`	Property that links a property to a class description or a data range
`sameAs`	States that two individuals are the same
`differentFrom`	States that an individual is different from other individuals
`AllDifferent`	States that a number of individuals are mutually distinct
`TransitiveProperty`	Defines a property as transitive
`SymmetricProperty`	Defines a property as symmetric
`FunctionalProperty`	A functional property is a property that can have only an unique value for each instance
`InverseFunctional-Property`	If a property is inverse functional then the inverse of the property is functional
`allValuesFrom`	This restriction requires that for every instance of the class that has instances of the specified property, the values of the property are all members of the class indicated by the *allValuesFrom* clause
`someValuesFrom`	In the case of *someValuesFrom* at least one of the properties of an instance of the class must point to an individual that is part of the class indicated in the restriction
`minCardinality`	Describes the class of all individuals that for a given property have at least N semantically distinct values (individuals or data values)
`maxCardinality`	Describes the class of all individuals that for a given property have at most N semantically distinct values (individuals or data values)
`cardinality`	States that a property on a class has both *minCardinality* and *maxCardinality* N
`oneOf`	Classes can be described by enumeration of the individuals that are part of the class
`hasValue`	A property can be required to have a certain data value or an individual as a value
`disjointWith`	Classes are defined as disjoint from each other
`unionOf`	Describes a class using a class list, the class extension contains those individuals that occur at least in one of the class extensions of the classes in the list
`intersectionOf`	Describes a class using a class list, the class extension contains those individuals that are members of the class extension of all classes in the list

We have not used constructs concerning ontology versioning, ontology inclusion and annotation properties. Those constructs will be treated in future work.

In order to provide efficient access to instance data, we must design an appropriate relational schema for the domain of the problem, which is encoded in the ontology itself. The intensional description of the ontology can be translated to a relational database schema. In order to capture as much of this intensional knowledge as possible, we must represent it using an appropriate semantic model that can be implemented in a relational system. In order to perform a seamlessly migration from OWL to relational databases, some authors [4] propose an intermediate transformation to an ER semantic model. This intermediate step is useful from the point of view of schema documentation, and the formalisation of the different transformations. However, the ER model presents certain representation limitations; entities cannot have another entity as instance, entities cannot have an instance of any entity type as a subtype and entity intersection is not easily representable.

Transformation of the domain information contained in the ontology to a ER semantic model allows to deploy the model as a schema in a relational database. As a direct consequence to this transformation, the database schema storage problem is reduced to an OWL domain ontology design problem, an its transformation to a semantic model. The ER semantic model can be transformed in a relational database model without too much effort; this process is thoroughly studied in [5, 21, 33].

After the ER to database schema transformation, instance data can be stored in the database tables, but the connection to the original ontology is lost. In order to perform rich semantic queries, it is necessary to preserve the original ontology.

At this point we have one problem which can be solved using two different approaches, ontology transformation to a database schema, and ontology structure storage. In order to decide which is the best way to solve this problem, we review some approaches dealing with ontology storage in databases.

In [23] a method to provide support to ontology-based semantic matching in a RDBMS is presented. The chapter proposes a method to store an OWL DL ontology in system-defined tables of a relational database. A set of *SQL* operators is defined to provide semantic matching operations, and a new indexing scheme is implemented to speed up these operations. In this approach, the ontology is stored in system-defined tables along with its instances avoiding information loss.

In [44] a relational database schema representation of an OWL domain ontology is presented. Instances are stored into tables and relations in a relational database schema. These tables and relations are obtained by the transformation of OWL classes and properties. Metadata regarding restrictions is stored in additional tables created for that purpose. This approximation lacks of a proper representation of the ontology itself, this can limit the reasoning capacity inside the database as some information is difficult to restore after the transformation.

In a similar fashion [2] proposes rules to transform OWL ontologies to database schemas. This general rules are oriented for human consumption and need an expert to apply them. An algorithmic formulation would be more appropriate, in order to discard any ambiguities in the processing. Moreover, a complete algorithmic definition can be used to implement the transformations in an automatic way.

These latter approaches deal with ontology data storage appropriately, obtaining benefits from database efficiency, but losing the original structure depicted in the ontology. The first approach covers the ontology storage problem in a more appropriate way, but lacks of an efficient access to instance data.

To avoid the information loss, we propose the storage of the original ontology in the database along with the instance data. Ontology storage must be performed in such a way that properties and relations are preserved, and the original structure can be recovered using queries. The intensional part of the ontology can be stored in a specific database schema that guarantees that no information is lost. With these considerations in mind, we propose a RDBMS hybrid schema to store ontologies and their instances, allowing a separate or combined use. A summarized version can be seen in [9].

In order to transform the conceptual model represented by the domain ontology to a physical implementation in a relational database, we propose an OWL to database schema transformation algorithm. This algorithm takes an OWL DL ontology as input, then generates a database schema for instance data, stores the ontology (TBox Structure) in the system catalog, inserts instance data (ABox assertions) in the schema previously created and relates instance schema data with their corresponding concepts in the stored ontology.

Using this approximation we preserve ontology information in the database and create an appropriate schema for instance handling. However, both parts are disconnected. There is no way to identify which ontology class corresponds to a specific database table. In order to establish a connection between both parts, we design a set of tables containing metadata, in order to link ontology descriptions and their instances in the database schema.

Figure 1 depicts the database structures used to store the ontology and its instances. The main OWL constructs are reflected in the schema.

The proposed schema is divided in two main parts:

- Individuals Schema: This schema is created by the transformation algorithm to store ontology individuals.
- System Catalog Ontology Tables: Store the ontology and the relationships between individuals and the ontology.

Each different OWL DL document would create a different Individuals Schema depending on the domain ontology described in it. In the database schema generation process the OWL ontology is transformed to a semantic model, ER in our case. Then, this is transformed into a database schema using known transformation rules [24]. This way the whole process allows to transform classes into tables, object properties into relations, datatype properties into table attributes and so on. The algorithm is depicted in more detail in a later section.

Table names and attribute names are automatically generated using the name of the class with a prefix with the name of the ontology the class belongs to.

The System Catalog of Ontology Tables contains two different structures:

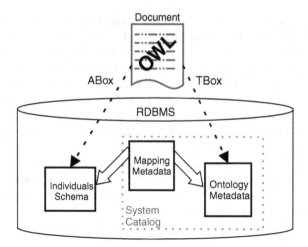

Fig. 1 Ontology and individuals schema

- Ontology Metadata: This set of tables stores the different elements of the ontology, terms, properties, relationships, restrictions...
- Mapping Metadata: These tables store the necessary data to establish a link between an individual in the Individuals Schema and the concept in the ontology to which the individual belongs.

Next sections detail each of the components of the proposed schema and define the necessary transformations from OWL ontology to relational database schema.

4.1 From Ontology to Semantic Model

This section deals with the formal definition of the transformation of OWL ontologies to the ER semantic model. We review the transformations necessary to express the ontology as an ER model.

The Entity-Relationship model was introduced in [21], although later other authors have contributed with some variations and extensions [11, 40, 41]. The ER model is the most used semantic model and it has become a standard for database conceptual design.

In order to better characterise the relation between ER and DLs we employ the formal description defined in [17].

An ER schema $\mathscr{S}$ is built starting from pairwise disjoint sets of entity symbols, relationship symbols, role symbols, attribute symbols and domain symbols.

$$\mathscr{S}\langle E, R, U, A, D \rangle$$

Each domain symbol D has an associated predefined basic domain $D^{\mathscr{B}\mathscr{D}}$ and it is assumed that basic domains are pairwise disjoint. A symbol with arity n has associated n role symbols, each with their respective entity symbol, and defines a relationship between those entities. We assume that each role belongs to just one relation, in such a way that determines an unique entity. Cardinality restrictions are represented by two functions applied on role symbols: $cmin_{\mathscr{S}}$ which returns a non negative integer and $cmax_{\mathscr{S}}$ which returns a value included in the set of natural positive values union the special symbol ∞. Specialization relationships between entities are modeled using a binary relation $\preceq_{\mathscr{S}}$.

As established previously, an ontology is an intensional definition. The ER schema is appropriate to represent this type of information, but it has no capacity to represent extensional information whatsoever. So, the relationship between an ontology an a schema is defined via an application.

We define the application of an ontology into an ER schema.

$$f : \mathscr{O} \longrightarrow \mathscr{S} \quad \text{or} \quad f : \mathscr{O}\langle \mathscr{T} \rangle \longrightarrow \mathscr{S}\langle E, R, U, A, D \rangle$$

We must remark that before performing the translation, a external reasoner is used to check the ontology and compute the inferred ontology to avoid redundancies.

Classes: Classes defined in the ontology are represented as entities in the ER model. A concept C_i maps to an ER model as an entity E_i.

$$C_i \longmapsto E_i$$

A particular case are subclasses. Subclasses in an ontology are shown in an ER model as entities that are specializations too. A concept C_i which is a sub-concept for a concept C_j maps to ER as an entity E_i which is a specialization of the general entity E_j.

$$C_i \sqsubseteq C_j \longmapsto E_i \preceq_{\mathscr{S}} E_j$$

Complex classes are translated to ER as normal entities, but we need to perform additional processing in the transformation as we will see later.

Datatype Properties: Datatype properties map to attributes of the entity in their domain in the ontology with basic datatypes.

Given a role R_i with domain the concept C_i defined in $\mathscr{O}\langle \mathscr{T} \rangle$ this maps to the ER schema $\mathscr{S}$ as an attribute $A_j \in A$ associated to the entity $E_i \in E$ with a domain $D_k \in D$ whose basic predefined domain is $D^{\mathscr{B}\mathscr{D}}$.

Object Properties: Object properties of an OWL ontology relate two classes through a property or role. In ER this is expressed as a relationship between the entities representing in the schema the ontology classes in the range and domain of the property. Cardinality of these properties in the ontology determines participation and cardinality of roles in the relationship modeled in the ER schema.

A property P_i of an ontology $\mathscr{O}$ with range the concept C_r and domain the concept C_d, is represented in the ER schema $\mathscr{S}$ as a relationship R_i between entities E_r and E_d (obtained by mapping concepts C_r y C_d), by the role U_j.

For a given object property, if minimum cardinality in the range of the property is 0, participation is partial, and for a value of 1 or greater, participation is total. Minimum cardinality of a property P_i denoted as $\geq nP_i$ maps to the minimum cardinality of role U_j, denoted as $cmin_{\mathscr{S}}(U_j) = n$. Analogously if maximum cardinality of property range is 1, cardinality is 1, while if it is greater than 1, the cardinality of the relationship in the ER schema is n. Property P_i cardinality is denoted as $\leq nP_i$ and maps to maximum cardinality of role U_j, denoted as $cmax_{\mathscr{S}}(U_j) = n$.

Different combinations of participation and cardinality generate relationships $1 : 1$, $1 : n$, $n : n$. Attending to the roles in each relationship and their cardinalities we can perform the translation to the relational model.

4.2 OWL to Relational Schema Algorithm

After the definition of the OWL domain ontology is finished, it is necessary to define a proper database schema for instance handling. The generation of this schema is performed using a transformation algorithm which takes the OWL ontology, transforms it to the ER semantic model and finally creates the schema structure in a relational database. Complex class definitions that cannot be represented in ER are directly translated to the schema.

The individuals schema presented earlier, is the result of the application of the transformation algorithm to the OWL ontology. Table and relation definitions are created in memory and then are translated to *SQL* sentences.

The algorithm is presented in pseudocode in order to show the transformation process in detail. Although the syntax employed is not very complex, we describe some details before moving to the listings.

Element `list` is a FIFO list where elements are included. Using the NEXT command on a list, the next object is recovered. DBSchema is a complex element containing table objects, it represents the relational schema to create. The behaviour of assignment operator $< -$, varies depending on if it is used before an ADD command or a CREATE command. In the first case it adds a new object, in the latter it creates it. Attributes of an element are referenced using dot syntax `element.attribute`. Command GET element id FROM complex_element searches for an object of type `element` with the specified `id`, inside a complex object. It can be used to retrieve tables from complex element DBSchema, using their identifiers.

The algorithm is presented as a succession of steps. In the first step DBSchema is empty, in next steps it contains the structures created during the algorithm execution.

The first step of the algorithm is the creation of tables. Starting from the root class, each time a class is processed all its subclasses are added to the processing list. Ontology classes are transformed into tables, establishing relations according to inheritance properties such as `rdfs:subClassOf`. Subclasses generate new tables referencing the tables obtained from their super-classes. Primary key (PK) columns are named after the class from which the table is generated. The whole process expressed in pseudocode can be seen in Algorithm 1.

Algorithm 1 Table Generation

```
 1:  list IS FIFO LIST;
 2:  list <- ADD Root Classes;
 3:
 4:  DBschema <- EMPTY;
 5:
 6:  WHILE list IS NOT EMPTY
 7:
 8:      class <- NEXT list;
 9:      table <- CREATE TABLE;
10:      table.name <- class.rdf:ID;
11:
12:      IF class IS ROOT THEN
13:          column <- CREATE PK COLUMN;
14:          column.name <- class.rdf:ID;
15:          table <- ADD column;
16:      ELSE
17:          column <- CREATE PK COLUMN;
18:          column.name <- class.rdf:ID;
19:          parentTable <- GET TABLE class.parentClass.rdf:ID FROM DBschema;
20:          column <- ADD FK REFERENCE TO parentTable.PK;
21:          table <- ADD column;
22:      END IF;
23:
24:      IF class HAS SUBCLASS THEN
25:          IF class.subClass IS owl:disjointWith ONE OR MORE SIBLINGS THEN
26:              list <- ADD subClass;
27:          END IF;
28:      END IF;
29:
30:      DBschema <- ADD table;
31:
32:  END WHILE;
```

Next step is the transformation of `owl:ObjectProperty` to represent relations between the already defined tables. Depending on factors such as cardinality, new tables or attributes are added to the schema. As a general rule multivalued object properties describing many-to-many relations are mapped as new tables with foreign keys (FK) to the `rdfs:domain` and `rdfs:range` tables with both identifiers as the primary key of the relation, whereas single-valued object properties are mapped in the `rdfs:domain` table as foreign keys to the `rdfs:range` table. There are some exceptions to this rule, as can be seen in the Algorithm 2.

Algorithm 2 Relation Generation

```
1:  list IS FIFO LIST;
2:  list <- ADD ALL Root owl:ObjectProperty;
3:  DBschema HAS CONTENT;
4:
5:  WHILE list IS NOT EMPTY
6:    property <- NEXT list;
7:    domain <- property.rdfs:domain;
8:
9:    IF domain.owl:cardinality == 1 OR domain.owl:maxCardinality == 1 THEN
10:       IF property IS SUBPROPERTY THEN
11:          column <- CREATE COLUMN;
12:          column.name <- property.rdf:ID;
13:
14:          IF property.rdf:type.rdf:resource IS owl:TransitiveProperty THEN
15:             column <- ADD UNIQUE CONSTRAINT;
16:          END IF;
17:
18:          IF property.owl:cardinality == 1 OR
19:             property.owl:minCardinality == 1 THEN
20:                column <- ADD NOT NULL CONSTRAINT;
21:          END IF;
22:
23:          tableParentRange <- GET TABLE property.parentProperty.rdfs:range.rdf:resource
24:                             FROM DBschema;
25:
26:          column <- ADD FK REFERENCE TO tableParentRange.FK
27:          tableRange <- GET TABLE property.rdfs:range.rdf:resource FROM DBschema;
28:          tableRange <- ADD column;
29:          DBschema <- UPDATE tableRange;
30:       ELSE
31:          column <- CREATE COLUMN;
32:          column.name <- property.rdf:ID;
33:
34:          IF property.rdf:type.rdf:resource IS owl:TransitiveProperty THEN
35:             column <- ADD UNIQUE CONSTRAINT;
36:          END IF;
37:
38:          IF property.owl:cardinality == 1 OR property.owl:minCardinality THEN
39:             column <- ADD NOT NULL CONSTRAINT;
40:          END IF;
41:
42:          tableRange <- GET TABLE property.rdfs:range.rdf:resource FROM DBschema;
43:          column <- ADD FK REFERENCE TO tableRange.FK
44:          tableDomain <- GET TABLE property.rdfs:domain.rdf:resource FROM DBschema;
45:          tableDomain <- ADD column;
46:          DBschema <- UPDATE tableDomain;
47:       END IF;
48:    ELSE
49:       IF property IS SUBPROPERTY THEN
50:
51:          table <- CREATE TABLE;
52:          table.name <- property.rdf:ID;
53:          columnParentRange <- CREATE PK COLUMN;
54:          tableParentRange <- GET TABLE property.parentProperty.rdfs:range.rdf:resource
55:                             FROM DBschema;
56:
57:          columnParentRange.name <- tableParentRange.FK.name;
58:          columnParentRange <- ADD FK REFERENCE TO tableParentRange.FK;
59:          columnRange <- CREATE PK COLUMN;
60:          tableRange <- GET TABLE property.rdfs:range.rdf:resource FROM DBschema;
61:
62:          columnRange.name <- tableRange.FK.name;
63:          columnRange <- ADD FK REFERENCE TO tableRange.FK;
64:          table <- ADD columnParentRange;
65:          table <- ADD columnRange;
66:          DBschema <- ADD table;
67:       ELSE
68:          table <- CREATE TABLE;
69:          table.name <- property.rdf:ID;
70:          columnDomain <- CREATE PK COLUMN;
71:          tableDomain <- GET TABLE property.rdfs:domain.rdf:resource FROM DBschema;
72:
73:          columnDomain.name <- tableDomain.FK.name;
74:          columnDomain <- ADD FK REFERENCE TO tableDomain.PK;
75:          columnRange <- CREATE PK COLUMN;
76:          tableRange <- GET TABLE property.rdfs:range.rdf:resource FROM DBschema;
77:
78:          columnRange.name <- tableRange.FK.name;
79:          columnRange <- ADD FK REFERENCE TO tableRange.PK;
80:          table <- ADD columnDomain;
81:          table <- ADD columnRange;
82:          DBschema <- ADD table;
83:       END IF;
84:    END IF;
85:
86:    IF property HAS SUBPROPERTIES THEN
87:       list <- ADD property.subProperties;
88:    END IF;
89: END WHILE;
```

Datatype properties are transformed to table attributes. The attribute type is the same as the one used in `rdfs:range` in the property `owl:DatatypeProperty`. These types are valid basic XML Schema types. These types map to *SQL* types in a straightforward way. XML serialization for OWL uses XML Schema (XSD) datatypes.

Algorithm 3 shows pseudocode for the transformation of OWL datatype properties to attributes in a relational schema table. SQLTYPE ( ) function uses XSD to SQL predefined mapping tables to determine the appropriate SQL type for a XSD type defined in the XML serialization of the OWL ontology.

Algorithm 3 Attribute Generation

```
 1: list IS FIFO LIST;
 2: list <- ADD ALL owl:DatatypeProperty;
 3:
 4: DBschema HAS CONTENT;
 5:
 6: WHILE list NOT EMPTY
 7:
 8:     property <- NEXT list;
 9:     domainTable <- GET TABLE property.rdfs:domain.name FROM DBschema;
10:     column <- CREATE COLUMN
11:             WITH TYPE SQLTYPE(property.rdfs:range.rdf:resource);
12:     column.name <- property.rdf:ID;
13:     domainTable <- ADD column;
14:
15: END WHILE;
```

Property restrictions are a special kind of class descriptions. They describe an anonymous class, namely a class of all individuals that satisfy the restriction. OWL distinguishes two kinds of property restrictions: value constraints and cardinality constraints. Cardinality constraints have been treated already, they are covered in the mapping of owl:ObjectProperty properties. Value constraints in OWL are defined with the constructs owl:someValuesFrom, owl:allValuesFrom and owl:hasValue.

The mapping of owl:allValuesFrom is done adding a foreign key constraint in the column generated previously by the property, pointing to the primary key column of the class specified by rdf:resource.

In the case of owl:someValuesFrom, the situation is more complex because if the property is multi-valued the consistency must be checked using a trigger. If it is single-valued, the value for the property must be one of the class in the restriction, making this case equivalent to the owl:allValuesFrom previously depicted.

Also owl:hasValue restriction is a complex case. The property links a restriction class to a value, which can be either an individual or a data value. In the case where the value is a data value it can be modeled as a *SQL* CHECK constraint. Problems arise when the value is an individual, then the use of a trigger is requested. All information needed by the trigger to operate is available in the Mapping Metadata table. Particular cases implementing triggers are obviated for now, but they are taking into account as future work. The algorithm performing the mapping of restrictions is depicted as Algorithm 4.

Algorithm 4 Property Restriction Generation

```
 1:  list IS FIFO LIST;
 2:  list <- ADD Root Properties;
 3:
 4:  DBschema IS HAS CONTENT;
 5:
 6:  WHILE list NOT EMPTY
 7:
 8:      property <- NEXT list;
 9:
10:      IF property HAS owl:Restriction THEN
11:
12:          CASE property.restriction OF
13:
14:            owl:allValuesFrom :
15:                tableDomain <- GET TABLE property.owl:domain FROM DBschema;
16:                tableRange <- GET TABLE property.owl:range FROM DBschema;
17:                column <- GET COLUMN property.rdf:ID FROM tableDomain;
18:                column <- ADD FK REFERENCE TO tableRange.PK;
19:                tableDomain <- UPDATE column;
20:                DBschema <- UPDATE tableDomain;
21:
22:            owl:someValuesFrom:
23:                IF property.owl:cardinality == 1 OR
24:                                property.owl:maxCardinality ==1 THEN
25:                    tableDomain <- GET TABLE property.owl:domain FROM DBschema;
26:                    tableRange <- GET TABLE property.owl:range FROM DBschema;
27:                    column <- GET COLUMN property.rdf:ID FROM tableDomain;
28:                    column <- ADD FK REFERENCE TO tableRange.PK;
29:                    tableDomain <- UPDATE column;
30:                    DBschema <- UPDATE tableDomain;
31:                END IF;
32:
33:            owl:hasValue:
34:                IF property.owl:hasvalue.rdfs:resource IS A VALUE
35:                    table <- GET TABLE property.owl:domain FROM DBschema;
36:                    column <- GET COLUMN property.rdf:ID FROM table;
37:                    column <- ADD CHECK CONSTRAINT
38:                                property.owl:hasvalue.rdfs:resource;
39:                    table <- UPDATE column;
40:                    DBschema <- UPDATE table;
41:                END IF;
42:      END IF;
43:
44:      IF property HAS SUBPROPERTIES THEN
45:          list <- ADD property.subProperties;
46:      END IF;
47:
48: END WHILE;
```

In the final step, instances are introduced in their corresponding tables. It is important to take into account that one instance can be separated between different tables as a result of the mapping performed.

When the ontology definition is finally stored in the database, individuals and ontology are related trough the Mapping Metadata. This relation allows the creation of procedures and operators to perform limited reasoning tasks such as class subsumption. The use of these operators and procedures in queries, must be explicit.

4.3 Ontology Storage Schema

In this section the ontology storage database schema is depicted. We use a vertical table schema where all objects identifiers are stored and several additional tables to provide additional information regarding the object, stored in the main table.

The OWL constructs previously depicted in Table 1 are translated into the schema. The Ontology Metadata tables are organized as Figure 2 shows. The description of each table is the following:

Table `Ont_NameSpaces` includes information about the namespaces defined and the prefixes used, a namespace can be used with different prefixes. Each term in the ontology has its corresponding namespace associated in order to deal with elements defined in different ontologies.

`Ont_Ontologies` stores general information about the different ontologies. Each ontology should also be described as a term in order to be associated with its corresponding namespace.

Table `Ont_Terms` contains the representation of the various concepts described in the ontology. Ontologies, classes and properties are stored here as basic elements to use in more complex constructs. Each term has an associate ontology and a namespace. In addition to these terms two complex elements can be defined here, `ClassList` that represents a list of classes, and `Group` which represents a group of individuals. The basic OWL constructs must be defined here before any ontology is inserted, in order to use them to create relations between terms.

`Ont_ObjectProperties` contains information about object properties and their characteristics. The domain and range of the property are stored as terms in the `Ont_Terms` table. The remaining attributes include the textual description of the property, and all its possible characteristics. Object properties should be included as terms in the table `Ont_Terms` in order to be used as part of relationships.

The table `Ont_Restrictions` includes information about property restrictions and the anonymous class that is created when restrictions are applied on the property. The `mincardinality` and `maxcardinality` attributes contain positive numeric values of cardinality. Attributes `somevaluesfrom` and `allvaluesfrom` contain a term identifier corresponding to a class term defined in the `Ont_Terms` table.

`Ont_Relationships` contains information about the existing relationships between two terms of the ontology. Basic OWL relationships as `SubClassOf` can be expressed here. The `termID` attributes point to terms defined in the `Ont_Terms` table. These terms can be class terms, property terms, group terms or class list terms.

`Ont_DataTypeProperties` contains information about datatype properties such as the identifier of the property, the ontology to which it belongs, the domain and the range datatype. The domain is a term in `Ont_Terms`. Datatype attribute specifies the name of the type, that could be the ones defined in the `xsd` XML Schema or fuzzy ones as we will describe later.

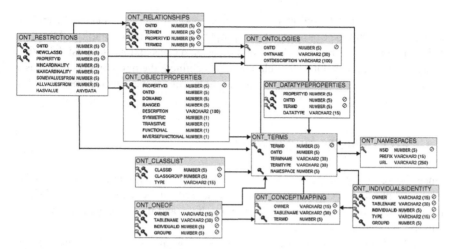

Fig. 2 System catalog metadata schema

The Mapping Metadata Tables relate the individuals in the Individuals Schema with the Ontology Metadata tables where the proper concepts are defined. The tables are defined as follows:

Ont_ConceptMapping relates a table and hence its tuples (individuals) in the Individuals Schema with its corresponding class term in Ont_Terms. A class is represented as a table to contain its individuals, and as a term to allow reasoning.

The table Ont_IndividualsIdentity contains information related to issues concerning individuals identity across ontologies. OWL defines a set of properties to deal with individuals identity owl:sameAs, owl:differentFrom and owl:AllDifferent. These properties relate individuals, not classes. Due to this, we deal with them in a separate system-defined table which links instances and definitions in an ontology. The type attribute describes the OWL property, one of the three mentioned before, and groupID is the identifier of the group containing the individuals that are related by the property. The group of individuals definition must be included as a term in the Ont_Terms table.

OWL provides the owl:oneOf construct to specify a class via a direct enumeration of its members. Ont_OneOf contains information about the individuals belonging to a particular enumeration. The owl:oneOf construct relates a class with the individuals describing it. In order to reflect them on the system catalog it is necessary to include a relationship between the class and a group of individuals. owl:oneOf should be present as an available OWL property in Ontology Metadata Schema and the class and group should be inserted on the Ont_Terms table, then the list of individuals belonging to the group is specified in Ont_OneOf.

OWL offers set operators like owl:intersectionOf and owl:unionOf to define complex classes based on operations performed on classes already defined. Classes constructed using those operators are like definitions, where their class extension consists of exactly the individuals obtained in the operation. Table Ont_Class

List contains the definition of class lists. owl:unionOf and owl:inter
sectionOf, should be present as OWL properties defined in the Ontology Metadata
Schema, in order to create a relationship between a class and its class list definition
using these properties. Once the relationship is defined, the new class is created as a
view over the classes in the group stored as tables, under the selected operation.

4.4 Fuzzy Datatype Management

Once the general process is depicted, it is important to explain how we deal with
fuzziness in our proposal. This proposal deals with fuzzy datatypes at two levels,
the expression of fuzzy datatypes at ontology level and their storage and operation
within a database.

Works dealing with fuzziness in ontologies usually center their attention in
describing fuzzy concepts and relations in ontologies, including fuzzy hierarchical
relationships. In [37] a fuzzy description logic is presented where concepts modeled
as fuzzy sets can be class attributes. Working with fuzziness at this level entails an
increment of complexity in the representation and management of the data.

Although fuzzy data handling is a hot topic in ontology research, there is not a
standard for representing fuzzy data in ontologies, especially in OWL ontologies.
Certain authors consider that a fuzzy extension of an ontology language is not likely
to become a standard in the foreseeable future. Thus, they propose Fuzzy OWL 2 [16],
a framework to represent fuzzy ontologies using current languages and resources.
Fuzzy OWL 2 provides a representation for fuzzy data in ontologies using the current
standard language OWL 2, by using annotation properties. This fuzzy data includes
fuzzy datatypes, fuzzy modifiers and weighted sum concepts.

The use of fuzziness at the datatype level allows to represent certain semantics,
approximate values, data ranges, etc, but does not affect complexity. Fuzzy data is
used at instance level and it is not necessary to deal with it at ontology level. As
we clearly distinguish between the treatment of the ontology structure from that of
the storage and management of instance data, it is possible to add semantics to data
using the capabilities provided by the underlying Fuzzy ORDBMS.

The use of fuzzy logic in queries helps the user to create meaningful queries
expressing flexible criteria. In [6] fuzzy data querying is used to improve query
results, particularly when using multiple conditions. Fuzzy queries allow the user
to express her/his information requirements in an intuitive and flexible way. Query
results provide additional related information that although is not what the user
specifically is looking for, may be of her/his interest.

Our approach starts from an OWL ontology and adds the possibility of defining
fuzzy datatypes as the range of owl:DatatypeProperty properties. The transformation shown in Sect. 4.2 can be adapted, just including the fuzzy datatypes as
attributes in the appropriate table in individuals schema. If the original ontology has
no instance data, we just specify in the datatype property the appropriate fuzzy type
for the range. These types must be defined beforehand as depicted in Figure 3.

If the initial ontology has instance data, it is necessary to provide the means to represent instance data using the fuzzy datatypes in the ontology. OWL ontologies do not support fuzzy types, so we must extend them. We create a new XML Schema Type named `NumericFuzzyData`; this datatype accepts values of type `CrispValue` for precise values, `ApproximateValue` for triangular possibility distributions, `UpperBoundValue` and `LowerBoundValue` for bound values, `IntervalValue` for range values, and `TrapezoidalValue` for trapezoidal possibility distributions.

To include these changes in the definition of datatype properties, we must change the OWL XML Schema. The problem with this change, is that no reasoner is capable of working with these ontologies, because it can not recognize the new datatypes. Instead of creating a new reasoner from scratch, we take advantage of the extension capacities offered by DBMS, in order to perform basic reasoning tasks inside the database where ontology data is stored.

In order to better understand the capabilities of the underlying FORDBS next section presents a brief summary of its functionality.

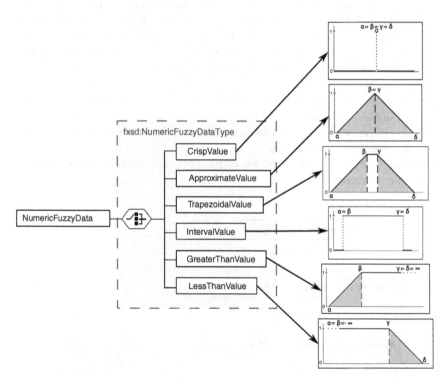

Fig. 3 Fuzzy datatypes

4.4.1 Fuzzy Datatypes in Databases

The best way of dealing with huge quantities of fuzzy data is the use of a Fuzzy DBMS. We include one in our proposal to store instance data with fuzzy attributes.

Several works on fuzzy databases [30, 34, 35, 43, 45] provide different ways of handling fuzzy information. We have decided to use a database data model [8, 22, 29] that allows extracted fuzzy data to be stored in a FORDBMS. This data model specifies a datatype hierarchy taking advantage of object oriented modeling benefits. Figure 4 shows this datatype hierarchy.

The FORDBMS proposal covers a wide variety of fuzzy datatypes:

- *FuzzyDataTypes* (FDT) is an abstract type which is the root ancestor of all supported fuzzy datatypes. This type declares abstract methods which must be implemented in its instantiable subtypes.
- *AtomicFuzzyTypes* (AFT) is an abstract type designed for collecting common behaviour of subtypes aimed to represent atomic data.
- *OrderedAFTs* (OAFT) datatype gives support for fuzzy numbers, which are atomic fuzzy data represented by a possibility distribution defined on an ordered domain. The order relation between domain members allows the definition of classical relational operators, from which extended relational operators are derived in order to obtain fuzzy comparators for this datatype. The extended relational operators for OAFT data are for instance FEQ (fuzzy equal to), FLEQ (fuzzy less than or equal to), etc.
- *NonOrderedAFTs* (NOAFT) datatype is designed to store fuzzy data defined on a scalar domain without any order between its elements.
- *FuzzyCollections* (FC) is an abstract datatype which extends the concept of classical collections to a fuzzy one.
- *DisjunctiveFCs* (DFC) datatype, a subtype of FC, supports fuzzy collections with disjunctive semantics.

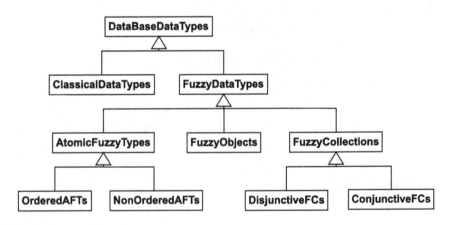

Fig. 4 Datatype hierarchy

- *ConjunctiveFuzzyCollections* (CFC) datatype is similar to DFC but supports collections with conjunctive semantics.
- *FuzzyObject* (FO) is an abstract datatype which sets a general framework for dealing with user defined complex fuzzy objects.

We focus on the use of OAFT datatypes as the range of properties of type `owl:DatatypeProperty`. This includes trapezoidal distributions, approximate values, data ranges, upper bound values, lower bound values and crisp data. This allows to express fuzzy attributes for concepts e.g. the age of a person can be expressed as approximate 25 years, between 20 and 25 years, more than 20 years, etc. The creation and handling of these new numeric datatypes can have effects on efficiency. In order to solver this drawback, [7] proposes an indexing mechanism for imprecise numerical data.

These datatypes have their counterpart in the ontology, a definition for all of them has been created in XML Schema, and it is included as part of a modified OWL XML Schema. We use OAFT datatypes because this type of fuzziness is enough to solve the kind of problems we are dealing with, where imprecision is only in the numeric attributes of data.

Queries in a FORDMBS are processed as standard SQL queries, and can be optimized. The processing of these queries is more complex, due to the use of the fuzzy datatypes and the greater amount of results recovered. Nevertheless, the main advantage of this approach is that results are ranked, and we can obtain values similar to the ones searched.

The data model is implemented in a Oracle® DBMS, using PL/SQL user defined datatypes. The extension of a commercial DBMS offers benefits such as good performance, scalability and reliability.

5 Conclusions

In this chapter we have presented a transformation algorithm and a storage schema for OWL DL ontologies which allows to use domain ontologies as a conceptual model for database schema design. This allows to create database schemas from scratch in order to store ontology instance data in the context of the Semantic Web. The proposed schema and algorithm cover the main constructs of OWL DL ontologies and set the groundwork necessary to develop internal reasoning in the database.

The use of an underlying FORDBMS adds the possibility to use fuzzy datatypes in the ontologies. These types provide richer semantics than traditional numerical datatypes, that can be exploited in querying. Whereas we focus on OAFT datatypes, future research will also deal with the inclusion of other datatypes defined in the FORDBMS into the ontology, the integration of fuzzy parameters at other levels, like fuzzy relations, fuzzy hierarchies, etc. analyzing the effect of these additions in the subsequent task of reasoning.

The benefit of the approach is the possibility of creating database schemas without the additional effort of designing an additional conceptual model, and the possibility to manage and query instance data in a semantic way. Data instances and the ontology can be queried together or separately, establishing a link between the objects defined and their definition, all these features supported by database technology which guarantees good performance and reliability. The whole process of transformation into a schema and the storage in the catalog can be automatized.

Future work will focus on the definition of a basic reasoner inside the database which will lead to the optimization of the schema and DML operations. A proper representation using triggers of complex semantic specifications is in the works.

Although the version of OWL used is compatible with OWL2, we plan to move to the new version of the language in order to take advantage of new features.

Another concern is ontology evolution inside our system, allowing changes in the ontology and propagate them to the schema. Finally, a graphical interface to import and export ontologies into the database will be designed.

Acknowledgments This work has been partially supported by the "Consejería de Economía, Innovación, Ciencia y Empleo de Andalucía" (Spain) under research projects P10-TIC-6109 and P11-TIC-7460.

References

1. The Protégé ontology editor: http://protege.stanford.edu/ (2012)
2. Astrova, I., Korda, N., Kalja, A.: Storing OWL ontologies in SQL relational databases. Int. J. Electr. Comput. Syst. Eng. **1**(4), 242–247 (2007)
3. Baader, F., Werner, N.: The Description Logic Handbook: Theory, Implementation, and Applications, chap. Basic Description Logics, pp. 47–100. Cambridge University Press, New York (2003)
4. Bagui, S.: Mapping OWL to the entity relationship and extended entity relationship models. Int. J. Knowl. Web Intell. **1**(1), 125–149 (2009)
5. Bagui, S., Earp, R.: Database design using entity-relationship diagrams. Auerbach Publications, Boca Raton (2003)
6. Barranco, C.D., Campaña, J.R., Medina, J.M.: Improving query expressiveness in product search interfaces using fuzzy logic. WSEAS Trans. Bus. Econ. **2**(2), 80–87 (2005)
7. Barranco, C.D., Campaña, J.R., Medina, J.M.: A B+-Tree based indexing technique for fuzzy numerical data. Fuzzy Sets Syst. (Advances in Intelligent Databases and Information Systems) **159**(12), 1431–1449 (2008)
8. Barranco, C.D., Campaña, J.R., Medina, J.M.: Handbook of Research on Fuzzy Information Processing in Databases, chap. Towards a Fuzzy Object-Relational Database Model, 1st edn., pp. 431–461. Hershey, Pennsylvania (2008)
9. Barranco, C.D., Campaña, J.R., Medina, J.M., Pons, O.: On storing ontologies including fuzzy datatypes in relational databases. Fuzzy systems conference, 2007. FUZZ-IEEE 2007. IEEE, international pp. 1–6, 23–26 July 2007
10. Barrasa, J., Corcho, O., Gómez-Pérez, A.: R2O, an extensible and semantically based database-to-ontology mapping language. In: Proceedings of the second workshop on semantic web and databases (SWDB2004) (2004)
11. Batini, C., Ceri, S., Navathe, S.B.: Conceptual Database Design: An Entity-Relationship Approach. Benjamin/Cummings, Redwood City (1992)

12. Berardi, D., Calvanese, D., De Giacomo, G.: Reasoning on UML class diagrams. Artif. Intell. **168**(1–2), 70–118 (2005)
13. Berners-Lee, T., Hendler, J., Lassila, O.: The semantic web. Sci. Am. **284**(5), 28–37 (2001)
14. Bizer, C., Seaborne, A.: D2RQ-treating non-RDF databases as virtual RDF graphs. In: Proceedings of the 3rd international semantic web conference (ISWC2004)
15. Blanco, I.J., Vila, M.A., Martinez-Cruz, C.: The use of ontologies for representing database schemas of fuzzy information. Int. J. Intell. Syst. **23**(4), 419–445 (2008)
16. Bobillo, F., Straccia, U.: Fuzzy ontology representation using OWL 2. Int. J. Approximate Reasoning **52**(7), 1073–1094 (2011)
17. Borgida, A., Lenzerini, M., Rosati, R.: The Description Logic Handbook: Theory, Implementation, and Applications, chap. Description Logics for Databases, pp. 462–484. Cambridge University Press, New York (2003)
18. Calvanese, D., De Giacomo, G., Lembo, D., Lenzerini, M., Rosati, R.: Tractable reasoning and efficient query answering in description logics: the DL-Lite family. J. Autom Reasoning **39**(3), 385–429 (2007)
19. Calvanese, D., De Giacomo, G., Lembo, D., Lenzerini, M., Rosati, R., Vetere, G.: DL-Lite: practical reasoning for rich DLs. In: Proceedings of the 2004 description logic workshop (DL 2004). CEUR electronic workshop proceedings (2004)
20. Calvanese, D., Lenzerini, M., Nardi, D.: Description Logics for Conceptual Data Modeling, chap. 8, pp. 229–264. Kluwer Academic Publisher, Dordrecht (1998)
21. Chen, P.: The entity-relationship model—toward a unified view of data. ACM Trans. Database Syst. (TODS) **1**(1), 9–36 (1976)
22. Cubero, J.C., Marín, N., Medina, J.M., Pons, O., Vila, M.A.: Fuzzy object management in an object-relational framework. In: X International conference of information processing and management of uncertainty in knowledge-based systems, pp. 1767–1774 (2004)
23. Das, S., Chong, E.I., Eadon, G., Srinivasan, J.: Supporting ontology-based semantic matching in RDBMS. In: Proceedings of the 30th VLDB conference. Toronto, Canada, pp. 1054–1065 (2004)
24. Date, C.: An Introduction to Database Systems. Addison-Wesley, Reading (1990)
25. Fahad, M.: Er2owl: generating owl ontology from er diagram. In: Intelligent information processing, pp. 28–37 (2008)
26. Jean, S., Pierra, G., Ait-Ameur: Domain ontologies: a database-oriented analysis. In: Web information systems and technologies (WEBIST 2006), pp. 238–254 (2006)
27. Kalyanpur, A., Parsia, B., Sirin, E., Grau, B.C., Hendler, J.: Swoop: a web ontology editing browser. Web Semant. Sci. Serv. Agents WWW **4**(2), 144–153 (2006)
28. Martínez-Cruz, C., Blanco, I.J., Vila, M.A.: Describing fuzzy DB schemas as ontologies: a system architecture view. In: Information Processing and Management of Uncertainty in Knowledge-Based Systems. Applications, Communications in Computer and Information Science, vol. 81, pp. 147–157. Springer, Berlin (2010)
29. Medina, J.M., Galindo, J., Berzal, F., Serrano, J.M.: Using object relational features to build a fuzzy database server. In: VIII international conference of information processing and management of uncertainty in knowledge-based systems (IPMU 2002), pp. 307–314 (2002)
30. Medina, J.M., Pons, O., Vila, M.A.: GEFRED: a generalized model of fuzzy relational databases. Inf. Sci. **76**(1–2), 87–109 (1994)
31. Motik, B., Horrocks, I., Sattler, U.: Bridging the gap between OWL and relational databases. J. Web Semant. **7**(2), 74–89 (2009)
32. Myroshnichenko, I., Murphy, M.C.: Mapping er schemas to owl ontologies. In: Proceedings of the 2009 IEEE international conference on semantic computing, ICSC '09, pp. 324–329 (2009)
33. Navathe, S., Elmasri, R.: Fundamentals of database systems. Addison Wesley, Reading (2007)
34. Pokorný, J., Vojtáš, P.: A data model for flexible querying. In Proceedings of ADBIS'01. Lecture Notes in Computer Science, vol. 2151, pp. 280–293 (2001)
35. Prade, H., Testemale, C.: Generalizing database relational algebra for the treatment of incomplete or uncertain information and vague queries. Inf. Sci. **34**, 115–143 (1984)

36. Roldán García, M., Navas Delgado, I., Aldana Montes, J.: A design methodology for semantic web database-based systems. In: ICITA (1), pp. 233–237 (2005)
37. Straccia, U.: A fuzzy description logic for the semantic web. In: Sanchez, E. (ed.) Fuzzy Logic and the Semantic Web, Capturing Intelligence, chap. 4, pp. 73–90. Elsevier, Amsterdam (2006)
38. Straccia, U.: SoftFacts: A top-k retrieval engine for ontology mediated access to relational databases. In: Proceedings of the 2010 IEEE International Conference on Systems, Man and, Cybernetics (SMC-10), pp. 4115–4122 (2010)
39. Sugumaran, V., Storey, V.C.: The role of domain ontologies in database design: an ontology management and conceptual modeling environment. ACM Trans. Database Syst. **31**(3), 1064–1094 (2006)
40. Teorey, T.J.: Distributed database design: A practical approach and example. SIGMOD Rec. **18**(4), 23–39 (1989)
41. Thalheim, B.: Foundations of entity—relationship modeling. Ann. Math. Artif. Intell. **7**(1–4), 197–256 (1993)
42. Trinkunas, J., Vasilecas, O.: A graph oriented model for ontology transformation into conceptual data model. Inf. Technol. Control **36**(1A), 126–132 (2007)
43. Umano, M.: Freedom-O: A fuzzy database system.In: Fuzzy Information and Decision Processes. North-Holland, New York (1982)
44. Vysniauskas, E., Nemuraite, L.: Transforming ontology representation from OWL to relational database. Inf. Technol. Control **35A**(3), 333–343 (2006)
45. Zemankova-Leech, M., Kandel, A.: Implementing imprecision in information systems. Inf. Sci. **37**, 107–141 (1985)
46. Zhang, F., Ma, Z.M., Yan, L., Wang, Y.: A description logic approach for representing and reasoning on fuzzy object-oriented database models. Fuzzy Sets Syst. **186**(1), 1–25 (2012)

Part III
Uncertain Databases

Chapter 7
Information Systems Uncertainty Design and Implementation Combining: Rough, Fuzzy, and Intuitionistic Approaches

Theresa Beaubouef and Frederick Petry

Abstract There are a number of alternative techniques for dealing with uncertainty. Here we discuss rough set, fuzzy rough set, and intuitionistic rough set approaches and how to incorporate uncertainty management using them in the relational database model. The impacts of rough set techniques on fundamental database concepts such as functional dependencies and information theory are also considered.

1 Introduction

The term "information system" here is meant to identify a system designed to model, store and retrieve effectively large amounts of information. From an historical point of view, the management of unstructured information (texts) on one side, and structured information (formatted data representing factual information, often for business) on the other side, gave rise to two different lines of research and products: information retrieval systems and database management systems, respectively. For both areas there have been significant research efforts for managing uncertainty. In this chapter we focus only on the database area.

There are different kinds of uncertainty in data and databases, and different techniques have been developed for managing the various types of uncertainty. Rough sets [35] are effective at managing uncertainty related to vagueness and to varying the granularity of discernment of items and in the representation of "certain" and "possible or partial" membership in a set based on the underlying partitioning of data items. Fuzzy sets [50] offer a more continuous approach by incorporating membership values to denote degrees of belonging to some sets. Intuitionistic fuzzy

T. Beaubouef
Southeastern Louisiana University, Hammond, Louisiana, USA

F. Petry (✉)
Naval Research Lab, Stennis Space Center, Mississippi, USA
e-mail: fred.petry@nrlssc.navy.mil

O. Pivert and S. Zadrożny (eds.), *Flexible Approaches in Data, Information and Knowledge Management*, Studies in Computational Intelligence 497, DOI: 10.1007/978-3-319-00954-4_7, © Springer International Publishing Switzerland 2014

sets [1, 2] take this one step further by also incorporating degrees of nonmembership in some set. Various combinations of rough, fuzzy, and intuitionistic sets have also been studied [23, 32, 47] for managing uncertainty. These take advantage of the benefits of each. Other uncertainty management approaches, which are not discussed in this chapter, include probabilistic reasoning [33], possibility theory [51] and interval-based logic [40].

A variety of approaches were taken by early researchers in use of fuzzy sets for uncertainty in database models [16, 38, 46, 52]. Of particular note are the significant research contributions by Patrick Bosc and his collaborators including among the most cited papers [13–15].

This chapter overviews alternative techniques such as rough sets, fuzzy rough sets, and intuitionistic rough sets for incorporating uncertainty management into the relational database model. More general aspects of various approaches to uncertainty than just the effects on database models can be found in [24]. Impacts of the rough set techniques on fundamental database concepts such as functional dependencies and information theory are discussed. In databases information theoretic measures can be used to measure the information content of data. Entropy is one such measure, and we discuss its use in fuzzy databases[18], intuitionistic fuzzy sets [45] and rough databases [9].

2 Rough Sets

Rough set theory, introduced and further developed mathematically by Pawlak [35], provides a framework for the representation of uncertainty. It has been used in various applications such as the rough querying of crisp data [3], uncertainty management in databases [5], the mining of spatial data [6], and improved information retrieval [43]. Rough set theory is a technique for dealing with uncertainty and for identifying cause-effect relationships in databases. It is based on a partitioning of some domain into equivalence classes and the defining of lower and upper approximation regions based on this partitioning to denote certain and possible inclusion in the rough set. An extensive theory for rough sets and their properties has been developed and become a well established approach for the management of uncertainty in a variety of applications. Rough sets involve the following:

U is the *universe* (non- empty)

R is the *indiscernibility relation*, or equivalence relation

$A = (U,R)$, an ordered pair, is called an *approximation space*

$[x]_R$ denotes the equivalence class of R containing x, for any element x of U

elementary sets in A - equivalence classes of R, *definable set* in A - finite union of elementary sets in A.

Given an approximation space defined on some universe U that has an equivalence relation R imposed upon it, U is partitioned into equivalence classes called elementary sets that may be used to define other sets in A. A rough set X, where $X \subseteq U$, can be defined in terms of the definable sets in A by the following:

lower approximation of X in A is the set $\underline{R}\,X = \{x \in U/[x]_R \subseteq X\}$

upper approximation of X in A is the set $\bar{R}X = \{x \in U/[x]_R \cap X \neq \emptyset\}$.

$POS_R(X) = \underline{R}X$ denotes the R-positive region of X, or those elements which certainly belong to the rough set. The R-negative region of X, $NEG_R(X) = U - \bar{R}X$, contains elements which do not belong to the rough set, and the boundary or R-borderline region of X, $BN_R(X) = \bar{R}X - \underline{R}X$, contains those elements which may or may not belong to the set. X is R-definable if and only if $\underline{R}X = \bar{R}X$. Otherwise, $\underline{R}X \neq \bar{R}X$ and X is rough with respect to R. A *rough set in A* is the group of subsets of U having both the same upper approximations and the same lower approximations. In Fig. 1 the universe U is partitioned into equivalence classes denoted by the squares. Those elements in the lower approximation of X, $POS_R(X)$, are denoted with the letter p and elements in the R-negative region by the letter n. All other classes belong to the boundary region of the upper approximation.

3 Fuzzy and Fuzzy Rough Sets

Fuzzy set theory is a very influential approach for managing uncertainty; proposed initially by Zadeh [50], it also has well developed theory, properties, and applications. Applications involving fuzzy logic are diverse and plentiful, ranging from fuzzy control systems in industry to fuzzy logic in databases.

Because there are advantages to both fuzzy set and rough set theories, several researchers have studied various ways of combining the two theories [23, 28, 32, 37] Others have investigated the interrelations between the two theories [19, 36, 47]. A similar approach is the fuzzy rough set in [29]. That approach is more in the spirit of functional analysis, however. Fuzzy sets and rough sets are not equivalent, but complementary.

Fig. 1 Example of a rough set X

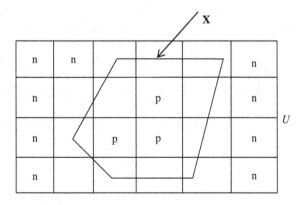

One specific approach we use to illustrate this [47] has rough sets expressed by a fuzzy membership function $\mu \rightarrow \{0, 0.5, 1\}$ to represent the negative, boundary, and positive regions. In this model, all elements of the lower approximation, or positive region, have a membership value of one. Those elements of the boundary region are assigned a membership value of 0.5. Elements not belonging to the rough set have a membership value of zero. Rough set definitions of union and intersection can be modified so that using this approach the fuzzy model can be made compatible with rough sets principles. [4, 5].

However there is a problem in general of truth-functionality in applying fuzzy operations to rough sets. This is discussed extensively by Yao [49] He describes how in general the theories of fuzzy and rough sets have similar qualitative aspects. Rough set upper and lower approximations can be seen to correspond to concepts of core and support of a fuzzy set. Methods for construction of a fuzzy set are described based on constraints that rough set theory induces on membership values, in particular constraints on membership values of related elements and also related sets.

We integrate fuzziness into the rough set model in order to quantify levels of roughness in boundary region areas through the use of fuzzy membership values. Therefore, we do not require membership values of elements of the boundary region to equal 0.5, but allow them to range from zero to one, non-inclusive. For example, if one were defining a fuzzy set on "people" whose membership function is "tall", then every person would have a membership value associated with him that indicated the degree to which he belongs to that fuzzy set. Most basketball players would have higher membership values (.9 or even 1) than people of average height. Additionally, the union and intersection operators for fuzzy rough sets are comparable to those for ordinary fuzzy sets, where MIN and MAX are used to obtain membership values of redundant elements.

Let U be a *universe*, X a rough set in U, and we have the following definitions:

A fuzzy rough set Y in U is a membership function $\mu_Y(x)$ which associates a grade of membership from the interval [0,1] with every element of U where

$$\mu_Y(\underline{R}X) = 1, \quad \mu_Y(U - \overline{R}X) = 0, \text{ and } 0 < \mu_Y(\overline{R}X) - \underline{R}X < 1.$$

The union of two fuzzy rough sets A and B is a fuzzy rough set C where

$$C = \{x | x \in A \text{ OR } x \in B\}, \text{ where } \mu_C(x) = \text{MAX}[\mu_A(x), \mu_B(x)].$$

The intersection of two fuzzy rough sets A and B is a fuzzy rough set C where

$$C = \{x | x \in A \text{ AND } x \in B\}, \text{ where } \mu_C(x) = \text{MIN}[\mu_A(x), \mu_B(x)].$$

4 Intuitionistic Sets

Intuitionistic sets extend the notion of fuzzy sets to include non-membership in a set. This double-sided fuzziness measure provides even greater management of uncertainty for many real world applications. An intuitionistic set [1, 2] (intuitionistic fuzzy set) is a generalization of the traditional fuzzy set. Let set X be fixed. An intuitionistic set A is defined by the following:

$$A = \{< x, \mu_A(x), \nu_A(x) > \mid x \in X\}, \text{ and where } \mu_A : X \to [0, 1], \text{ and } \nu_A : X \to [0, 1].$$

The degree of membership of element $x \in X$ to the set A is denoted by $\mu_A(x)$, and the degree of non-membership of element $x \in X$ to the set A is denoted by $\nu_A(x)$. A is a subset of X.

Additionally, for all $x \in X$,

$$0 \leq \mu_A(x) + \nu_A(x) \leq 1.$$

A hesitation margin,
$$\pi_A(x) = 1 - (\mu_A(x) + \nu_A(x)),$$

expresses a degree of uncertainty about whether x belongs to X or not, or uncertainty about the membership degree. This hesitancy may cater toward membership or non-membership.

Example.

A shopper buys electronic products online. Using traditional 2-valued logic the person may be *satisfied* or *dissatisfied* with the purchase. There is no continuum between satisfied and dissatisfied; nor is there any uncertainty involved.

In rough sets many things may be considered in the realm of product satisfaction, and some of them will be grouped together in equivalence classes. Some of these classes are entirely included in the set *satisfied*: [delighted, overjoyed] or [accepting, pleased], for example. Some are not in the rough set *satisfied* at all [furious, disgusted] or [disappointed], [dissatisfied], for example. Lastly, there are some that involve uncertainty about the belonging to the rough set *satisfied*. These may include equivalence classes such as [ambivalent, noncommittal, indifferent]. These would belong to the boundary, or uncertain region of the rough set.

In fuzzy sets a person could be pleased to a certain degree. The degree of membership of an element to the fuzzy set of "pleased" is represented by a membership value between zero and one. For example, one could be "pleased" with the purchase to a degree of .8. This implies dissatisfaction to a degree of .2. However, a person could be pleased to a degree of .8, but not dissatisfied at all, or at least not to that extent. This cannot be represented in fuzzy sets.

In intuitionistic fuzzy sets, however, there are measures for both the degree of membership and the degree of non-membership. A person could be pleased with the product to a degree of .8, but only dissatisfied to a degree of .05, resulting in a hesitancy of .15. This two sided fuzziness in the intuitionistic set provides greater management of uncertainty for many real world cases.

For the fuzzy rough set case, the situation would be similar. However, with the inclusion of indiscernibility providing equivalences, both "accepting" or "pleased" would be treated as equivalent during query evaluation, and results would include matches with either or both of these. So "pleased" and "accepting" are indiscernible from each other through roughness, and fuzziness is applied with that in mind. In a similar manner, with intuitionistic rough sets, there is an underlying indiscernibility relation, and then also the values denoting degrees of membership and non-membership.

5 Intuitionistic Rough Sets

Both fuzzy sets and intuitionistic sets can be enhanced with rough set techniques to provide versatile uncertainty management. Intuitionistic rough sets are generalizations of fuzzy rough sets that give more information about the uncertain, or boundary region. They follow the definitions for partitioning of the universe into equivalence classes as in traditional rough sets, but instead of having a simple boundary region, there are basically two boundaries formed from the membership and non-membership functions.

Let U be a *universe*, Y a rough set in U, defined on some approximation space which partitions U into equivalence classes, then an intuitionistic rough set Y in U is $< Y, \mu_Y(x), \nu_Y(x) >$, where $\mu_Y(x)$ is a membership function which associates a grade of membership from the interval [0,1] with every element (equivalence class) of U, and $\nu_Y(x)$ associates a degree of non- membership from the interval [0,1] with every element (equivalence class) of U, where

$$0 \leq \mu_Y(x) + \nu_Y(x) \leq 1,$$

where x denotes the equivalence class containing x.

A hesitation margin,

$$\pi_Y(x) = 1 - (\mu_Y(x) + \nu_Y(x))$$

Let Y′ denote the complement of Y. Then the intuitionistic set having $< \mu_Y(x)$ $\mu_{Y'}(x) >$ is same as fuzzy rough set. The last two cases in Table 1, $<0, q>$ and $<x, y>$, cannot be represented by fuzzy sets, rough sets, or fuzzy rough sets. These are the situations which show that intuitionistic rough sets provide greater uncertainty management than the others alone. Note, however, that with the intuitionistic set we do not lose the information about uncertainty provided by other set theories, since from the first few cases we see that they are special cases of the intuitionistic rough set.

Table 1 Special cases $< \mu, v >$ for some element of Y

Case	Description
$<1, 0>$	Denotes total membership; corresponds to elements found in $\underline{R}Y$
$<0, 1>$	Denotes elements that do not belong to Y; same as $U - \bar{R}Y$.
$<0.5, 0.5>$	Corresponds to traditional rough set boundary region
$<p, 1-p>$	Corresponds to fuzzy rough set in that there is a single boundary. In this case we assume that any degree of membership has a corresponding complementary degree of non-membership.
$<p, 0>$	Corresponds to fuzzy rough set; in this case there is no complement to what p shows membership in.
$<0, q>$	This case cannot be modeled by fuzzy rough sets; denoting things not a member of $\underline{R}Y$ or $\bar{R}Y$. It falls somewhere in region $U - \bar{R}Y$.
$<x, y>$	Intuitionistic set general case , uncertain double boundary, one for membership and one for non-membership.

We may also perform operations on the intuitionistic rough sets such as union and intersection. We define these operations next. The definition of these operators is necessary for applications such as the intuitionistic rough relational database model, and is similar to the fuzzy rough operations, but they include the concept of non-membership v.

The <u>union</u> of two intuitionistic rough sets A and B is an intuitionistic rough set C where

$$C = \{x \mid x \in A \text{ OR } x \in B\} : \mu_C(x) = MAX[\mu_A(x), \mu_B(x)],$$
$$v_C(x) = MIN[v_A(x), v_B(x)].$$

The <u>intersection</u> of two intuitionistic rough sets A and B is an intuitionistic rough set C where

$$C = \{x \mid x \in A \text{ AND } x \in B\}; \mu_C(x) = MIN[\mu_A(x), \mu_B(x)],$$
$$v_C(x) = MAX[v_A(x), v_B(x)].$$

In this section we defined intuitionistic rough sets and compared them with rough sets and fuzzy sets. Although there are several various ways of combining rough and fuzzy sets, we focused on those fuzzy rough sets as defined in [5] and used for fuzzy rough databases, since our intuitionistic rough relational database model follows from this. The intuitionistic rough relational database model has an advantage over the rough and fuzzy rough database models in that the non-membership uncertainty of intuitionistic set theory will also play a role, providing even greater uncertainty management than the original models.

6 Rough Relational Database

The rough relational database model [11] is an extension of the standard relational database model of Codd [20]. It captures all the essential features of rough sets theory including indiscernibility of elements denoted by equivalence classes and lower and upper approximation regions for defining sets which are indefinable in terms of the indiscernibility.

Every attribute domain is partitioned by some equivalence relation designated by the database designer or user. Within each domain, those values that are considered indiscernible belong to an equivalence class. This information is used by the query mechanism to retrieve information based on equivalence with the class to which the value belongs rather than equality, resulting in less critical wording of queries.

Recall is also improved in the rough relational database because rough relations provide *possible* matches to the query in addition to the *certain* matches which are obtained in the standard relational database. This is accomplished by using set containment in addition to equality of attributes in the calculation of lower and upper approximation regions of the query result.

The rough relational database has several features in common with the ordinary relational database. Both models represent data as a collection of *relations* containing *tuples*. These relations are sets. The tuples of a relation are its elements, and like elements of sets in general, are unordered and non-duplicated. A tuple t_i takes the form $(d_{i1}, d_{i2}, \ldots, d_{im})$, where d_{ij} is a *domain value* of a particular *domain set* D_j. In the ordinary relational database, $d_{ij} \in D_j$. In the rough database, however, as in other non-first normal form extensions to the relational model [30], $d_{ij} \subseteq D_j$, and although it is not required that d_{ij} be a singleton, $d_{ij} \neq \emptyset$. Let $P(D_i)$ denote the powerset (D_i)–$\emptyset$. and A rough relation R is a subset of the set cross product $P(D_1) \times P(D_2) \times \cdots \times P(D_m)$.

A rough tuple t is any member of R, which implies that it is also a member of $P(D_1) \times P(D_2) \times \cdots \times P(D_m)$. If t_i is some arbitrary tuple, then $t_i = (d_{i1}, d_{i2}, \ldots, d_{im})$ where $d_{ij} \subseteq D_j$. A tuple in this model differs from that of ordinary databases in that the tuple components may be sets of domain values rather than single values. The set braces are omitted from singletons for notational simplicity.

Let $[d_{xy}]$ denote the equivalence class to which d_{xy} belongs. When d_{xy} is a set of values, the equivalence class is formed by taking the union of equivalence classes of members of the set; if $d_{xy} = \{c_1, c_2, \ldots, c_n\}$, then $[d_{xy}] = [c_1] \cup [c_2] \cup \ldots \cup [c_n]$. Tuples $t_i = (d_{i1}), d_{i2}, \ldots, d_{im})$ and $t_k = (d_{k1}, d_{k2}, \ldots, d_{km})$ are redundant if $[d_{ij}] = [d_{kj}]$ for all $j = 1, \ldots, m$. Redundant tuples are removed in the merging process since duplicates are not allowed in sets, the structure upon which the relational model is based.

In the rough relational database, relations are rough sets as opposed to ordinary sets. Therefore, new rough operators $(-, \cup, \cap, \sigma, \pi, \bowtie)$, which are comparable to the standard relational operators, are required for the rough relational database. Moreover, a mechanism must exist within the database to mark tuples of a rough relation as belonging to the lower or upper approximation of that rough relation. Rough relational operator definitions and their properties can be found in [11]. We extend these operators for fuzzy rough relations in the next section.

7 Fuzzy Rough Relational Database

The fuzzy rough relational database is an extension of the rough database where a relation tuple t_i takes the form $(d_{i1}, d_{i2}, \ldots, d_{im}, d_{i\mu})$, where $d_{i\mu} \in D_\mu$, and where D_μ is the interval $[0,1]$, the domain for fuzzy membership values.

For a specific relation, R, membership is determined semantically. Given that D_1 is the set of names of manufacturers, D_2 is the set of electronics products, and assuming that T-VAL is the only electronics manufacturer for the distributor that produces EZ-MONITORs, and this is the only product of T-VAL, then

$$(\text{T-VAL, EZ-MONITOR, } 1.0)$$

$$(\text{T-VAL, QMONITOR, } 0.7)$$

$$(\text{T-VAL, KEYB1, } 1.0)$$

$$(\text{NTEL, EZ-MONITOR, } 0.3)$$

are all elements of $P(D_1) \times P(D_2) \times D_\mu$. However, only the element (T-VAL, EZ-MONITOR, 1.0) of those listed above is a member of the relation $R(\text{MANUFACTURER, PRODUCT, } \mu)$, which associates each manufacturer with the products that it manufactures. A fuzzy rough tuple t is any member of R. If t_i is some arbitrary tuple, then $t_i = (d_{i1}, d_{i2}, \ldots, d_{im}, d_{i\mu})$ where $d_{ij} \subseteq D_j$ and $d_{i\mu} \in D_\mu$. An $\underline{\text{interpretation }} \alpha = (a_1, a_2, \ldots, a_m, a_\mu)$ of a fuzzy rough tuple $t_i = (d_{i1}, d_{i2}, \ldots, d_{im}, d_{i\mu})$ is any value assignment such that $a_j \in d_{ij}$ for all j.

The interpretation space is the cross product $D_1 \times D_2 \times \cdots \times D_m \times D_\mu$, but is limited for a given relation R to the set of those tuples which are valid according to the underlying semantics of R. In an ordinary relational database, because domain values are atomic, there is only one possible interpretation for each tuple t_i. Moreover, the interpretation of t_i is equivalent to the tuple t_i. In the fuzzy rough relational database, this is not always the case.

Let $[d_{xy}]$ denote the equivalence class to which d_{xy} belongs. When d_{xy} is a set of values, the equivalence class is formed by taking the union of equivalence classes of members of the set; if $d_{xy} = \{c_1, c_2, \ldots, c_n\}$, then $[d_{xy}] = [c_1] \cup [c_2] \cup \ldots \cup [c_n]$. Fuzzy rough tuples $t_i = (d_{i1}, d_{i2}, \ldots, d_{in}, d_{i\mu})$ and $t_k = (d_{k1}, d_{k2}, \ldots, d_{kn}, d_{k\mu})$ are $\underline{\text{redundant}}$ if $[d_{ij}] = [d_{kj}]$ for all $j = 1, \ldots, n$.

If a relation contains only those tuples of a lower approximation, i.e., those tuples having a μ value equal to one, the interpretation α of a tuple is unique. This follows immediately from the definition of redundancy. In fuzzy rough relations, there are no redundant tuples. The merging process used in relational database operations removes duplicate tuples since duplicates are not allowed in sets, the structure upon which the relational model is based.

Tuples may be redundant in all values except μ. As in the union of fuzzy rough sets where the maximum membership value of an element is retained, it is the convention of the fuzzy rough relational database to retain the tuple having the higher μ value when removing redundant tuples during merging. If we are supplied with identical data from two sources, one certain and the other uncertain, we would want to retain the data that is certain, avoiding loss of information.

Another definition, which will be used for upper approximation tuples, is necessary for some of the alternate definitions of operators to be presented. This definition captures redundancy between elements of attribute values that are sets:

Two sub-tuples $X = (d_{x1}, d_{x2}, \ldots, d_{xm})$ and $Y = (d_{y1}, d_{y2}, \ldots, d_{ym})$ are *roughly-redundant*, $\approx_R$, if for some $[p] \subseteq [d_{xj}]$ and $[q] \subseteq [d_{yj}]$, $[p] = [q]$ for all $j = 1, \ldots, m$.

In order for any database to be useful, a mechanism for operating on the basic elements and retrieving specified data must be provided. The concepts of redundancy and merging play a key role in the operations defined.

We next formally define the fuzzy rough relational database operators.. We may view indiscernibility as being modeled through the use of the indiscernibility relation, imprecision through the use of non-first normal form constructs, and degree of uncertainty and fuzziness through the use of tuple membership values, which are given as the value for the μ attribute in every fuzzy rough relation.

8 Fuzzy Rough Relational Operators

In [11], we defined several operators for the rough relational algebra. We now define similar operators for the fuzzy rough relational database as in [5]. Recall that for all of these operators the indiscernibility relation is used for equivalence of attribute values rather than equality of values.

Difference: The fuzzy rough relational difference operator is very much like the ordinary difference operator in relational databases and in sets in general. In the fuzzy rough relational database, the difference operator is applied to two fuzzy rough relations and, as in the rough relational database, indiscernibility, rather than equality of attribute values, is used in the elimination of redundant tuples. Hence, this difference operator is somewhat more complex. Let X and Y be two union compatible fuzzy rough relations. The <u>fuzzy rough difference</u>, **X - Y**, between X and Y is a fuzzy rough relation T where

$$T = \{t(d_1, \ldots, d_n, \mu_i) \in X \,|\, t(d_1, \ldots, d_n, \mu_i) \notin Y\} \cup \{t(d_1, \ldots, d_n, \mu_i) \in X \,|$$
$$t(d_1, \ldots, d_n, \mu_j) \in Y \text{ and } \mu_i > \mu_j\}$$

The resulting fuzzy rough relation contains all those tuples which are in the lower approximation of X, but not redundant with a tuple in the lower approximation of

Y. It also contains those tuples belonging to upper approximation regions of both X and Y, but which have a higher μ value in X than in Y. For example, let X contain the tuple (SATISFIED, 1.0) and Y contain the tuple (SATISFIED, 0.02). It would not be desirable to subtract out certain information with possible information, so X – Y yields (SATISFIED, 1.0). Moreover, less certain information will not cause tuples that are more certain to be eliminated, even if their membership values are close in value. So if X contains the tuple (SATISFIED, .55) and Y contains (SATISFIED, .53), the result of the difference operation X – Y will contain (SATISFIED, .55) as one of its tuples.

Union: Because relations in databases are considered as sets, the union operator can be applied to any two union-compatible relations to result in a third relation which has as its tuples all the tuples contained in either or both of the two original relations. The union operator can be extended to apply to fuzzy rough relations. Let X and Y be two union compatible fuzzy rough relations. The fuzzy rough union of X and Y, **X ∪ Y** is a fuzzy rough relation T where

$$T = \{t | t \in X \text{ OR } t \in Y\} \quad \text{and} \quad \mu_T(t) = \text{MAX}[\mu_X(t), \mu_Y(t)].$$

The resulting relation T contains all tuples in either X or Y or both, merged together and having redundant tuples removed. If X contains a tuple that is redundant with a tuple in Y except for the μ value, the merging process will retain only that tuple with the higher μ value.

Intersection: The fuzzy rough intersection, another binary operator on fuzzy rough relations, can be defined similarly. The fuzzy rough intersection of X and Y, **X ∩ Y** is a fuzzy rough relation T where

$$T = \{t \mid t \in X \text{ AND } t \in Y\} \quad \text{and} \quad \mu_T(t) = \text{MIN}[\mu_X(t), \mu_Y(t)].$$

In intersection, the MIN operator is used in the merging of equivalent tuples having different μ values and the result contains all tuples that are members of both of the original fuzzy rough relations.

Select: The select operator for the fuzzy rough relational database model, σ, is a unary operator which takes a fuzzy rough relation X as its argument and returns a fuzzy rough relation containing a subset of the tuples of X, selected on the basis of values for a specified attribute. The operation $\sigma_{A=a}(X)$, for example, returns those tuples in X where attribute A is equivalent to the class [**a**]. In general, select returns a subset of the tuples that match some selection criteria.

Let R be a relation schema, X a fuzzy rough relation on that schema, A an attribute in R, **a** = {a_i} and **b** = {b_j}, where $a_i, b_j \in$ dom(A), and $\cup_x$ is interpreted as "the union over all x". The fuzzy rough selection, $\sigma_{A=a}(X)$, of tuples from X is a fuzzy rough relation Y having the same schema as X and where

$$Y = \{t \in X \mid \cup_i [a_i] \subseteq \cup_j [b_j]\},$$

and $a_i \in \mathbf{a}$, $b_j \in t(A)$, and where membership values for tuples are calculated by multiplying the original membership value by card($\mathbf{a}$)/card($\mathbf{b}$) where *card(x)* returns the cardinality, or number of elements, in x.

Assume we want to retrieve those elements where PRODUCT = "QMON" from the following fuzzy rough tuples:

$$(QMON \qquad\qquad\qquad 1.0)$$
$$(\{QMON, EZ\text{-}MONITOR, N14\}\ 0.7)$$
$$(KEYB1 \qquad\qquad\qquad 1.0)$$
$$(\{QMON, T2\text{-}JACK\} \qquad\qquad 0.9)$$

The result of the selection is the following:

$$(QMON \qquad\qquad\qquad 1.0)$$
$$(\{QMON, EZ\text{-}MONITOR, N14\}\ 0.23)$$
$$(\{QMON, T2\text{-}JACK\} \qquad\qquad 0.45).$$

where the μ for the second tuple is the product of the original membership value 0.7 and 1/3.

Project: Project is a unary fuzzy rough relational operator. It returns a relation that contains a subset of the columns of the original relation. Let X be a fuzzy rough relation with schema A, and let B be a subset of A. The fuzzy rough projection of X onto B is a fuzzy rough relation Y obtained by omitting the columns of X which correspond to attributes in A - B, and removing redundant tuples. Recall the definition of redundancy accounts for indiscernibility, which is central to the rough sets theory and that higher μ values have priority over lower ones. The fuzzy rough projection of X onto B, $\pi_{\mathbf{B}}(\mathbf{X})$, is a fuzzy rough relation Y with schema Y(B) where

$$Y(B) = \{t(B) \mid t \in X\}.$$

Join : Join is a binary operator that takes related tuples from two relations and combines them into single tuples of the resulting relation. It uses common attributes to combine the two relations into one, usually larger, relation. Let $X(A_1, A_2, \ldots, A_m)$ and $Y(B_1, B_2, \ldots, B_n)$ be fuzzy rough relations with $\mathbf{m}$ and $\mathbf{n}$ attributes, respectively, and AB = C, the schema of the resulting fuzzy rough relation T.

The fuzzy rough join, $\mathbf{X} \bowtie_{<\text{JOIN CONDITION}>} \mathbf{Y}$, of two relations X and Y, is a relation $T(C_1, C_2, \ldots, C_{m+n})$ where $T = \{t \mid \exists\ t_X \in X,\ t_Y \in Y$ for $t_X = t(A)$, $t_Y = t(B)\}$, and where

(1) $t_X(A \cap B) = t_Y(A \cap B)$, $\mu=1$
(2) $t_X(A \cap B) \subseteq t_Y(A \cap B)$ or $t_Y(A \cap B) \subseteq t_X(A \cap B)$, $\mu = MIN(\mu_X, \mu_Y)$

<JOIN CONDITION> is a conjunction of one or more conditions of the form $\mathbf{A} = \mathbf{B}$.

Only those tuples which resulted from the "joining" of tuples that were both in lower approximations in the original relations belong to the lower approximation of the resulting fuzzy rough relation. All other "joined" tuples belong to the upper approximation only (the boundary region), and have membership values less than one. The fuzzy membership value of the resultant tuple is simply calculated as in [16] by taking the minimum of the membership values of the original tuples. Taking the minimum value also follows the logic of [34], where in joins of tuples with different levels of information uncertainty, the resultant tuple can have no greater certainty than that of its least certain component.

9 Intuitionistic Rough Relational Database Model

The intuitionistic rough relational database [8] is an extension of the fuzzy rough model above using the intuitionistic membership and non-membership values where D_μ is the interval [0,1], the domain for intuitionistic membership values, and D_v is the interval [0,1], the domain for intuitionistic non-membership values.

For a specific relation, R, membership is determined semantically. Given that D_1 is the set of names of electronics manufacturers, D_2 is the set of products, and assuming that T-VAL is the only manufacturer to produce the EZ-MONITOR, with EZ-MONITOR its only product, then

$$(\text{T-VAL, EZ-MONITOR}, 1.0, 0.0)$$
$$(\text{T-VAL, QMONITOR}, 0.7, 0.3)$$
$$(\text{T-VAL,KEYB1}, 1.0, 0.0)$$
$$(\text{NTEL, EZ-MONITOR}, 0.3, 0.2)$$

are all elements of $P(D_1) \times P(D_2) \times D_\mu \times D_v$. However, only the element (T-VAL, EZ-MONITOR, 1.0, 0.0) of those listed above is a member of the relation R(MANUFACTURER, PRODUCT, μ, v), which associates products with their manufacturers. An *intuitionistic rough tuple* $\mathbf{t}$ is any member of R. If $\mathbf{t_i}$ is some arbitrary tuple, then $\mathbf{t_i} = (d_{i1}, d_{i2}, \ldots, d_{im}, d_{i\mu}, d_{iv})$ where $d_{ij} \subseteq D_j$ and $d_{i\mu} \in D_\mu$, $d_{iv} \in D_v$. Definitions for interpretation, redundancy, and rough redundancy are similar to those for the fuzzy rough database with the addition of v.

The intuitionistic rough relational operators extend those of the fuzzy rough database by inclusion of v, for non-membership. There are two cases of note here. For the *intuitionistic rough selection*, the non-membership value v remains the same as in the intuitionistic rough relation X, since the result of performing this operation does not give us any additional information about non-membership. Also for the *intuitionistic rough join* if $\mu_X = \mu_Y$, then $v = \text{MAX}(v_X, v_Y)$.

10 Functional Dependencies

A functional dependency can be defined as in [25] through the use of a universal database relation concept. Let $R = \{A_1, A_2, \ldots, A_n\}$ be a universal relation schema describing a database having n attributes. Let X and Y be subsets of R. A functional dependency between the attributes of X and Y is denoted by $X \rightarrow Y$. This dependency specifies the constraint that for any two tuples of an instance r of R, if they agree on the X attribute(s) they must agree on their Y attributes(s): if $t_1[X] = t_2[X]$, then it must be true that $t_1[Y] = t_2[Y]$. Tuples that violate the constraint cannot be inserted into the database. The rough functional dependency is based on the rough relational database model. The classical notion of functional dependency for relational databases does not naturally apply to the rough relational database, since all the "roughness" would be lost. In the rough querying of crisp data [3], however, the data is stored in the standard relational model having ordinary functional dependencies imposed upon it and rough relations result only from querying; they are not a part of the database design in which the designer imposes constraint upon relation schemas. Rough functional dependencies for the rough relational database model are defined as follows [7]:

A rough functional dependency, $X \rightarrow Y$, for a relation schema R exists if for all instances $\underline{T(R)}$,

(1) for any two tuples t, $t' \in \underline{R}T$,

   $redundant(t(X), t'(X) \Rightarrow redundant(t(Y), t'(Y))$

(2) for any two tuples s, $s' \in \bar{R}T$,

   $roughly\text{-}redundant(s(X), s'(X)) \Rightarrow roughly\text{-}redundant(s(Y), s'(Y))$.

Y is roughly functional dependent on X, or X roughly functionally determines Y, whenever the above definition holds. This implies that constraints can be imposed on a rough relational database schema in a rough manner that will aid in integrity maintenance and the reduction of update anomalies without limiting the expressiveness of the inherent rough set concepts.

It is obvious that the classical functional dependency for the standard relational database is a special case of the rough functional dependency; indiscernibility reduces to simple equality and part (2) of the definition is unused since all tuples in relations in the standard relational model belong to the lower approximation region of a similar rough model.

The first part of the definition of rough functional dependency compares with that of fuzzy functional dependencies discussed in [42], where adherence to Armstrong's axioms was proven. The results apply directly in the case of rough functional dependencies when only the lower approximation regions are considered. It is also necessary to show that axioms hold for upper approximations.

Rough functional dependencies satisfy Armstrong's axioms.

PROOF:

(1) **Reflexive**
    If $Y \subseteq X \subseteq U$, then
    redundant(t(X), t'(X)) $\Rightarrow$ *redundant*(t(Y), t'(Y)), and
    roughly-redundant(t(X), t'(X)) $\Rightarrow$ *roughly-redundant*(t(Y), t'(Y)).
    Hence, $X \rightarrow Y$.

(2) **Augmentation**
    If $Z \subseteq U$ and the rough functional dependency $X \rightarrow Y$ holds, then
    redundant(t(XZ), t'(XZ)) $\Rightarrow$ *redundant*(t(YZ), t'(YZ)), and
    roughly-redundant(t(XZ), t'(XZ)) $\Rightarrow$ *roughly-redundant*(t(YZ), t'(YZ)).
    Hence, $XZ \rightarrow YZ$.

(3) **Transitive**
    If the rough functional dependencies $X \rightarrow Y$ and $Y \rightarrow Z$ hold, then
    redundant(t(X), t'(X)) $\Rightarrow$ *redundant*(t(Z), t'(Z)), and
    roughly-redundant(t(X), t'(X)) $\Rightarrow$ *roughly-redundant*(t(Z), t'(Z)).
    Hence, $X \rightarrow Z$.

Hence, rough functional dependencies satisfy Armstrong's axioms. Given a set of rough functional dependencies, the complete set of rough functional dependencies can be derived using Armstrong's axioms. The rough functional dependency, therefore, is an important formalism for design in the rough relational database.

Fuzzy and rough set techniques integrated into the underlying data model result in databases that can more accurately represent real world enterprises since they incorporate uncertainty management directly into the data model itself. This is useful as is for obtaining greater information through the querying of rough and fuzzy databases..

One practical topic of great concern is the issue of security in databases. Aspects of security violations by query inferencing in statistical and probabilistic databases have been studied [22, 31]. Similar approaches to database security for databases using fuzzy and rough set representations [9, 17] have also been considered in particular using information measures such as introduced in the next section.

11 Information Theory

In communication theory, Shannon [41] introduced the concept of entropy which was used to characterize the information content of signals. Since then, variations of these information theoretic measures have been successfully applied to applications in many diverse fields. In particular, the representation of uncertain information by entropy measures has been applied to all areas of databases, including fuzzy database querying [18], data allocation [26], classification in rule-based systems [39], and measuring uncertainty in rough and fuzzy rough relational databases [10].

In fuzzy set theory the representation of uncertain information measures has been extensively studied [12, 21, 27, 48]. For fuzzy databases fuzzy entropy may be measured as a function of a domain value or as a function of a relation. Intuitively, the uncertainty of a domain value increases as its cardinality $| d_{ij} |$ increases or when the similarity $s_j(x,y)$ decreases. So if a domain value in a relational scheme, d_{ij}, consisting of a single element represents exact information and multiple elements are a result of fuzziness, then this uncertainty can be represented by entropy. DeLuca and Termini [21] have devised formulas for uncertainty based on fuzzy measures. Adapting their result to a fuzzy database, the entropy $H_{fz}(d_{ij})$, for a domain value $d_{ij} \subseteq D_j$ would be

$$E_{fz}(d_{ij}) = - \sum_{\{x,y\} \in d_{ij}} [s_j(x, y) \log_2 (s_j(x, y)) + (1 - s_j(x, y))$$

$$\log_2(1 - s_j(x, y))]$$

This definition cannot be directly extended to tuples, so a probabilistic entropy measure after Shannon is needed for an entire tuple. Based on the concept of interpretation of a tuple, for the ith tuple, t_i, there are p_i possible interpretations, i.e., the cardinality of the cross product of the domain values, $| d_{i1} \times d_{i2} \times \ldots \times d_{im}|$. Viewing all interpretations as a priori equally likely, the entropy of tuple t_i can be defined as [18]

$$E_{pb}(t_i) = - \sum_{k=0}^{p_i}(1/p_i) \log_2 (1/p_i) = \log_2(p_i) \sum_{k=1}^{p_i}$$

For a non-fuzzy database, clearly $p_i = 1$ and $E_{pb}(t_i) = 0$.

If the choice of a tuple in a relation r is independent of the interpretation of the tuple, the joint probabilistic entropy $E_{pb}(r,t)$ of a relation can be expressed as

$$E_{pb}(r, t) = - \sum_{i=1}^{n} \sum_{k=1}^{p_i}(n\,p_i)^{-1} \log_2[(n\,p_i)^{-1}]$$

where there are n tuples.

An entropy measure in [44] is based on a geometric interpretation of intuitionistic fuzzy sets [45] and the distances between such sets. The entropy measure is based on the ratio of the largest cardinalities (max Σ Counts) involving only the set F and its complement F^c. So the form of the entropy of intuitionstic fuzzy set F with n elements is

$$E(F) = \frac{1}{n} \sum_{i=1}^{n} (\text{max Count}(F_i \cap F_i^c))/(\text{max Count}(F_i \cup F_i^c))$$

Next information theory measures can be applied to rough sets and compared to established rough set metrics of uncertainty. The measures are then applied to the rough relational database model [11]. Information content of both stored relational schemas and rough relations are expressed as types of rough entropy.

Rough set theory [35] inherently models two types of uncertainty. The first type of uncertainty arises from the indiscernibility relation that is imposed on the universe, partitioning all values into a finite set of equivalence classes. If every equivalence class contains only one value, then there is no loss of information caused by the partitioning. In any coarser partitioning, however, there are fewer classes, and each class will contain a larger number of members. Our knowledge, or information, about a particular value decreases as the granularity of the partitioning becomes coarser.

Uncertainty is also modeled through the approximation regions of rough sets where elements of the lower approximation region have total participation in the rough set and those of the upper approximation region have uncertain participation in the rough set. Equivalently, the lower approximation is the *certain* region and the boundary area of the upper approximation region is the *possible* region.

Pawlak [36] discusses two numerical characterizations of imprecision of a rough set X: *accuracy* and *roughness*. Accuracy, which is simply the ratio of the number of elements in the lower approximation of X, $\underline{R}X$, to the number of elements in the upper approximation of the rough set X, $\bar{R}X$, measures the degree of completeness of knowledge about the given rough set X. It is defined as a ratio of the two set cardinalities as follows:

$$\alpha_R(X) = \text{card}(\underline{R}X)/\text{card}(\bar{R}X), \text{ where } 0 \leq \alpha_R(X) \leq 1.$$

The second measure, roughness, represents the degree of incompleteness of knowledge about the rough set. It is calculated by subtracting the accuracy from 1: $\rho_R(X) = 1 - \alpha_R(X)$.

These measures require knowledge of the number of elements in each of the approximation regions and are good metrics for uncertainty as it arises from the boundary region, implicitly taking into account equivalence classes as they belong wholly or partially to the set. However, accuracy and roughness measures do not necessarily provide us with information on the uncertainty related to the granularity of the indiscernibility relation for those values that are totally included in the lower approximation region. For example,

Let the rough set X be defined as follows:

$$X = \{A11, A12, A21, A22, B11, C1\}$$

with lower and upper approximation regions defined as

$$\underline{R}X = \{A11, A12, A21, A22\} \text{ and}$$
$$\bar{R}X = \{A11, A12, A21, A22, B11, B12, B13, C1, C2\}$$

These approximation regions may result from one of several partitionings. Consider, for example, the following indiscernibility relations:

$$A_1 = \{[A11, A12, A21, A22], [B11, B12, B13], [C1, C2]\},$$
$$A_2 = \{[A11, A12], [A21, A22], [B11, B12, B13], [C1, C2]\},$$
$$A_3 = \{[A11], [A12], [A21], [A22], [B11, B12, B13], [C1, C2]\}.$$

All three of the above partitionings result in the same upper and lower approximation regions for the given set X, and hence the same accuracy measure (4/9 = .444) since only those classes belonging to the lower approximation region were re-partitioned. It is obvious, however, that there is more uncertainty in A_1 than in A_2, and more uncertainty in A_2 than in A_3. Therefore, a more comprehensive measure of uncertainty is needed.

We derive such a measure from techniques used for measuring entropy in classical information theory. Countless variations of the classical entropy have been developed, each tailored for a particular application domain or for measuring a particular type of uncertainty. Our rough entropy is defined such that we may apply it to rough databases. We define the entropy of a rough set X as follows: The rough entropy $E_r(X)$ of a rough set X is calculated by

$$E_r(X) = -(\rho_R(X))[\Sigma\, Q_i \log(P_i)] \text{ for } i = 1, \dots n \text{ equivalence classes.}$$

The term $\rho_R(X)$ denotes the roughness of the set X. The second term is the summation of the probabilities for each equivalence class belonging either wholly or in part to the rough set X. There is no ordering associated with individual class members. Therefore the probability of any one value of the class being named is the reciprocal of the number of elements in the class. If c_i is the cardinality of, or number of elements in, equivalence class i and all members of a given equivalence class are equal, $P_i = 1/c_i$ represents the probability of one of the values in class i. Q_i denotes the probability of equivalence class i within the universe. Q_i is computed by taking the number of elements in class i and dividing by the total number of elements in all equivalence classes combined. The entropy of the sample rough set X, $E_r(X)$, is given below for each of the possible indiscernibility relations A_1, A_2, and A_3.

Using A_1 : $-(5/9)[(4/9)\log(1/4) + (3/9)\log(1/3) + (2/9)\log(1/2)] = .274$
Using A_2 : $-(5/9)[(2/9)\log(1/2) + (2/9)\log(1/2) + (3/9)\log(1/3)+$
$\qquad (2/9)\log(1/2)] = .20$
Using A_3 : $-(5/9)[(1/9)\log(1) + (1/9)\log(1) + (1/9)\log(1) + (1/9)\log(1)+$
$\qquad (3/9)\log(1/3) + (2/9)\log(1/2)] = .048$

From the above calculations it is clear that although each of the partitionings results in identical roughness measures, the entropy decreases as the classes become smaller through finer partitionings.

12 Entropy and the Rough Relational Database

The basic concepts of rough sets and their information-theoretic measures carries over to the rough relational database model [11]. Recall that in the rough relational database all domains are partitioned into equivalence classes and relations are not restricted to first normal form. We therefore have a type of rough set for each attribute of a relation. This results in a rough relation, since any tuple having a value for an attribute that belongs to the boundary region of its domain is a tuple belonging to the boundary region of the rough relation.

There are two things to consider when measuring uncertainty in databases: uncertainty or entropy of a rough relation that exists in a database at some given time and the entropy of a relation schema for an existing relation or query result. We must consider both since the approximation regions only come about by set values for attributes in given tuples. Without the extension of a database containing actual values, we only know about indiscernibility of attributes. We cannot consider the approximation regions.

We define the entropy for a rough relation schema as follows: The rough schema entropy for a rough relation schema S is

$$E_s(S) = -\Sigma_j[\Sigma\, Q_i \log(P_i)] \text{ for } i = 1, \ldots n; j = 1, \ldots, m$$

where there are n equivalence classes of domain j, and m attributes in the schema $R(A_1, A_2, \ldots, A_m)$.

This is similar to the definition of entropy for rough sets without factoring in roughness since there are no elements in the boundary region (lower approximation = upper approximation). However, because a relation is a cross product among the domains, we must take the sum of all these entropies to obtain the entropy of the schema. The schema entropy provides a measure of the uncertainty inherent in the definition of the rough relation schema taking into account the partitioning of the domains on which the attributes of the schema are defined.

We extend the schema entropy $E_s(S)$ to define the entropy of an actual rough relation instance $E_R(R)$ of some database D by multiplying each term in the product by the roughness of the rough set of values for the domain of that given attribute.

The rough relation entropy of a particular extension of a schema is

$$E_R(R) = -\Sigma_j D\rho_j(R)\, [\Sigma\, DQ_i \log(DP_i)] \quad \text{for } i = 1, \ldots n; j = 1, \ldots, m$$

where $D\rho_j(R)$ represents a type of database roughness for the rough set of values of the domain for attribute j of the relation, m is the number of attributes in the database relation, and n is the number of equivalence classes for a given domain for the database.

We obtain the $D\rho_j(R)$ values by letting the non-singleton domain values represent elements of the boundary region, computing the original rough set accuracy and subtracting it from one to obtain the roughness. DQ_i is the probability of a tuple in the database relation having a value from class i, and DP_i is the probability of a value for class i occurring in the database relation out of all the values which are given.

Information theoretic measures again prove to be a useful metric for quantifying information content. In rough sets and the rough relational database, this is especially useful since in ordinary rough sets Pawlak's measure of roughness does not seem to capture the information content as precisely as our rough entropy measure.

In rough relational databases, knowledge about entropy can either guide the database user toward less uncertain data or act as a measure of the uncertainty of a data set or relation. As rough relations become larger in terms of the number of tuples or attributes, the automatic calculation of some measure of entropy becomes a necessity. Our rough relation entropy measure fulfills this need.

13 Summary

Uncertainty in information systems has been extensively and successfully modeled using fuzzy set theory. There are a number of aspects of uncertainty that can also be represented with other approaches. In this chapter we have introduced some of these approaches including rough set theory, fuzzy rough sets and intuitionistic set theory. These were applied to a relational database model and extensions to the standard relational operators developed. The basics of functional dependencies for the rough relational database were also described. Lastly the application of information was used to develop formulations for entropy to be applicable to the rough relational database.

Acknowledgments This work was supported by the Naval Research Laboratory's Base Program, Program Element No. 0602435N

References

1. Atanassov, K.: Intuitionistic Fuzzy Sets. Fuzzy Sets Syst. **20**, 87–96 (1986)
2. Atanassov, K.: On Intuitionistic Fuzzy Sets Theory. Springer-Verlag (2012)
3. Beaubouef, T., Petry F.: Rough Querying of Crisp Data in Relational Databases. Proceedings of Third International Workshop on Rough Sets and Soft Computing (RSSC'94), pp. 368–375, San Jose, California (1994)
4. Beaubouef, T., Petry, F.: Fuzzy Set Quantification of Roughness in a Rough Relational Database Model. Proceedings of Third IEEE International Conference on Fuzzy Systems, pp. 172–177, Orlando, Florida (1994)
5. Beaubouef, T., Petry, F.: Fuzzy rough set techniques for uncertainty processing in a relational database. Int. J. Intell. Syst. **15**, 389–424 (2000)
6. Beaubouef, T., Petry F.: A rough set foundation for spatial data mining involving vague regions. Proceedings of FUZZ-IEEE'02, pp. 767–772, Honolulu, Hawaii (2002)
7. Beaubouef, T., Petry F.: Rough Functional Dependencies, 2004 Multiconferences: International Conference On Information and Knowledge Engineering (IKE'04), pp. 175–179, Las Vegas (2004)

8. Beaubouef, T., Petry, F.: Uncertainty modeling for database design using intuitionistic and rough set theory. Int. J. Intell. Fuzzy Syst. **20**(3), 105–117 (2009)

9. Beaubouef, T., Petry F.: Imprecise Database Security and Information Measures., International J. Comput. Intell.: Theory Pract. **5**(2), 61–7 (2010)

10. Beaubouef, T., Petry, F., Arora, G.: Information-theoretic measures of uncertainty for rough sets and rough relational databases. Inf. Sci. **109**, 185–195 (1998)

11. Beaubouef, T., Petry, F., Buckles, B.: Extension of the relational database and its algebra with rough set techniques. Comput. Intell. **11**, 233–245 (1995)

12. Bhandari, D., Pal, N.R.: Some new information measures for fuzzy sets. Inform. Sci. **67**, 209–228 (1993)

13. Bosc, P., Gailbourg, M., Hamlin, G.: Fuzzy querying with SQL: extensions and implementation aspects. Fuzzy Sets Syst. **28**(3), 333–339 (1988)

14. Bosc, P., Pivert, O.: Some approaches for relational databases flexible querying. J. Intell. Inf. Syst. **1**, 323–354 (1992)

15. Bosc, P., Pivert, O.: SQLf : a relational database language for fuzzy querying. IEEE Trans. Fuzzy Syst. **3**, 1–17 (1995)

16. Buckles, B., Petry, F.: A fuzzy model for relational databases. Int. J. Fuzzy Sets Syst. **7**, 213–226 (1982)

17. Buckles, B., Petry, F.: Security and Fuzzy Databases, Proceedings of 1982 IEEE International Conference on Cybernetics and Society, pp. 622–625, Seattle WA (1982)

18. Buckles, B., Petry, F.: Information-theoretical characterization of fuzzy relational databases. IEEE Trans. Syst. Man Cybern. **13**, 74–77 (1983)

19. Chanas, S., Kuchta, D.: Further remarks on the relation between rough and fuzzy sets. Fuzzy Sets Syst. **47**, 391–394 (1992)

20. Codd, E.: A relational model of data for large shared data banks. Commun. ACM **13**(6), 377–387 (1970)

21. de Luca, A., Termini, S.: A definition of a non-probabilistic entropy in the setting of fuzzy set theory. Inf. Control **20**, 301–312 (1972)

22. Denning, D.: Secure statistical databases with random sample queries. Trans. Database Syst. **5**(3), 291–315 (1980)

23. Dubois, D., Prade, H.: Putting rough sets and fuzzy sets together. In: Slowinski, R. (ed.) Intelligent Decision Support: Handbook of Applications and Advances of the Rough Sets Theory. Kluwer Academic Publishers, Boston (1992)

24. Dubois, D., Godo, L., Prade, H., Esteva, F.: An information-based discussion of vagueness. In: Cohen, H., Lefebvre, C. (eds.) Handbook of Categorization in Cognitive Science, Chap. 40, pp. 892–913 , Elsevier (2005)

25. Elmasri, R., Navathe, S.: Fundamentals of Database Systems, 5th edn. Pearson/Addison Wesley (2007)

26. Fung, K., Lam, C.: The database entropy concept and its application to the data allocation problem. INFOR **18**(4), 354–363 (1980)

27. Klir, G., Folger, T.: Fuzzy Sets, Uncertainty, and Information. Prentice Hall, Englewood Cliffs NJ (1988)

28. Ligeza, A.: Granular Sets and Granular Relation. Intelligent Information Systems, pp. 331–340, Physica Verlag (2002)

29. Lin, T.Y.: Topological and fuzzy rough sets. In: Slowinski, R. (ed.) Intelligent Decision Support: Handbook of Applications and Advances of the Rough Sets Theory, pp. 287–304. Kluwer Academic Publishers, Boston (1992)

30. Makinouchi, A.: A Consideration on normal form of not-necessarily normalized relation in the relational data model. Proceedings of the 3rd International Conference on VLDB, pp. 447–453 (1977)

31. Motro, A., Marks, D., Jajodia, S.: Aggregation in relational databases: controlled disclosure of sensitive information. In: Proceedings of ESORICS 94, Third European Symposium on Research in Computer Security. Lecture Notes in Computer Science, vol. 875, pp. 431–445, Brighton, UK, Springer-Verlag (1994)

32. Nanda, S., Majumdar, S.: Fuzzy rough sets. Fuzzy Sets Syst. **45**, 157160 (1992)
33. Nilsson, N.: Probabilistic Logic. Artif. Intell. **28**(1), 71–87 (1986)
34. Ola, A., Ozsoyoglu, G.: Incomplete relational database models based on intervals. IEEE Trans. Knowl. Data Eng. **5**, 293–308 (1993)
35. Pawlak, Z.: Rough sets. Int. J. Comput. Inform. Sci. **11**, 341–356 (1982)
36. Pawlak, Z.: Rough sets and fuzzy sets. Fuzzy Sets Syst. **17**, 99–102 (1985)
37. Pawlak, Z.: Rough Sets: Theoretical Aspects of Reasoning About Data. Kluwer Academic Publishers, Norwell, MA (1991)
38. Prade, H., Testemale, T.: Generalizing database relational algebra for the treatment of incomplete/uncertain information and vague queries. Inform. Sci. **34**, 115–143 (1984)
39. Quinlan, J.: Induction of decision trees. Mach. Learn. **1**, 81–106 (1986)
40. Randell, D., Cui, Z., Cohn, A.: An interval logic for space based on connection. In: Proceedings of ECAI, pp. 394–398 (1992)
41. Shannon, C.: The mathematical theory of communication. Bell Syst. Tech. J. **27**, 379–423 (1948)
42. Shenoi, S., Melton, A., Fan, L.: Functional dependencies and normal forms in the fuzzy relational database model. Inf. Sci. **60**, 1–28 (1992)
43. Srinivasan, P.: The importance of rough approximations for information retrieval. Int. J. Man Mach. Stud. **34**, 657–671 (1991)
44. Szmidt, E., Kacprzyk, J.: On distances between intuitionistic fuzzy sets. Fuzzy Sets Syst. **114**, 505–518 (2000)
45. Szmidt, E., Kacprzyk, J.: Entropy for intuitionistic fuzzy sets. Fuzzy Sets Syst. **118**, 467–477 (2001)
46. Umano, M.: FREEDOM-O: a fuzzy database system. In: Gupta, M., Sanchez, E. (eds.) Fuzzy Information and Studies in Fuzziness Series, pp. 339–347. Physica-Verlag, Heidelberg, Decision Processes, North-Holland (1982)
47. Wygralak, M.: Rough sets and fuzzy sets-some remarks on interrelations. Fuzzy Sets Syst. **29**, 241–243 (1989)
48. Yao, P.: Fuzzy rough set and information entropy based feature selection for credit scoring. Proc. 6th Int. Conf. Fuzzy Syst. Knowl. Disc. **6**, 247–251 (2009)
49. Yao, Y.: Semantics of Fuzzy Sets in Rough Set Theory. T. Rough Sets **II**, 297–318 (2004)
50. Zadeh, L.: Fuzzy sets. Inf. Control **8**, 338–353 (1965)
51. Zadeh, L.A.: Possibility theory and soft data analysis. In: Cobb, L., Thrall, R.M. (eds.) Mathematical Frontiers of the Social and Policy Sciences. Westview, Boulder, CO., pp. 69–129 (1981)
52. Zemankova, M., Kandel, A.: Implementing Imprecision in Information Systems. Inf. Sci. **37**, 107–141 (1985)

Chapter 8
Flexible Bipolar Querying of Uncertain Data Using an Ontology

Patrice Buche, Sébastien Destercke, Valérie Guillard, Ollivier Haemmerlé and
Rallou Thomopoulos

Abstract In this chapter, we propose an approach to query a database where the
user preferences can be bipolar (i.e., express both constraints and wishes about the
desired result) and the data stored in the database can be uncertain. Query results are
then completely ordered with respect to these bipolar preferences, giving priority to
constraints over wishes. Furthermore, we consider user preferences expressed on a
domain of values which is not "flat", but contains values that are more specific than
others according to the "kind of" relation. These preferences are represented by specific fuzzy sets, called "Hierarchical Fuzzy Sets" and defined over a simple ontology.
We propose a use of "Hierarchical Fuzzy Sets" for query enlargement purposes. The
approach is illustrated on a real-world problem concerning the selection of optimal
packaging material for fresh fruits and vegetables.

P. Buche (✉) · R. Thomopoulos
INRA IATE, 2, place Pierre Viala, F-34060 Montpellier Cedex 02, France
e-mail: buche@supagro.inra.fr

P. Buche
LIRMM/CNRS-UM2/INRIA GRAPHIK, F-34392 Montpellier, France

S. Destercke
CNRS HEUDYASIC, Centre de recherches de Royallieu, F-60205 Compiegne Cedex, France
e-mail: sebastien.destercke@hds.utc.fr

V. Guillard
UM2 IATE, cc 023 Pl. E. Bataillon, F-34095 Montpellier, France
e-mail: guillard@univ-montp2.fr

O. Haemmerlé
IRIT-Melodi, Université Toulouse le Mirail, 5, Allées Antonio Machado,
F-31058 Toulouse Cedex 9, France
e-mail: ollivier.haemmerle@univ-tlse2.fr

R. Thomopoulos
LIRMM/CNRS-UM2/INRIA GRAPHIK, F-34392 Montpellier, France
e-mail: rallou@supagro.inra.fr

O. Pivert and S. Zadrożny (eds.), *Flexible Approaches in Data, Information
and Knowledge Management*, Studies in Computational Intelligence 497,
DOI: 10.1007/978-3-319-00954-4_8, © Springer International Publishing Switzerland 2014

1 Introduction

In some applications, there may be a need to differentiate, within queries, between negative preferences and positive ones. Negative preferences correspond to constraints, as they specify which values or objects have to be rejected (i.e., those that do not satisfy constraints), while positive preferences correspond to wishes, as they specify which objects are more desirable than others (i.e., satisfy user wishes) without rejecting those that do not meet the wishes. Indeed, while the first type of preferences should be satisfied by query results, satisfying the second type of preferences can be considered as optional, as the user does not consider them to be necessary requirements.

Also, preferences may be expressed over elements organized into a hierarchy rather than on a 'flat' domain. This kind of hierarchy is typically modeled as a simple ontology in which concepts are partially-ordered by the 'kind of' relation. Considering these two extensions (i.e., allowing bipolar preferences expressed over hierarchies) answers a bipolar query enlargement purpose, as the resulting query will send back more results than classical bipolar ones.

Finally, there may be uncertainty in the available data, and there is a need to integrate this uncertainty in the query processing. In this chapter, we propose to consider these three problems in a common framework, using the notion of bipolar information and of fuzzy pattern matching.

The notions of bipolar preferences and of bipolar information in general have recently received increasing attention [1, 21]. Roughly speaking, information is said to be bipolar when there is a positive and a negative part of the information. These negative and positive parts of the information may have different natures, and should therefore be processed in parallel, and not as a single piece of information. This kind of bipolarity [11], coined as asymmetric, is the one we are concerned with. For example, we may feel both positive and negative about something, without being able to fuse these two feelings in a unique one (for example, eating ice cream gives a gustative pleasure, but one can also feel guilty about it).

In the case of database queries, asymmetric bipolarity is useful to distinguish negative preferences or constraints (i.e. criteria that a good answer **must** satisfy) from positive preferences or wishes (i.e. criteria that a good answer **should** satisfy, if possible). For example, in the query "a new car not too expensive and if possible red", "not too expensive" is clearly a requirement while "red" merely expresses a wish.

Some preliminary studies of this work have already been published in [16] and [28]. In this chapter, we provide a synthetic overview of a method to treat bipolar preferences in databases where data can be uncertain and expressed on a hierarchical domain. In particular, this method uses the bipolar nature of preferences to induce an (pre-)ordering between query results, so that priority is given to those instances that are the most likely to satisfy all expressed constraints and wishes. Section 2 describes the method, while Sect. 3 illustrates the approach on a use case coming from a new decision support system (DSS) currently developed in IATE laboratory where a (industrial/researcher) user wants to select a packaging material that best suits his/her needs. Finally, we give some elements of comparison with previous works in Sect. 4.

2 Method

This section first recalls some basic tools that will be used in the method, before describing the method itself.

2.1 Preliminaries: Fuzzy Pattern Matching

In this chapter, we use fuzzy sets [33] to represent preferences in our queries and possibility distributions [19] to represent uncertainty in the data. A normalized fuzzy set μ over a variable X assuming its value on D_X is a mapping $\mu : D_X \rightarrow [0, 1]$ with at least one $x \in D_X$ such that $\mu(x) = 1$. Here, we assume that D_X is either a finite set of elements (e.g., the colour of a car), possibly partially ordered by the 'kind of' relation (see Sect. 2.3), or a subset of the real line (e.g., the maximal speed of a car).

Here, fuzzy sets are used to express preferences provided by a user in a query. That is, for a given variable X, the fuzzy value $\mu(x)$ expresses to what degree the value x satisfies the preference represented by μ, with $\mu(x) = 1$ meaning that the preference is fully satisfied and $\mu(x) = 0$ that it is completely unsatisfied.

Example 1 Consider again our car example "a new car not too expensive and if possible red". Assume the user has specified that "not too expensive" means that any price over 18,000 \$ is unacceptable, while any price lower than 14,000 \$ can be considered as totally satisfactory. The corresponding preference is represented by the fuzzy set μ_{NTE} in Fig. 1. Given this representation, we have, for example, that a price of 15,000 \$ fulfils the user preferences at a degree $\mu_{NTE}(15, 000) = 0.75$.

Possibility distributions, on the other hand, are simple uncertainty representations allowing to model the ill-known value of some variable. A possibility distribution π over a variable X is also a mapping $\pi : D_X \rightarrow [0, 1]$ with at least one $x \in D_X$ such

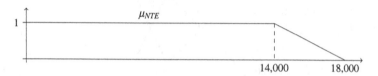

Fig. 1 Fuzzy set μ_{NTE} describing "Not Too Expensive"

that $\pi(x) = 1$. They are therefore equivalent to fuzzy sets from a formal point of view, but possess different semantics. Indeed, they describe our knowledge about the potential value of X. Two measures or set-functions can be derived from a possibility distribution, namely the necessity and possibility measures, which are such that, for every event $A \subseteq D_X$,

$$\Pi(A) = \sup_{x \in A} \pi(x); \quad N(A) = \inf_{x \in A^c} (1 - \pi(x)) = 1 - \Pi(A^c),$$

where $\Pi(A)$ and $N(A)$ express to what extent it is respectively plausible and certain that the actual value of X lies in A.

Note that possibility distributions can model both precisely known values ($X = x$ corresponds to the distribution $\pi(x) = 1$ and zero everywhere else) and set-valued variables ($X \in A$ corresponds to the distribution $\pi(x) = 1$ if $x \in A$, zero otherwise). In the same way, fuzzy sets can model crisp preferences (i.e., those used in classical queries).

In the rest of the chapter, we consider that each query (or preference) P on an attribute X assuming its value on D_X is expressed by a fuzzy set μ_P (possibly degenerated into a crisp preference). Our knowledge D about the attribute value for a particular tuple is given by a possibility distribution π_D (also possibly degenerated in a crisp set). Our knowledge about the imprecise evaluation of P given uncertainty D is summarised by the following lower and upper values [19, 20]:

$$\Pi(P; D) = \sup_{x \in D_X} \min(\mu_P(x), \pi_D(x)), \tag{1}$$

$$N(P; D) = \inf_{x \in D_X} \max(\mu_P(x), 1 - \pi_D(x)).$$

In the following, we will speak of evaluations of a fuzzy preference when talking about the interval $[N(P; D), \Pi(P; D)]$.

Example 2 Consider the preference of Example 1, and a car for which the price is known to belong to the interval $[14, 500; 16, 000]$, with $15, 500$ the most likely value. Figure 2 illustrates both the preference and the knowledge about the price. From this information, we have (using Eq. (1)) that

$$\Pi(P; D) = 0.7 \quad \text{and} \quad N(P; D) = 0.55.$$

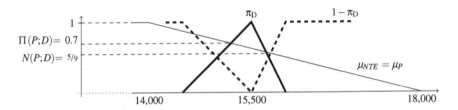

Fig. 2 Evaluation of a fuzzy preference with uncertain data

2.2 Notations and Problem

The problem we consider is as follows: we assume that we have a database consisting of a set $\mathscr{T}$ of T objects o_t, $t = 1, \ldots, T$, with each object taking its values on the Cartesian product $\times_{i=1}^{N} D_{X_i}$ of N domains $D_{X_1}, \ldots, D_{X_N}$. An object o_t is here described by a set of N possibility distributions π_t^i, $i = 1, \ldots, N$, where $\pi_t^i : D_{X_i} \to [0, 1]$ is the possibility distribution describing our knowledge about the value of the ith attribute of object t. When D_{X_i} is finite, its elements are partially ordered in an ontology according to the 'kind of' relation (classical finite sets are retrieved when all elements are incomparable w.r.t. this order, see Sect. 2.3). We also assume that the user provides the following information:

- a set $\mathscr{C} = \{C_1^{i_1}, \ldots, C_{N_c}^{i_{N_c}}\}$ of N_c constraints ($N_c \leq N$) to be satisfied by the retrieved objects, where $C_j^{i_j} : D_{X_{i_j}} \to [0, 1]$ is a normalised fuzzy set defined on the attribute i_j ($1 \leq i_j \leq N$).
- a set $\mathscr{W} = \{W_1^{i_1}, \ldots, W_{N_w}^{i_{N_w}}\}$ of N_w wishes ($N_w \leq N$) that the retrieved objects should satisfy if possible, where $W_j^{i_j} : D_{X_{i_j}} \to [0, 1]$ is a normalised fuzzy set defined on the attribute i_j ($1 \leq i_j \leq N$).
- complete pre-orderings $\leq_c$ and $\leq_w$ between the constraints to be satisfied and between the wishes, respectively. These pre-orderings take account of the fact that some constraints may be considered as more important to satisfy than others (and similarly for wishes). In the following, we denote by $\mathscr{C}_{(i)}$ (resp. $\mathscr{W}_{(i)}$) the constraints (resp. the wishes) that have rank i w.r.t. to the pre-ordering[1] $\leq_c$ (resp. $\leq_w$). We denote by $| \leq_c |$ and $| \leq_w |$ the total number of ranks (i.e., of equivalence classes) induced by the two orderings.

Note that constraints and wishes may well be defined on the same attribute. For example, having an acceptable price may be considered as a constraint, but since a lower price (all other things being equal) is always preferable, lowering the price may become a wish for prices lower than completely satisfying prices (in Example 1, one can define a wish that would start from 14, 000 $).

The problem we consider now is how to retrieve from a set $\mathscr{T}$ of objects, those that primarily satisfy the constraints, and among the latter, those that fulfill the most wishes. Of course, the querying approach has to take account of the bipolar nature of the information, of the possible uncertainty in the data, and of the user's preferences among the constraints and wishes. The next section presents how user preferences are handled when defined over a domain of elements partially ordered by the "kind of" relation. In this latter case, a special kind of fuzzy sets, called *hierarchical fuzzy sets*, will be used.

---

[1] Note that since $\leq_c$ and $\leq_w$ are complete pre-orderings, each constraint/wish has a well-defined rank.

2.3 Fuzzy Sets Defined on a Hierarchical Domain

The notion of hierarchical fuzzy set rose from our need to express fuzzy values in the case where elements receiving a membership value are part of an ontology domain (e.g., packaging material components). First (Sect. 2.3.1), a fuzzy set is created directly by the user and defined on a part of the hierarchy (only some elements are given membership values). Second, for reasons explained in Sect. 2.3.2, we extend the fuzzy set to the whole hierarchy, thus obtaining the *closure* of the fuzzy set. Section 2.3.3 defines how we extend the evaluation of fuzzy preferences when classic fuzzy sets are extended to hierarchical fuzzy sets.

2.3.1 Presentation

The definition domains of the fuzzy sets that we define below are subsets of hierarchies composed of elements partially ordered by the "kind of" relation, i.e. they are defined over a subset $B \subseteq D_X$ of the domain of attribute X. An element $x \in D_X$ is more general than an element $x' \in D_X$ (denoted $x' \prec x$), if x' is a predecessor of x in the partial order induced by the "kind of" relation (denoted $\prec$) of the hierarchy. An example of such a hierarchy is given in Fig. 3. A hierarchical fuzzy set is then defined as follows.

Definition 1 A **hierarchical fuzzy set** *is a fuzzy set whose definition domain $B \subseteq D_X$ is a subset of the elements of a finite hierarchy partially ordered by the "kind of" relation $\prec$.*

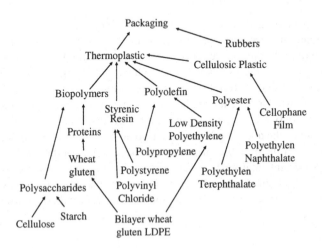

Fig. 3 Example of a hierarchy

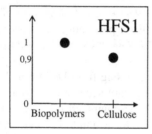
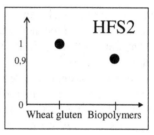

Fig. 4   Fuzzy sets *HFS1* and *HFS2*

For example, the fuzzy sets *HFS1* and *HFS2* represented in Fig. 4 conform to Definition 1. Their definition domains are subsets of the hierarchy given in Fig. 3.

We can note that no restriction has been imposed concerning the elements that compose the definition domain of a hierarchical fuzzy set and their membership values. Therefore, the user may associate a given degree d with an element x and another degree d' with an element x' more specific than x. $d' \leq d$ represents a semantic of restriction for x' compared to x, whereas $d' \geq d$ represents a semantic of reinforcement for x' compared to x.

For example, if there is particular interest in wheat gluten because the user is studying the properties of wheat chain by-products to design packaging, but also wants to retrieve complementary information about other kinds of biopolymers, these preferences can be expressed using for instance the following fuzzy set[2]: $\{(Wheat gluten, 1), (Biopolymers, 0.9)\}$. In this example, the element *Wheat gluten* has a larger degree than the more general element *Biopolymers*, which corresponds to a semantic of reinforcement for *Wheat gluten* compared to *Biopolymers*. On the contrary, if the user is interested in all kinds of biopolymers to design packaging, but to a lesser extent in *Cellulose* because of its higher value to make bioethanol rather than packaging, the preferences can be expressed using the following fuzzy set: $\{(Biopolymers, 1), (Cellulose, 0.9)\}$. In this case, the element *Cellulose* has a smaller degree than the more general element *Biopolymers*, which corresponds to a semantic of restriction for *Cellulose* compared to *Biopolymers*.

2.3.2 Closure of a Hierarchical Fuzzy Set

We can make two remarks concerning the use of hierarchical fuzzy sets:

- the first one is semantic. Let $\{(Polysaccharides, 1), (Biopolymers, 0.9)\}$ be an expression of preferences in a query. We can note that this hierarchical fuzzy set implicitly gives information about elements of the hierarchy other than *Polysaccharides* and *Biopolymers*. For instance, it can be deduced that the user does not

---

[2] Here, we adopt the usual notation (x, y) for specifying fuzzy sets over symbolic variables, where (x, y) means that modality x has membership value y.

expect results concerning packagings like *Rubber* or *Polyolefin*, even if the degree 0 has not explicitly been associated with these packagings. It is also possible to assume that any kind of *Polysaccharides* (*Cellulose* and *Starch* for example) interests the user with the degree 1;

- the second one is operational. The problem rising from Definition 1 is that two different fuzzy sets on the same hierarchy do not necessarily have the same definition domain, which means they cannot be compared using the classic comparison operations of fuzzy set theory (see for example Eq. (1)). For instance, {(*Wheat gluten*, 1), (*Biopolymers*, 0.9)} and {(*Biopolymers*, 1), (*Cellulose*, 0.9)} are defined on two different subsets of the hierarchy of Fig. 3, respectively {*Wheat gluten*, *Biopolymers*} and {*Biopolymers*, *Cellulose*}, and thus are not comparable.

From these remarks can be defined the concept of *closure* of a hierarchical fuzzy set, which is a developed form of the hierarchical fuzzy set defined on the whole hierarchy. The closure of a hierarchical fuzzy set is computed by propagating the degree of an element according to the "kind of" relation: the degree associated with an element is propagated to its sub-elements (more specific elements) in the hierarchy, provided the latter have no degree yet. For instance, in a query, if the user is interested in the element *Biopolymers*, we consider that all kinds of *Biopolymers*—*Polysaccharides*, *Proteins*, etc.—are of interest. On the other hand, we consider that the super-elements (more general elements) of *Biopolymers* in the hierarchy— *Thermoplastic*, *Packaging*, ...—are too general to be relevant for the user's query.

Definition 2 *Let F be a hierarchical fuzzy set defined on a subset B of the elements of a hierarchy D_X. Its membership function is denoted μ_F. The closure$^+$ and the closure$_-$ of F, denoted $\mu_{clos^+(F)}$ and $\mu_{clos_-(F)}$, are two hierarchical fuzzy sets defined on the whole set of elements D_X.*

For each element x of D_X, let $E_x = \{x_1, \ldots, x_n\}$ be the set of the smallest super-elements of x in B, i.e. elements such that for any $i = 1, \ldots, n$, $x \preceq x_i$ ($x \in E_x$ if $x \in B$) and there is no $x' \in B$ such that $x \prec x' \prec x_i$. Then:

- *if E_x is not empty,*

$$\mu_{clos^+(F)}(x) = \max_{1 \le i \le n} (\mu_F(x_i)) \tag{2}$$

and

$$\mu_{clos_-(F)}(x) = \min_{1 \le i \le n} (\mu_F(x_i)); \tag{3}$$

- *otherwise $\mu_{clos^+(F)}(x) = \mu_{clos_-(F)}(x) = 0$.*

In other words, the *closure*$^+$ and the *closure*$_-$ of a hierarchical fuzzy set F are built according to the following rules. For each element x of D_X:

1. if $x \in B$, then x keeps the same degree in both closures of F, i.e., $\mu_{clos^+(F)}(x) = \mu_{clos_-(F)}(x) = \mu_F(x)$ (case where $E_x = \{x\}$);

2. if E_x has a unique smallest super-element x_1 in B, then the degree associated with x_1 is propagated to x in both closures of F, i.e., $\mu_{clos^+(F)}(x) = \mu_{clos_-(F)}(x) = \mu_F(x_1)$ (case where $E_x = \{x_1\}$ with $x \prec x_1$);
3. if x has several smallest super-elements $\{x_1, \ldots, x_n\}$ in B, with different degrees, the proposition made in Definition 2 consists in choosing the maximal degree associated with $x_1, \ldots, x_n$ in the $closure^+$, and the minimal degree in the $closure_-$;
4. all the other elements of D_X, i.e., those that are more general than, or not comparable with the elements of B according to $\prec$ are considered as non-relevant. The degree 0 is associated with them (case where $E_x = \emptyset$).

Example 1 *Figure 5 shows the two closures of the hierarchical fuzzy set* $\{(Wheatgluten, 1), (Biopolymers, 0.8), (Cellulose, 0.3), (LowDensity Polyethylene, 0.2)\}$.

The use of both a permissive ($closure^+$) and a restrictive ($closure_-$) closure is due to the bipolar nature of the preferences involved. In the case of a wish, Eq. (2) ensures a semantic of reinforcement by the use of the max operator (i.e., an element outside B is at least as desirable as its most desirable super-element in B), while the use of min operator in Eq. (3) ensures a semantic of restriction for constraints (i.e., an element outside B is at most as desirable as its least desirable super-element in B).

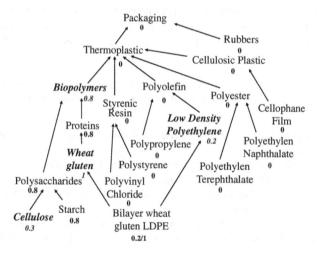

Fig. 5 Closures of a hierarchical fuzzy set: $closure^+$ and $closure_-$ only differ for the element *Bilayer wheat gluten LDPE* for which $\mu_{clos^+(F)}(x) = 1$ and $\mu_{clos_-(F)}(x) = 0.2$

2.3.3 Pattern Matching for Hierarchical Fuzzy Sets

Using the concept of closure, all fuzzy sets defined on a given hierarchy can be extended to the same definition domain (the whole hierarchy D_X) and thus can be compared using the classical comparisons and operations between fuzzy sets (e.g., those presented in Eq. (1)).

Similarly to preferences, our knowledge about data will usually be expressed on a subset B of D_X, here by a possibility distribution π (note that here, we assume that the ontology structure and concepts are certain, only the actual value of some data on this ontology is uncertain). Computing the closure of π over D_X is slightly different, as we do not consider bipolarity in the information (only negative information in the form of π is given) and as the semantic of possibility distributions is different.

Let us define, for an element $x \in D_X$, the set $\underline{E}(x) = \{y_1, \ldots, y_m\}$ of the biggest sub-elements of x in B, i.e. elements such that for any $i = 1, \ldots, m$, $y_i \prec x$ and there is no $y \in B$ such that $y_i \prec y \prec x$. The closure $clos(\pi)(x)$ of π is defined as follows:

- if $x \in B$, then $clos(\pi)(x) = \pi(x)$;
- if $E_x = \{x_1, \ldots, x_n\}$ is not empty, then $clos(\pi)(x) = \max_{1 \leq i \leq n}(\pi(x_i))$;
- if E_x is empty and $\underline{E}(x) = \{y_1, \ldots, y_m\}$ is not, then $clos(\pi)(x) = \max_{1 \leq i \leq m}(\pi(y_i))$;
- else $clos(\pi)(x) = 0$.

This procedure may give quite imprecise possibilities, but it corresponds to the desire not to miss any interesting data. It is also consistent with usual procedures modifying uncertainty models in the case of refinement or coarsening of an initial non-hierarchical space (in the example of Fig. 4, *Biopolymers* can be seen as a coarsening of the elements *Polysaccahrides, Proteins* and as an element of the refinement of *Thermoplastic*).

Definition 3 *Let π and F be two hierarchical fuzzy sets defined on the same hierarchy, respectively defining some knowledge about the variable value and some preferences about these values. Then:*

1. *the possibility degree of matching between π and F a positive preference (resp. a negative one) $\Pi(\pi; F)$ is defined as*
   $\Pi(clos(\pi); clos^+(F)) = \sup_{x \in D_X} \min(clos(\pi)(x), \mu_{clos^+(F)}(x))$
   (resp. $\Pi(clos(\pi); clos_-(F)) = \sup_{x \in D_X} \min(clos(\pi)(x), \mu_{clos_-(F)}(x)))$;
2. *the necessity degree of matching between π and F a positive preference (resp. a negative one), $N(\pi; F)$, is defined as*
   $N(clos(\pi); clos^+(F)) = \inf_{x \in D_X} \max(\mu_{clos^+(F)}(x), 1 - clos(\pi)(x))$
   (resp. $N(clos(\pi); clos_-(F) = \inf_{x \in D_X} \max(\mu_{clos_-(F)}(x), 1 - clos(\pi)(x))))$.

We will see in the next section how bipolar preferences, including positive and negative preferences defined by hierarchical fuzzy sets, are used to query uncertain data.

2.4 From Bipolar Querying with Imprecise Data to Answer Ordering

Previous sections have dealt with the problem of modeling and specifying bipolar preferences and uncertain data over hierarchies defined by simple ontologies. We now detail methods allowing the retrieval and ordering of answers from these preferences and propose some elements explaining this ordering to the users.

2.4.1 Ordering Answers

As underlined by [1], when bipolar information concerns preferences, satisfying constraints should be a primary aim, while satisfying wishes remains secondary. To do this, a solution is to first retain all the objects that may satisfy the constraints, order them w.r.t. the degree to which they satisfy these constraints, and then refine this order by using degrees to which objects satisfy those wishes. If the user has specified preferences between constraints (resp. between wishes)[3], we also provide a means to take these preferences into account.

We propose, for constraints $\mathscr{C}_{(i)}$ of rank i, to summarise the way an object o_t satisfies these constraints by an aggregated interval $[N_t^{(i)}, \Pi_t^{(i)}]_c$ given by the following formula:

$$N_t^{(i)} = \underset{C_k^{jk} \in \mathscr{C}_{(i)}}{\mathsf{T}} N(C_k^{jk}; \pi_t^{jk}), \quad \text{and} \quad \Pi_t^{(i)} = \underset{C_k^{jk} \in \mathscr{C}_{(i)}}{\mathsf{T}} \Pi(C_k^{jk}; \pi_t^{jk}), \quad (4)$$

with $N(C_k^{jk}; \pi_t^{jk}), \Pi(C_k^{jk}; \pi_t^{jk})$ given by Eq. (1) or definition 3 if the domain associated with C_k^{jk} is a hierarchy, and T a t-norm[4] [23]. T-norms are conjunctive aggregation operators and are chosen here for the reason that ALL constraints have to be satisfied simultaneously. Here, we take $\mathsf{T} = \min$, the minimum operator.

Similarly, we build, for each $\mathscr{W}_{(i)}$ and object o_t satisfying the constraints, the interval $[N_t^{(i)}, \Pi_t^{(i)}]_w$, such that

$$N_t^{(i)} = \underset{W_k^{jk} \in \mathscr{W}_{(i)}}{\oplus} N(W_k^{jk}; \pi_t^{jk}), \quad \text{and} \quad \Pi_t^{(i)} = \underset{W_k^{jk} \in \mathscr{W}_{(i)}}{\oplus} \Pi(W_k^{jk}; \pi_t^{jk}), \quad (5)$$

where $\oplus$ is an aggregation operator that can be a t-norm, an averaging operator such as an OWA [32] operator or a t-conorm, depending on the behaviour we want to adopt w.r.t. the satisfaction of wishes. Indeed, since satisfying wishes is not compulsory, we can adopt different attitudes [1]. For instance, using a t-conorm means that we are satisfied as soon as one wish is fulfilled, while using a t-norm means that we still

---

[3] No preferences means here that all constraints (or wishes) have the same rank, i.e., are of equal importance.

[4] A T-norm $\mathsf{T} : [0, 1]^2$ to $[0, 1]$ is an associative, commutative operator that has 1 for neutral element and 0 for absorbing element.

want all the wishes to be satisfied to increase our overall satisfaction. In this chapter, we consider the latter case, and will take $\oplus = \min$.

It is necessary then to order objects that could satisfy the constraints and some wishes, according to the previous evaluations. To do so, we will use a lexicographic order and a dominance relation $\leq_{[N^{(i)}, \Pi^{(i)}]}$ between objects such that, for two interval evaluations $[N_t^{(i)}, \Pi_t^{(i)}]$, $[N_{t'}^{(i)}, \Pi_{t'}^{(i)}]$ related to objects o_t and $o_{t'}$ and to a group of constraints $\mathscr{C}_{(i)}$ or a group of wishes $\mathscr{W}_{(i)}$, $o_t \leq_{[N^{(i)}, \Pi^{(i)}]} o_{t'}$ if $N_t^{(i)} \leq N_{t'}^{(i)}$ and $\Pi_t^{(i)} \leq \Pi_{t'}^{(i)}$ (with $o_t <_{[N^{(i)}, \Pi^{(i)}]} o_{t'}$ if at least one inequality is strict). That is, an object $o_{t'}$ dominates another one o_t if its satisfaction bounds are pair-wise higher than the satisfaction bounds of o_t. The lexicographic order is then used to take account of the difference between negative and positive preferences and of the orderings $\leq_c$ and $\leq_w$ (i.e. objects are first ordered using constraints of rank one, then two, …).

Note that, although $\leq_{[N^{(i)}, \Pi^{(i)}]}$ is a partial order, we will induce from it a complete pre-order that refines $\leq_{[N^{(i)}, \Pi^{(i)}]}$, for the reason that users are more at ease with complete orderings. However, we will use the fact that $\leq_{[N^{(i)}, \Pi^{(i)}]}$ is a partial order to differentiate negative and positive preferences. The procedure consists in building iteratively an ordered partition $\{\mathscr{T}_0, \ldots, \mathscr{T}_M\}$ of $\mathscr{T}$. Rejected objects that do not satisfy all constraints are put in $\mathscr{T}_0$, while objects in $\mathscr{T}_M$ can be considered as the most satisfactory.

In a preliminary step, Algorithm 1 rejects those objects of $\mathscr{T}$ that do not at all satisfy some constraints.

Algorithm 1: Determination of $\mathscr{T}_0$, the set of rejected objects which will not belong to the query result

---

    Input: The set of objects $\mathscr{T} = \{o_1, \ldots, o_T\}$
    Output: Ordered partition$\{\mathscr{T}_0, \mathscr{T} \setminus \mathscr{T}_0\}$ of $\mathscr{T}$
1    $\mathscr{T}_0 = \emptyset$;
2   **foreach** $o_t \in \mathscr{T}$ **do**
3   **if** $\Pi_t^{(i)} = 0$ *for some* $i = 1, \ldots, |\leq_c|$ **then**
4         $\mathscr{T}_0 = \mathscr{T}_0 \cup \{o_t\}$;

Algorithm 2 describes how results are ordered within a subset of $\mathscr{T} \setminus \mathscr{T}_0$ (called $\mathscr{T}'$), according to constraints of a given rank. The whole procedure consists in building a partition of $\mathscr{T} \setminus \mathscr{T}_0$. The partition is refined iteratively by applying, at every rank i ($i \in [1, |\leq_c|]$), Algorithm 2 within each equivalence class of objects obtained at the previous rank $i - 1$. When $i = 1$, the unique initial equivalence class $\mathscr{T}'$ is $\mathscr{T} \setminus \mathscr{T}_0$. In every run of Algorithm 2, equivalence classes $\{\mathscr{T}_1', \ldots, \mathscr{T}_n'\}$ are incrementally built, starting from the worst ($\mathscr{T}_1'$) and ending with the best ($\mathscr{T}_n'$). At each step, the objects included and then suppressed from $\mathscr{T}'$ are those objects that do not dominate other objects (line 4), in the sense of $\leq_{[N^{(i)}, \Pi^{(i)}]}$. This means that objects with imprecise evaluations (i.e., $[N_t^{(i)}, \Pi_t^{(i)}]$ with larger width) will be in lower classes, along with objects having low evaluations (i.e., low $\Pi_t^{(i)}$). This corresponds to a pessimistic attitude towards imprecision, since imprecise evaluations are associated with poorly satisfying objects. Such an attitude is coherent with negative preferences, as the possibility of not satisfying a constraint is penalised.

Algorithm 2: Query result ordering for constraints of rank (i)

Input: $\mathscr{T}' \subseteq \mathscr{T} \setminus \mathscr{T}_0$ with $\mathscr{T}'$ an element of the partition issued from rank $(i-1)$,
$[N_t^{(i)}, \Pi_t^{(i)}]_c$ for each $o_t \in \mathscr{T}'$
Output: Ordered partition$\{\mathscr{T}_1', \ldots, \mathscr{T}_n'\}$ of $\mathscr{T}'$

1   $K = \mathscr{T}'$; j=1;
2   **while** $K \neq \emptyset$ **do**
3   **foreach** $o_t \in K$ **do**
4   **if** $\nexists o_j \in K$ *s.t.* $o_t \geq_{[N^{(i)}, \Pi^{(i)}]} o_j$ **then**
5   Put o_t in $\mathscr{T}_j'$
6      $K = K \setminus \mathscr{T}_j'$;
7      $j = j + 1$;

After having applied Algorithm 1 once and Algorithm 2 $| \leq_c |$ times, the complete pre-order is further refined according to wishes by using Algorithm 3. There are two main differences with Algorithm 1 and Algorithm 2. First, no objects are rejected, as we are dealing with positive preferences (satisfying them is not compulsory). Second, we start here from the best equivalence class and finish with the worst[5], and at each step the objects included and then suppressed from $\mathscr{T}'$ are those objects that are not dominated by other objects (line 7), in the sense of $\leq_{[N^{(i)}, \Pi^{(i)}]}$. Contrary to Algorithm 2, objects with imprecise evaluations will be in the upper classes. This corresponds to an optimistic attitude towards imprecision, which is coherent with positive preferences, as it promotes the possibility of satisfying more wishes. Note that inconsistency problems between positive and negative information [1] do not occur here, since constraints and wishes are treated separately and lexicographically.

The knowledge uncertainty is fully acknowledged through the use of the partial order $\leq_{[N^{(i)}, \Pi^{(i)}]}$ (which considers both end-points of intervals $[N^{(i)}, \Pi^{(i)}]$) in algorithms 2 and 3 which allow us to make a clear distinction in the treatment of negative and positive aspects of bipolar preferences. However, a possible drawback for huge databases is the complexity that the use of these algorithms represents. Indeed, each run of Algorithms 2 and 3 requires comparing each object with all the other objects of a same equivalence class. If n objects have to be ordered, then in the worst case $(| \leq_c | + | \leq_w |) n^2$ comparisons are performed, assuming that no object strictly dominates another for any rank of constraints or wishes. In the best case, i.e. when objects are completely ordered after a first run, n^2 comparisons have to be made. It must be noted that n is reduced to $|\mathscr{T} \setminus \mathscr{T}_0|$ thanks to Algorithm 1. Such complexities are quite acceptable for most databases, but could be problematic for databases counting billions of objects. In such a case, it is possible to use other propositions presenting

---

[5] The shift loop (Lines 3-5) is there to keep the same indexing of subsets $\mathscr{T}_j$

Algorithm 3: Query result ordering for wishes of rank (i)

Input: $\mathcal{T}' \subseteq \mathcal{T} \setminus \mathcal{T}_0$ with $\mathcal{T}'$ an element of a partition issued from rank $(i-1)$,
      $[N_t^{(i)}, \Pi_t^{(i)}]_w$ for each $o_t \in \mathcal{T}'$
Output: Ordered partition$\{\mathcal{T}'_1, \ldots, \mathcal{T}'_m\}$ of $\mathcal{T}'$

1   $K = \mathcal{T}'$; j=0;
2   **while** $K \neq \emptyset$ **do**
3   **for** $i = j, \ldots, 1$ *(skip if $j = 0$)* **do**
4         $\lfloor\ \mathcal{T}'_{j+1} = \mathcal{T}'_j$
5      $\mathcal{T}'_1 = \emptyset$;
6   **foreach** $o_t \in K$ **do**
7   **if** $\not\exists o_j \in K$ *s.t.* $o_t \leq_{[N^{(i)}, \Pi^{(i)}]} o_j$ **then**
8            $\lfloor$ Put o_t in $\mathcal{T}'_1$
9      $K = K \setminus \mathcal{T}'_1$;
10    $\lfloor\ j = j + 1$;

---

a lower complexity where object ordering is solely based on one of the two numbers $N^{(i)}$ or $\Pi^{(i)}$ [20]. However, using orderings based on single numbers means that the imprecision in $[N^{(i)}, \Pi^{(i)}]$ is not fully taken into account and some of the information contained in the interval is lost.

Example 3 Let us consider a set $\mathcal{T}$ of six objects $o_1, \ldots, o_6$, two ranks of constraints and only one rank of wish. The intervals $[N_t^{(i)}, \Pi_t^{(i)}]_c$ $(i = \{1, 2\})$ and $[N_t^{(1)}, \Pi_t^{(1)}]_w$ are summarized in Table 1.

Running Algorithm 1 gives $\mathcal{T}_0 = \{o_4\}$. o_4 is the only rejected object, because $\Pi_4^{(2)} = 0$, even if it satisfies rank one constraints necessarily to a high degree. After a first run of Algorithm 2, we obtain the following partition:

$$\mathcal{T}_0 = \{o_4\} < \mathcal{T}_1 = \{o_1, o_6\} < \mathcal{T}_2 = \{o_2, o_3\} < \mathcal{T}_3 = \{o_5\}.$$

All elements potentially satisfy constraints in $\mathscr{C}_{(1)}$ (although o_6 does not necessarily satisfy them). Note that o_6, for which information is fully imprecise, is at the end of the ordering (whereas it would have been at the front if we used Algorithm 3). Since there are two ranks of constraints, a second run of Algorithm 2 gives

$$\mathcal{T}_0 = \{o_4\} < \mathcal{T}_1 = \{o_6\} < \mathcal{T}_2 = \{o_1\} < \mathcal{T}_3 = \{o_2, o_3\} < \mathcal{T}_4 = \{o_5\}.$$

Table 1 Example 3 evaluations for constraints and wishes

	$[N_t^{(1)}, \Pi_t^{(1)}]_c$	$[N_t^{(2)}, \Pi_t^{(2)}]_c$	$[N_t^{(1)}, \Pi_t^{(1)}]_w$
o_1	[0.1,0.4]	[0.8,1]	[1,1]
o_2	[0.5,0.8]	[0.5,0.6]	[0.6,0.9]
o_3	[0.3,1]	[0.4,0.8]	[0.2,0.5]
o_4	[0.8,1]	[0,0]	[0.5,0.7]
o_5	[1,1]	[0.2,0.4]	[0,0]
o_6	[0,1]	[0.6,0.9]	[0.3,0.7]

This second run refined the ordering between o_1 and o_6. Also note that the bad scores of o_5 w.r.t. constraints of rank two do not change its order, due to the constraint preferences and the use of a lexicographic order. Finally, a run of Algorithm 3 gives

$$\mathscr{T}_0 = \{o_4\} < \mathscr{T}_1 = \{o_6\} < \mathscr{T}_2 = \{o_1\} < \mathscr{T}_3 = \{o_3\} < \mathscr{T}_4 = \{o_2\} < \mathscr{T}_5 = \{o_5\}.$$

Note that o_5 is not rejected, since satisfying wishes is not a requirement.

2.4.2 Explaining the Ordering

Answers provided by DSS, expert systems or multi-criteria decision making methods can be hard to interpret for end-users. It is therefore useful to provide them with simple and understandable (e.g., expressed in natural language) elements of explanation [24].

We therefore propose such explanations of our ordering. As Algorithms 1- 3 use a lexicographic ordering implicitly based on pair-wise comparisons, such explanations can only concern a single rank of constraint or wish and will therefore remain simple. These explanations can be stored in an $n \times n$ matrix $Expl$ where the element $Expl(\ell, k)$ will contain the explanation of why object ℓ has been judged better/worse than object k. This matrix is somehow anti-symmetric, as the reason $Expl(\ell, k)$ will be the opposite of $Expl(k, \ell)$. Note that we do not need to consider objects in $\mathscr{T}_0$, as such objects will not be part of the answer received by the user.

Consider first Algorithm 2 and assume that we are running it on the ith rank of constraints and that loop of lines 3-4 has just ended for the jth time (i.e., the set $\mathscr{T}_j'$ has just been built). Then, for each o_ℓ, o_k such that $o_\ell \in \mathscr{T}_j'$ and $o_k \in K \setminus \mathscr{T}_j'$, we propose the following explanation in $Expl(\ell, k)$:

- if $N_k^{(i)} > \Pi_\ell^{(i)}$, then $Expl(\ell, k) = \{o_\ell$ is judged worse than o_k because it is certainly worse on constraints of priority i, and they are indistinguishable on more important constraints$\}$;
- else, $Expl(\ell, k) = \{o_\ell$ is judged worse than o_k because it is possibly worse on constraints of priority i, and they are indistinguishable on more important constraints$\}$.

Note that explanations make a distinction between the relation $\geq_{[N^{(i)}, \Pi^{(i)}]}$ and the more constraining (but stronger) relation (known as interval dominance) that consists in saying that $o_k > o_\ell$ if and only if $N_k^{(i)} > \Pi_\ell^{(i)}$.

Proposed explanations are similar for Algorithm 3, except that users should be informed that wishes are now considered. Assume that we look at the ith rank of wishes and that loop of lines 3-9 has just finished (the new set $\mathscr{T}_1'$ has just been built). Then, for each o_ℓ, o_k such that $o_\ell \in K \setminus \mathscr{T}_1'$ and $o_k \in \mathscr{T}_1'$, we propose the following explanation in $Expl(\ell, k)$:

- if $N_k^{(i)} > \Pi_\ell^{(i)}$, then $Expl(\ell, k) = \{o_\ell$ is judged worse than o_k because it is certainly worse on wishes of priority i, and both satisfy constraints in an indistinguishable way$\}$;
- else, $Expl(\ell, k) = \{o_\ell$ is judged worse than o_k because it is possibly worse on wishes of priority i, and both satisfy constraints in an indistinguishable way$\}$.

For instance, in Example 3, the element $Expl(6, 1)$ would have been "Object 6 is judged worse than Object 1 because it is possibly worse on constraints of priority 2, and they are indistinguishable on more important constraints". In practical applications, the names of attributes concerned by the constraints or wishes separating two objects can be explicitly cited rather than giving ranks, as they will be more meaningful to the user.

A possible inconvenience of this method is that values of Eq. (4) and (5) are aggregated on many attributes, meaning that a detailed explanation on each attribute of rank i cannot be given. Possible solutions to solve this issue are (1) to consider complete orderings for $\leq_c$ and $\leq_w$ (i.e., $| \leq_c | = N_c$ and $| \leq_w | = N_w$) or (2) to use decision strategies not based on aggregated values (e.g., a voting rule on each constraint/wish of the same rank).

3 A New Decision Support System for Food Packaging Design

In this section, we present a new decision support system (DSS) for fresh fruit and vegetable packaging design in which the flexible bipolar querying approach plays a central role. To the best of our knowledge, only one DSS for fresh fruits and vegetables packaging already exists (see [25]), but it does not take into account the criteria ensuring a sustainable design (a critical issue in food science). Such a sustainable design must satisfy, at least, three kinds of criteria: economic, environmental and societal. An example of the economic aspect may be the cost of the packaging material. Concerning environmental aspects, important criteria are the biodegradability of the package or the optimization of product preservation at ambient temperature (in order to decrease the use of the energy-greedy cold chain). Societal aspects can concern the fact that consumers may reject the use of some additives or of nano-technology in the packaging material because of the unknown consequences on their health, or more simply they may prefer transparent rather than opaque packaging.

In our DSS, starting from a given fruit or vegetable, the user specifies his/her needs in terms of several criteria (e.g., conservation temperature, transparency, material cost, ...) in order to determine a list of packaging. These types of packaging are ordered according to their degree of satisfaction of the criteria. The bipolar approach gives the user the possibility to specify in a flexible way what criteria must be considered as constraints and what other criteria will be used to refine the ranking of packaging satisfying the constraints. Starting from the user specifications, a flexible bipolar query is executed against a database containing information about packaging materials. This information has been collected from different sources which may be technical descriptions of commercial packaging materials or data extracted from scientific publications concerning new packaging materials. This information may be uncertain, due to the variability of engineered packaging and the biological variability of vegetables. The bipolar approach proposed in this chapter deals with this uncertainty. In Sect. 3.1, we present the global architecture of the DSS. A use case concerning endive packaging will be presented in Sect. 3.2.

3.1 Decision Support System Architecture

Starting from the name of the vegetable/fruit of interest specified by the user (see Fig. 6), the system scans in the first step the vegetable/fruit database in order to retrieve the O_2 respiration rate (and associated parameters) of the studied vegetable/fruit. In the second step, the optimal O_2 permeance[6] of the targeted packaging is computed thanks to a model of gas exchanges inside the package called PassiveMap (see [12] for more details about the model). In the third step, using the targeted optimal O_2 permeance and the other user requirements about criteria of various types (economical, environmental or societal), a query is executed against the packaging database using the flexible bipolar querying engine, which is the central part of the DSS. A list of packaging materials ordered according to the method presented in the previous sections is finally presented to the user. The use case presented in the next section focuses on the DSS flexible bipolar querying engine.

Fig. 6 Global architecture of the DSS

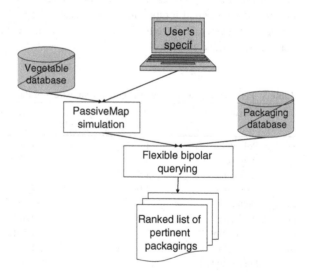

---

[6] A measure of the ability of a package to conduct gas fluxes.

3.2 Endive Packaging Use Case

In this section, we present a use case of the DSS concerning the choice of a packaging material for endives. The user has to specify a set of parameters needed by the DSS to determine the optimal O_2 permeance of the targeted packaging: the mass of the vegetable (500 grams), the surface, the volume and the thickness of the targeted packaging (respectively 0.14 m^2, 0.002 m^3 and 5e-5 m), the shelf life of the vegetable (7 days) and the storage temperature (20 °C). Using the O_2 respiration rate (and associated parameters) retrieved from the vegetable database, an optimal O_2 permeance of 3.65E-11 $mol.m^{-2}.s^{-1}.Pa^{-1}$ is computed. The optimal permeance and the temperature will be considered as criteria to scan the package database.

We consider in this use case that the user is also interested in two other criteria: the biodegradability and the transparency of the package. An extract of the packaging database content is presented in Tables 2 and 3 and will be used to illustrate the flexible bipolar querying process. Note that imprecise data are here reduced to degenerated possibility distributions (given by the min–max permeance span), since currently there is no possibilistic uncertainty in the database (however, such uncertainty will be integrated in future evolutions of the DSS including robust design methods [17]).

We will consider two examples of queries expressed by the user (in the current case, they were given by one of the co-authors, V. Guillard). In the first one, the user specifies one constraint and two wishes. The user first requires the package to be transparent in order to be accepted by the consumer who wants to see the endive through the package. It will be expressed as the first and unique constraint. Concerning his/her wishes, the user would like to maximize the shelf life of the product at an ambient temperature (and consequently to select a packaging whose oxygen permeance is close to the optimal one). It will be expressed as the wishes, here of equal rank.

Table 2 Permeance at a given temperature for an extract of the packaging database

o_{id}	Packaging type	Permeance$_{min}$ ($mol.m^{-2}.s^{-1}.Pa^{-1}$)	Permeance$_{max}$ ($mol.m^{-2}.s^{-1}.Pa^{-1}$)	Temperature (°C)
o_1	Polyolefin	1,29E-13	1,29E-13	23
o_2	Polyolefin	4,05E-11	4,05E-11	23
o_3	Cellophane	1,55E-14	1,55E-14	23
o_4	Polyolefin	1,96E-11	2,39E-11	20
o_5	Cellulose	1,55E-14	1,55E-14	23
o_6	Polyester	4,46E-12	4,46E-12	23
o_7	Polyolefin	1,50E-11	1,50E-11	23
o_8	Polyester	1,55E-13	1,55E-13	23
o_9	Polystyrene	1,03E-12	1,03E-12	23
o_{10}	Polyester	6,23E-12	6,23E-12	23
o_{11}	Wheat gluten	1,55E-11	1,67E-11	25
o_{12}	PolyVinyl Chloride	7,47E-11	7,47E-11	25

Table 3 Transparency and biodegradability for the same extract of the packaging database

o_{id}	PackagingType	Transparency	Biodegradability
o_1	Polyolefin	Transparent	no
o_2	Polyolefin	Transparent	no
o_3	Cellophane	Transparent	yes
o_4	Polyolefin	Transparent	no
o_5	Cellulose	Transparent	yes
o_6	Polyester	Transparent	yes
o_7	Polyolefin	Transparent	no
o_8	Polyester	Translucent	yes
o_9	Polystyrene	Translucent	no
o_{10}	Polyester	Translucent	yes
o_{11}	Wheat gluten	Translucent	yes
o_{12}	PolyVinyl Chloride	Transparent	no

In the second query, the user specifies two constraints and two wishes. To design a sustainable package, the user expresses that the packaging must be biodegradable as a first constraint (rank one) and must be made of renewable resources (i.e. constraint of rank two on the packaging type). Then, the user expresses as first wish that the packaging should be transparent in order to be accepted by the consumer and as second wish that it should maximize the shelf life of the product at an ambient temperature for economic reasons.

As already said in Sect. 2.1, the user preferences are, for each criterion, expressed by a fuzzy set used as a general formalism which enables the representation of fuzzy, interval or crisp values. Concerning the permeance criterion, 60 % of variation is authorized around the optimal value computed by the PassiveMap subsystem, with decreasing degrees of preferences. For the temperature, a total variation of 100 % is authorized, with no preference for the different values. The fuzzy sets associated with the permeance and temperature preferences are presented in Fig. 7.

The fuzzy set associated with the transparency (resp. biodegradability) criterion is $Pref_{transparency}=\{(\text{transparent},1),(\text{translucent},0),(\text{opaque},0)\}$ (resp. the fuzzy set $Pref_{biodegradability}=\{(\text{yes},1),(\text{no},0)\}$). They correspond to crisp requirements provided by the user, as the concept of graded biodegradability made little sense to the user, while translucency is not graded in our current data. The hierarchical fuzzy set associated with the packaging type is $Pref_{packagingType}=\{(\text{Biopolymers},1)\}$.

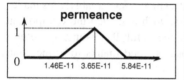 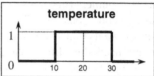

Fig. 7 Preferences for permeance and temperature

Table 4 Evaluations for the constraint and the wishes of the first query

	$[N_t^{(1)}, \Pi_t^{(1)}]_c$	$[N_t^{(1)}, \Pi_t^{(1)}]_w$
o_1	[1,1]	[0,0]
o_2	[1,1]	[0,817,0,817]
o_3	[1,1]	[0,0]
o_4	[1,1]	[0,228,0,427]
o_5	[1,1]	[0,0]
o_6	[1,1]	[0,0]
o_7	[1,1]	[0,021,0,021]
o_8	[0,0]	[0,0]
o_9	[0,0]	[0,0]
o_{10}	[0,0]	[0,0]
o_{11}	[0,0]	[0,043,0,098]
o_{12}	[1,1]	[0,0]

It expresses that the user preferences are for renewable resources but without specifying a specific type of biopolymer.

Using the notations introduced in Sect. 2.1, the first query is built as follows: $\mathscr{C}_{(1)} = \{Pref_{transparency}\}$ and $\mathscr{W}_{(1)} = \{Pref_{permeance}, Pref_{temperature}\}$.

Let us consider the set $\mathscr{T} = \{o_1, \ldots, o_{12}\}$ of the twelve packages whose characteristics are given in Tables 2 and 3 and whose evaluations for the constraint and wishes of query 1 are given in Table 4 (as the two wishes are of the same rank, they have been aggregated in $[N_t^{(1)}, \Pi_t^{(1)}]_w$ according to Eq. (5)). After running Algorithm 1, we obtain $\mathscr{T}_0 = \{o_8, o_9, o_{10}, o_{11}\}$. After running Algorithm 2 with $\mathscr{C}_{(1)}$, we obtain the following partition:

$$\mathscr{T}_0 = \{o_8, o_9, o_{10}, o_{11}\} < \mathscr{T}_1 = \{o_1, o_2, o_3, o_4, o_5, o_6, o_7, o_{12}\}.$$

After running Algorithm 3 with $\mathscr{W}_{(1)}$, we obtain the following partition:

$$\mathscr{T}_0 = \{o_8, o_9, o_{10}, o_{11}\} < \mathscr{T}_1 = \{o_1, o_3, o_5, o_6, o_{12}\} <$$
$$\mathscr{T}_2 = \{o_7\} < \mathscr{T}_3 = \{o_4\} < \mathscr{T}_4 = \{o_2\}.$$

The second query is built as follows:

$\mathscr{C}_{(1)} = \{Pref_{biodegradability}\}$, $\mathscr{C}_{(2)} = \{Pref_{packagingType}\}$, $\mathscr{W}_{(1)} = \{Pref_{transparency}\}$, $\mathscr{W}_{(2)} = \{Pref_{permeance}, Pref_{temperature}\}$. The first constraint is judged more important than the second one: one wants biodegradable packaging to preserve the environment (first constraint) which is sustainable, thus made of renewable resource (second constraint). The first wish is judged more important than the second one: one wants transparent packaging to fulfill consumers' preferences (first wish) and optimized shelf life of the packed food thanks to a fine control of O_2 permeance (second wish).

Table 5 Evaluations for the constraints and the wish of the second query

	$[N_t^{(1)}, \Pi_t^{(1)}]_c$	$[N_t^{(2)}, \Pi_t^{(2)}]_c$	$[N_t^{(1)}, \Pi_t^{(1)}]_w$	$[N_t^{(2)}, \Pi_t^{(2)}]_w$
o_1	[0,0]	[0,0]	[1,1]	[0,0]
o_2	[0,0]	[0,0]	[1,1]	[0,817,0,817]
o_3	[1,1]	[0,0]	[1,1]	[0,0]
o_4	[0,0]	[0,0]	[1,1]	[0,228,0,427]
o_5	[1,1]	[1,1]	[1,1]	[0,0]
o_6	[1,1]	[0,0]	[1,1]	[0,0]
o_7	[0,0]	[0,0]	[1,1]	[0,021,0,021]
o_8	[1,1]	[0,0]	[0,0]	[0,0]
o_9	[0,0]	[0,0]	[0,0]	[0,0]
o_{10}	[1,1]	[0,0]	[0,0]	[0,0]
o_{11}	[1,1]	[1,1]	[0,0]	[0,043,0,098]
o_{12}	[0,0]	[0,0]	[1,1]	[0,0]

Consider again the set $\mathcal{T}$ of packages described in Tables 2 and 3 and whose evaluations for the constraints and the wish of query 2 are given in Table 5. After running Algorithm 1, we obtain $\mathcal{T}_0 = \{o_1, o_2, o_3, o_4, o_6, o_7, o_8, o_9, o_{10}, o_{12}\}$. Packaging which are not biodegradable have been discarded. Moreover, the hierarchical fuzzy set associated with the packaging type, $Pref_{packagingType}$, permits to express a generic constraint in a simple way: packaging which are not bio-sourced have been discarded too. It must be noticed that the use of a classical fuzzy set for $Pref_{packagingType}$ instead of a hierarchical fuzzy set would have delivered an empty set of answers (all the objects in $\mathcal{T}_0$) after running Algorithm 1. After the first run of Algorithm 2 with $\mathcal{C}_{(1)}$, we obtain the following partition:

$$\mathcal{T}_0 = \{o_1, o_2, o_3, o_4, o_6, o_7, o_8, o_9, o_{10}, o_{12}\} < \mathcal{T}_1 = \{o_5, o_{11}\}.$$

The second run of Algorithm 2 with $\mathcal{C}_{(2)}$ ($[N_t^{(2)}, \Pi_t^{(2)}]_c$) keeps the partition unchanged. After the first run of Algorithm 3 with $\mathcal{W}_{(1)}$, we obtain the following partition:

$$\mathcal{T}_0 = \{o_1, o_2, o_3, o_4, o_6, o_7, o_8, o_9, o_{10}, o_{12}\} < \mathcal{T}_1 = \{o_{11}\} < \mathcal{T}_2 = \{o_5\}.$$

The second run of Algorithm 3 with $\mathcal{W}_{(2)}$ keeps the partition unchanged.

We can see with the result obtained for the second query, from which only two results are retrieved, that the constraints may be very restrictive compared to the content of the database. In those cases where no answer is found, we have proposed in [16] an approach to provide to the users "best" answers among all the rejected ones (i.e., answers that are the closest to satisfying the constraints).

4 Related Works

There exist many works that propose to use fuzzy sets to introduce graded preferences and possibility distributions to handle uncertainty in databases. Our work can be related to these two complementary propositions.

The fuzzy set framework has been shown to be a sound scientific choice to model flexible queries [4]. It is a natural way of representing the notion of preference using a gradual scale. In [7], the semantics of a language called SQLf has been proposed to extend the well-known SQL language by introducing fuzzy predicates processed on crisp information. Other approaches have also been proposed to introduce preferences into queries in the database community [8, 13, 22]. However, in all these approaches, preferences are of the same nature. It is only recently that the concept of bipolarity and its potential use in flexible queries has been studied [18, 21]. This extended approach discriminates between two types of preferences, one acting as compulsory constraints, the other acting as optional wishes. Several works have recently been proposed in order to extend the relational algebra with this concept of bipolarity [6, 5]or to propose a framework to deal with bipolarity in regular relational databases [30]. It should be noticed that, to the best of our knowledge, the introduction in bipolar flexible querying of preferences expressed on a hierarchical domain is an original point of our approach.

The second proposition is to use possibility distributions (whose formalism is mathematically equivalent to that of fuzzy set) to represent uncertain values [34]. Several authors have developed this approach in the context of databases [2, 3, 10, 26, 27, 29]. To the best of our knowledge, the only other work dealing with the concept of bipolarity in flexible querying of databases including uncertain values, outside some research perspectives in [21], is that of G. De Tré et al. [31]. However, they deal with a different aspect of bipolar preferences, as they mainly consider the use of interval-valued fuzzy sets (or similar models) to cope with imprecisely defined preferences, and treat positive and negative preferences in a common framework, rather than considering them separately (as we do here).

5 Conclusion and Perspectives

In this Chapter, we have introduced a method for querying a database when preferences are bipolar (contains both constraints and wishes), data are uncertain and can be expressed on a hierarchical domain. We use fuzzy sets and possibility distributions to model preferences and uncertainty, respectively.

Using basic tools to evaluate query satisfaction, we have proposed methods allowing us to (1) extend fuzzy sets to hierarchical fuzzy sets which put in adequacy two order relations (the preference order relation and the 'kind of' relation) to permit a query enlargement (2) consider orderings between constraints or wishes and

(3) pre-order the results according to the bipolar preferences, thus presenting a list of equivalence classes to the user.

The proposed approach is applied in a real-case problem, and is included in a new support decision tool aiming at designing (optimal) packages for fresh fruits and vegetables.

Concerning the method, perspectives include the handling of more generic kinds of uncertainty models [14, 15] that could be included in the database, as well as methods that would allow to extract information concerning packages from the web automatically [9], since manually entering this information is time-consuming and can only be done by an expert.

Concerning the support decision tool, we are planning to link it with a preliminary step which will combine preferences expressed by the actors of the food packaging chain, which can be potentially in conflict, using argumentation methods.

Acknowledgments The research leading to these results has received funding from the European Community's Seventh Framework Programme (FP7/ 2007-2013) under the grant agreement FP7-265669-EcoBioCAP project.

References

1. Benferhat, S., Dubois, D., Kaci, S., Prade, H.: Bipolar possibility theory in preference modelling: representation, fusion and optimal solutions. Inf. Fusion **7**, 135–150 (2006)
2. Bordogna, G., Pasi, G.: A fuzzy object oriented data model managing vague and uncertain information. Int. J. Intell. Syst. **14**(6), SCI 3495 (1999)
3. Bosc, P., Lietard, L., Pivert, O.: Fuzzy theory techniques and applications in data-base management systems. In: Bosc, P., Kacprzyk, J. (eds.) Fuzziness in Database Management Systems, pp. 666–671. Academic Press, New York (1999)
4. Bosc, P., Lietard, L., Pivert, O.: Soft querying, a new feature for database management system. In: Proceedings DEXA'94 (Database and EXpert system Application), Lecture Notes in Computer Science, vol. 856, pp. 631–640. Springer-Verlag (1994)
5. Bosc, P., Pivert, O., Mokhtari, A., Lietard, L.: Extending relational algebra to handle bipolarity. In: Proceedings of the 2010 ACM Symposium on Applied Computing (SAC), pp. 1718–1722. Sierre, Switzerland, ACM, 22–26 March 2010
6. Bosc, P., Pivert, O.: About bipolar division operators. In: Flexible Query Answering Systems, 8th International Conference, FQAS 2009, Roskilde, Denmark, October 26–28, 2009. Proceedings. Lecture Notes in Computer Science, vol. 5822, pp. 572–582. Springer (2009)
7. Bosc, P., Pivert, O.: SQLf: a relational database language for fuzzy querying. IEEE Trans. Fuzzy Syst. **3**(1), 1–17 (1995)
8. Bruno, N., Chaudhuri, S., Gravano, L.: Top-k selection queries over relational databases: Mapping strategies and performance evaluation. ACM Trans. Database Syst. **27**(2), 153–187 (2002)
9. Buche, P., Dibie-Barthelemy, J., Ibanescu, L., Soler, L.: Fuzzy web data tables integration guided by an ontological and terminological resource. IEEE Trans. Knowl. Data Eng. **24**(4), 805–819 (2011)
10. Buche, P., Haemmerlé, O.: Towards a unified querying system of both structured and semi-structured imprecise data using fuzzy views. In: Proceedings of the 8th International Conference on Conceptual Structures, Lecture Notes in Artificial Intelligence, vol. 1867. pp. 207–220. Darmstadt, Germany, Springer-Verlag (August 2000)

11. Cacioppo, J.T., Gardner, W.L., Berntson, G.G.: Beyond bipolar conceptualizations and measures: the case of attitudes & evaluative space. Pers. Soc. Psychol. Rev. **1**, 3–25 (1997)
12. Charles, F., Sanchez, J., Gontard, N.: Modeling of active modified atmosphere packaging of endives exposed to several postharvest temperatures. J. Food Sci. **8**, 443–448 (2005)
13. Chomicki, J.: Preference formulas in relational queries. ACM Trans. Database Syst. **28**(4), 427–466 (2003)
14. Destercke, S., Dubois, D., Chojnacki, E.: Unifying practical uncertainty representations: I generalized p-boxes. Int. J. Approximate Reasoning **49**(3), 664–677 (2008)
15. Destercke, S., Dubois, D., Chojnacki, E.: Unifying practical uncertainty representations: II clouds. Int. J. Approximate Reasoning **49**(3), 649–663 (2008)
16. Destercke, S., Buche, P., Guillard, V.: A flexible bipolar querying approach with imprecise data and guaranteed results. Fuzzy Sets Syst. **169**(1), 51–64 (2011)
17. Destercke, S., Guillard, V.: Interval analysis on non-linear monotonic systems as an efficient tool to optimise fresh food packaging. Comput. Electron. Agric. **79**(2), 116–124 (2011)
18. Dubois, D., Prade, H.: Bipolarity in flexible querying. In: Flexible Query Answering Systems, 5th International Conference, FQAS 2002, Copenhagen, Denmark, October 27–29, 2002, Proceedings. Lecture Notes in Computer Science, vol. 2522, pp. 174–182. Springer (2002)
19. Dubois, D., Prade, H.: Possibility Theory: An Approach to Computerized Processing of Uncertainty. Plenum Press, New York (1988)
20. Dubois, D., Prade, H.: Tolerant fuzzy pattern matching: an introduction. In: Bosc, P., Kacprzyk, J. (eds.) Fuzziness in Database Management Systems. Physica-Verlag (1995)
21. Dubois, D., Prade, H.: An overview of the asymmetric bipolar representation of positive and negative information in possibility theory. Fuzzy Sets Syst. **160**, 1355–1366 (2009)
22. Kießling, W., Köstler, G.: Preference SQL—design, implementation, experiences. In: VLDB Proceedings of 28th International Conference on Very Large Data Bases. pp. 990–1001. Morgan Kaufmann, Hong Kong, China, 20–23 August 2002
23. Klement, E., Mesiar, R., Pap, E.: Triangular Norms. Kluwer Academic Publisher, Dordrecht (2000)
24. Labreuche, C.: A general framework for explaining the results of a multi-attribute preference model. Artif. Intell. **175**(7–8), 1410–1448 (2011)
25. Mahajan, P., Oliveira, F., Montanez, J., Frias, J.: Development of user-friendly software for design of modified atmosphere packaging for fresh and fresh-cut produce. Innovative Food Sci. Emerg. Technol. **8**, 84–92 (2007)
26. Prade, H.: Lipski's approach to incomplete information data bases restated and generalized in the setting of Zadeh's possibility theory. Inf. Syst. **9**(1), 27–42 (1984)
27. Prade, H., Testemale, C.: Generalizing database relational algebra for the treatment of incomplete or uncertain information and vague queries. Inf. Sci. **34**, 115–143 (1984)
28. Thomopoulos, R., Buche, P., Haemmerlé, O.: Fuzzy sets defined on a hierarchical domain. IEEE Trans. Knowl. Data Eng. **18**(10), 1397–1410 (2006)
29. Tré, G.D., Caluwe, R.D.: A generalized object-oriented database model. In: Bordogna, G. Pasi, G. (eds.) Recent Research Issues on the Management of Fuzziness in Databases. Studies in Fuzziness and Soft computing, vol. 53, pp. 155–182. Physica-Verlag, Heidelberg, Germany (2000)
30. Tré, G.D., Zadrozny, S., Matthé, T., Kacprzyk, J., Bronselaer, A.: Dealing with positive and negative query criteria in fuzzy database querying. In: Flexible Query Answering Systems, 8th International Conference, FQAS 2009, Proceedings. Lecture Notes in Computer Science, vol. 5822, pp. 593–604. Roskilde, Denmark, Springer, 26–28 October 2009
31. Tré, G.D., Zadrozny, S., Bronselaer, A.: Handling bipolarity in elementary queries to possibilistic databases. IEEE Trans. Fuzzy Syst. **18**, 599–612 (2010)
32. Yager, R.: On ordered weighted averaging aggregation operators in multicriteria decision making. IEEE Trans. Syst. Man Cybern. **18**, 183–190 (1988)
33. Zadeh, L.: The concept of a linguistic variable and its application to approximate reasoning-i. Inf. Sci. **8**, 199–249 (1975)
34. Zadeh, L.: Fuzzy sets as a basis for a theory of possibility. Fuzzy Sets Syst. **1**, 3–28 (1978)

Chapter 9
Aspects of Dealing with Imperfect Data in Temporal Databases

José Pons, Christophe Billiet, Olga Pons and Guy De Tré

Abstract In reality, some objects or concepts have properties with a time-variant or time-related nature. Modelling these kinds of objects or concepts in a (relational) database schema is possible, but time-variant and time-related attributes have an impact on the consistency of the entire database. Therefore, temporal database models have been proposed to deal with this. Time itself can be at the source of imprecision, vagueness and uncertainty, since existing time measuring devices are inherently imperfect. Accordingly, human beings manage time using temporal indications and temporal notions, which may contain imprecision, vagueness and uncertainty. However, the imperfection in human-used temporal indications is supported by human interpretation, whereas information systems need extraordinary support for this. Several proposals for dealing with such imperfections when modelling temporal aspects exist. Some of these proposals consider the basis of the system to be the conversion of the specificity of temporal notions between used temporal expressions. Other proposals consider the temporal indications in the used temporal expressions to be the source of imperfection. In this chapter, an overview is given, concerning the basic concepts and issues related to the modelling of time as such or in (relational) database

J. Pons (✉) · O. Pons
Department of Computer Science and Artificial Intelligence, Escuela Técnica Superior de Ingeniería Informática, Universidad de Granada, C/Periodista Daniel Saucedo Aranda s/n, 18071 Granada, Spain
e-mail: jpons@decsai.ugr.es

O. Pons
e-mail: opc@decsai.ugr.es

C. Billiet · G. De Tré
Department of Telecommunications and Information Processing, Ghent University, St.-Pietersnieuwstraat 41,9000 Gent, Belgium
e-mail: Christophe.Billiet@telin.ugent.be

G. De Tré
e-mail: Guy.DeTre@telin.ugent.be

O. Pivert and S. Zadrożny (eds.), *Flexible Approaches in Data, Information and Knowledge Management*, Studies in Computational Intelligence 497, DOI: 10.1007/978-3-319-00954-4_9, © Springer International Publishing Switzerland 2014

189

models and the imperfections that may arise during or as a result of this modelling. Next to this, a novel and currently researched technique for handling some of these imperfections is presented.

1 Introduction

The concept of time itself is very complex to handle and interpret [52, 74], though it is very natural and omnipresent. As information systems often attempt the modelling of natural objects, concepts or processes, they often require modelling temporal aspects or concepts. Thus, several proposals have been concerned with the obtaining of theoretical models that allow the modelling or representation of time [7, 16].

A very specific type of information systems are database systems, which are computer systems designed to manage databases. A database contains data representing real objects or concepts. Each (atomic) part of these data is a result value of a measurement of a property or a description of a property of a real object or concept. In reality, some aspects or properties of objects or concepts are time-variant or time-related. e.g., the moment of a bank transaction is traditionally a moment in time and thus a time-related notion, the function of an employee in a company can change through recorded history and is thus time-variant. A temporal database schema is a database schema that models real objects or concepts with time-related or time-variant properties. However, the modelling of temporal aspects has a direct impact on the consistency of the temporal database, because the temporal nature of these aspects imposes extra integrity constraints. An example. Consider a relation in a relational library database, modelling the presence of books in the library. Every physical book is represented by a unique identifier. Every record in the relation contains such an identifier, a date on which the corresponding book was loaned and a date on which it was subsequently returned (if it was returned). Without further precautions, a library employee could add several records with the same book identifier, different 'loaned'-dates and no 'returned'-dates. This group of records would represent the same physical book being loaned several times on different dates and never returned, which is of course impossible. A temporal database model will typically constrain record insertion and prevent similar modelling inconsistencies.

A lot of research concerns temporal database models and their approaches to the modelling of time. The first efforts were towards the representation of historical information related to objects represented by records in a database [14]. Some proposals tried to extend the Entity Relationship Model [53], without impact on any database standards like SQL [72].

Notably, in 1994, "A Consensus Glossary of Temporal Database Concepts" was published [36]. For this publication, 44 temporal database researchers, among which some of the main researchers in this field, cooperated to reach a consensus on the nature and definitions of some of the main temporal database concepts and terminology. This glossary was subsequently updated in 1998 [48].

An interesting issue in temporal modelling concerns relationships between temporal notions. Notably, Allen [1] studied temporal relationships between time intervals (and as a special case time points). Among others, the querying of temporal databases has greatly profited from these temporal relationships, because they allowed for richer and more complex user-specified temporal query demands, by allowing to express more complex relationships between the temporal notions in the temporal expressions in the query and the temporal indications in the database. e.g., in a relation modelling who was department head of an institution during which periods of time, a query like 'who were the department heads when Thomas worked for the institution' can be evaluated using similar relationships.

Humans handle temporal information using certain temporal notions like time intervals or time points [36], and they often have to deal with imperfections like imprecision's, vagueness, uncertainties or inconsistencies possibly contained in the descriptions of these temporal notions. Among many others, these possible imperfections in descriptions of temporal notions determine an important issue in temporal modelling. e.g., the description of the temporal notion in a sentence like 'The Belfry of Bruges was finished on a day somewhere between 01/01/1201 and 31/12/1300 A.D.' contains imperfection because of the uncertainty in the used time-related expression. It is known that the building was finished on a single day, but it is not known precisely which day this was.

To allow information systems to cope with these and similar data imperfections, many approaches adopt fuzzy sets [59] for the representation of temporal information [5, 26, 61, 62]. The temporal relationships studied by Allen were fuzzified by several authors [62, 65, 73]. Garrido et al. [43] present different temporal operators, defined by a combination of regular fuzzy comparisons. Both [43, 68] deal with uncertainty in temporal expressions concerning time intervals. Other approaches, like [70], use rough sets [66] to represent time intervals.

Next to temporal modelling, some attention has been spent on temporal reasoning [1]. Although temporal reasoning is not discussed in this chapter, it should be noted that, among others, Dubois and Prade et al. [26, 30] have dealt with fuzziness and uncertainty in temporal reasoning.

The aim of this chapter is to present and explain some main concepts regarding time in information systems and to present and discuss some issues and techniques concerned with handling data imperfections related to time. The rest of this chapter is structured as follows. Section 2 presents some basic concepts and terminology about temporal modelling and discusses some of its important aspects and issues, while Sect. 3 presents some important issues concerning the combination of data imperfections and temporal modelling. In Sect. 4, some basic concepts and terminology about temporal databases are presented, followed by an overview of some interesting issues concerning temporal databases and a survey of some commercial temporal database systems. Finally, in Sect. 5, an approach to querying temporal databases containing imperfect temporal information is presented, followed by some conclusions and some suggestions for future work in Sect. 6.

2 Basic Concepts and Issues in Time Modelling

Before considering the introduction of temporal modelling to information systems, it is necessary to define and explain some main concepts concerning temporal modelling and their corresponding terminology, to situate these and to discuss some properties and issues related to these concepts. In this section, several basic concepts and their corresponding terminology will be defined, explained and situated. Most of these concepts are widely used in the community of temporal databases and their definitions have been agreed upon in the context of [36]. For these concepts, in the entire chapter, the contents of [36] are followed (and often cited).

2.1 Basic Concepts and Properties

In information systems, time itself is usually perceived as a linear or cyclic concept [50]. Therefore, a time domain modelling time is usually represented by a set with an imposed partial order. In general, two main types of time models can be discerned: a *linear* model [4] and a *cyclic* model [58]. In the linear model, a total order is imposed on the set and the progress of time is seen as a linear matter, while cyclic models are mainly used in the modelling of recurrent processes. It should be noted that the majority of proposals use linear time models.

Data models used by information systems (and in specific, temporal database systems) may represent an underlying time axis using *chronons* [36], which can informally be described as the smallest distinguishable time units that can be used in the system. However, to explain what chronons are, an explanation of some other temporal concepts is necessary.

Definition 1 Instant [36]
An *instant* is a time point on an underlying timeaxis.

Thus, an instant is basically an instantaneous time point on the time axis underlying a time model. The term is used in the context of the time model too.

Orthogonal to the classification of time models as linear or cyclic, they can be classified as *discrete*, *dense* or *continuous* models [36, 48]. In a discrete model [14], the notion exists that every instant has a unique successor and the set of (modelled) instants is seen as a discrete one. Here, intuitively, the set of instants can be seen as isomorphic to the set of natural numbers $\mathbb{N}$. In a dense model, the notion exists that between any two instants always lies another. Here, intuitively, the set of instants can be seen as isomorphic to the set of rational numbers $\mathbb{Q}$ (when the set of (modelled) instants is a discrete one) or the set of real numbers $\mathbb{R}$ (when the set of (modelled) instants is a continuous one). In a continuous model, the notion also exists that between any two instants always lies another one, but the set of (modelled) instants is always seen as continuous and there are no "gaps" between successive instants.

Some other necessary concepts are:

Definition 2 Time Interval [36]
A *time interval* is the time between two instants.

Definition 3 Duration [36]
A *duration* is an amount of time with known length, but no specific starting or ending instants.

A time interval as such is bounded by two instants, whereas a duration is not. Also, it should be noted that an instant is in fact a singular case of a time interval.

Definition 4 Temporal Element [36]
A *temporal element* is a finite union of time intervals.

Definition 5 Event [36]
An *event* is an instantaneous fact, i.e. something occurring at an instant.

Definition 6 'Temporal' as Modifier [36]
The modifier *'temporal'* is used to indicate that the modified concept concerns some aspect of time.

Data models used for time modelling might now represent an underlying time axis using chronons:

Definition 7 Chronon [36]
In a data model, a *chronon* is a non-decomposable time interval of some fixed, minimal duration.

A time model contained in a data model may now represent an underlying time axis by a sequence of consecutive chronons. These chronons have identical durations. A data model will typically not specify the exact chronon duration, so it can be fixed later by applications implementing the data model.

The fact that chronons are actually time intervals has a particular effect on the representation of instants and time intervals. In a time model using chronons, an instant is of course represented by a chronon. A time interval may be represented by a set of contiguous chronons, depending on the amount of time the time interval comprises.

Another classification of time models concerns the use of points or intervals to model time. The equivalence between interval-based and point-based time models is demonstrated in [6].

Restrictions on time range may exist, as time may be bounded orthogonally in the past and in the future [50].

2.2 Granularities

An important issue in time modelling concerns the concept of *temporal granularities*. A formal definition for this concept is given in [57]:

Definition 8 Granularity [81]
A *granularity* is an ordered set of non-overlapping and continuous temporal elements called *granules*.

Definition 9 Granule [81]
A *granule* is the basic time unit in a granularity.

A temporal granularity is in fact a partitioning of the time line (time model) used by a system, usually dependent on the application. For example, the age of an adult human being is usually expressed in years: one will use sentences like 'Laura is 21 years old' instead of sentences like 'Laura is 21 years, 3 months and 4 days old'. In this example any duration shorter than a year needs no representation and thus the used granularity allows no specification for durations shorter than a year. The granules are years.

As a granularity G is an ordered set, each granule may be represented by an integer. In this representation, to keep track of the granularity a granule is an element of, the corresponding granularity name is added in subscript:

$$G = \{i_G \mid i \in \mathbb{Z}\} \tag{1}$$

In a system, the granularity with the shortest granules is the *chronons granularity*, which is denoted by '$\perp$'. It is the granularity of which the granules are chronons.

Definition 10   Mapping function [57]

A mapping function f is a function that maps a given granule i_G, $i \in \mathbb{Z}$, in a given granularity G, to a set of corresponding chronons:

$$f : G \to \mathcal{P}(\bot)$$
$$i_G \mapsto \{c_{\bot} \mid (c_{\bot} \text{ is contained in } i_G) \wedge (c_{\bot} \in \bot)\}$$

Note that a mapping function f always maps from a given granularity G to the powerset of the set of chronons $\bot$. Therefore, the output for a mapping function is an element of $\mathcal{P}(\bot)$ and thus a subset of $\bot$.

A mapping function f requires that the following properties hold [57]:

- G *is an ordered set.*
- G *is a set of continuous granules.*
- *The granules in G do not overlap.*

The existence of mapping functions between granularities and the chronons granularity also allows comparing granularities with respect to the length of their granules. In this context, two important concepts can be discerned.

Definition 11   Finner Than [57]

Consider a mapping function f and let i_G and j_H be elements of granularities G and H respectively. Granularity G is now said to be *finner than* granularity H if:
$$|f(i_G)| < |f(j_H)|$$

Definition 12   Coarser Than [57]

Consider a mapping function f and let i_G and j_H be elements of granularities G and H respectively. Granularity G is now said to be *coarser than* granularity H if:

$$|f(i_G)| > |f(j_H)|$$

It is also possible to describe the relation between different granularities. This is called a casting function:

Definition 13 Casting function [57]

Consider two different granularities G and H. A granularity-to-granularity *casting function* *cast* is then a function mapping granules from G to granules from H:

$$cast : G \times \mathbb{G} \times \mathbb{G} \to H$$
$$: (i_G, G, H) \to j_H$$

where $i_G \in G$ and $j_H \in H$ and where $\mathbb{G}$ denotes the set of all granularities.

Thus, the function *cast* associates a granule i_G in G to a corresponding granule j_H in H. Two kinds of granularity-to-granularity mappings can now be discerned: an *upwards mapping* is a mapping from a granularity G to a coarser granularity H, whereas a downwards mapping is a mapping from a granularity K to a finer granularity L. Orthogonal to this classification, mappings between two granularities may be classified as *regular mappings, irregular mappings* or *congruent mappings* [37, 57].

- *Regular mapping.* A regular mapping is a granularity-to-granularity mapping where the mapping function value is calculated by means of multiplications and/or divisions and (maybe) an anchor adjustment. e.g., the mapping value of the mapping between hours and minutes is calculated using a multiplication by 60.
- *Irregular mapping.* An irregular mapping is a granularity-to-granularity mapping where the mapping function value can not be calculated by means of multiplications and/or divisions. e.g., the mapping value of the mapping between months and days is dependent on the exact month or day.
- *Congruent mapping.* A congruent mapping is a granularity-to-granularity mapping where the two granularities involved in the mapping have the same granules but a different anchor. e.g., the mapping between the days (Gregorian calendar days) and the academic days is a congruent mapping.

Different granularity-to-granularity mappings between several granularities can be represented using a granularity graph, which is a directed graph indicating the mapping conversions. The above is illustrated in the following example.

Example 1 Consider a system that models both Gregorian calendar dates as well as academic calendar dates. In this system, the chronons granularity is a set of milliseconds. Figure 1 shows the complete granularity graph corresponding to this example. The transition between the chronons granularity and the seconds granularity is an example of a regular mapping. Regular mappings are represented by thin arrows in the visualisation of the graph. The transition between the days granularity and the months granularity is an example of an irregular mapping. In the graph visualisation, irregular mappings are represented by a bold arrow. Finally, the transition between

Fig. 1 The granularity graph
corresponding to Example 1

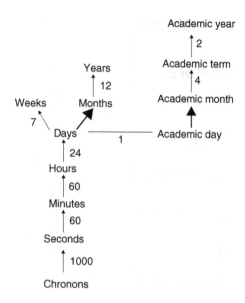

the Gregorian calendar day granularity and the academic day granularity is an example of a congruent mapping. Both concern the same days, but the academic year starts on October 1st, whereas the Gregorian calendar year starts on January 1st. In the graph visualisation, congruent mappings are visualised as straight lines without arrow heads.

2.3 Temporal Relationships

In this section, a brief introduction can be found, concerning *temporal relationships*, sometimes also called 'temporal relations' [5]. Temporal relationships can be seen as relationships between temporal elements belonging to the same time domain. These relationships express how the temporal elements are related to one another, with respect to temporal precedence and overlap.

Several (collections of) operators have been proposed in order to compare temporal elements and model the temporal relationships between them. Allen [1] most notably described such relationships between time intervals and as a special case, between instants. Figure 2 shows the temporal relationships Allen discerned. Some proposals can be applied to both crisp and other time intervals [43, 62, 65, 73].

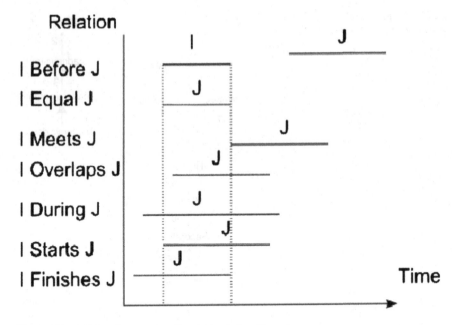

Fig. 2 Allen relations between two time intervals I and J

3 Data Imperfections in Time Modelling

As explained in the introduction, humans handle temporal information using temporal notions like time intervals or time points [36]. While the used temporal notions may contain imperfections [22, 26, 30, 62], humans often gracefully deal with these, as their inherent interpretation capability accounts for a lot of them. This phenomenon has been studied a.o. in the field of artificial intelligence [19, 35] and language understanding [17, 22, 62]. An information system, however, cannot appeal to a similar interpretation functionality. Thus, many proposals have been concerned with the combination of time and imperfections in the context of information systems [62]. In this section, some main concepts and issues concerning this combination are presented.

3.1 Types of Imperfections in Temporal Modelling

Generally, in temporal modelling, a distinction is made between the following types of imperfections [62].

- *Uncertainty*. Temporal information or data may contain uncertainty. This means that the exact temporal value is (partially) unknown, however, generally some knowledge is present anyhow, possibly describing the value [26, 30, 61, 62]. E.g., the temporal notion described in a sentence like 'The Belfry of Bruges was finished on a day somewhere between 01/01/1201 and 31/12/1300 A.D.' contains uncertainty: it is known that the belfry of Bruges was finished on a single day and that this day lays somewhere between 01/01/1201 and 31/12/1300 A.D., but it is not known exactly which day it is.

- *Vagueness.* Temporal information or data might contain inherent vagueness, as a precise instant or time interval may be intended, but the description of it is certainly vague [22, 62, 73]. E.g., the temporal notion described in a sentence like 'It happened during summer.' contains vagueness, as even the boundaries of the mentioned temporal notion are not clearly expressed.
- *Subjectivity or ambiguity.* Temporal notions might be subject to subjectivity or ambiguity. In certain cases, the temporal notion concerns a historical period like 'late romanticism' or 'the early middle ages' and thus contains subjectivity [62]. In other cases, the interpretation of the temporal notion depends on extra factors. E.g., consider a person saying to another person 'Let's meet each other at six.' The person hearing these words doesn't now if 6 a.m. or 6 p.m. is intended, though the person saying the words does.

As to the sources of the imperfections in temporal information, most proposals consider no specific source [30, 35, 61, 62, 65, 73, 80]. Some proposals, however, deal with the imperfections specifically resulting from aspects of language [22] and other proposals consider transitions between different granularities to be the source of imperfection in temporal information [57]. Therefore, some proposals consider granularity as the base of the temporal model [16].

In an information system, temporal information is usually related to facts or events [13]. In light of this, a classification of temporal information can be considered, in which the following types of temporal information may be found:

- *Definite temporal information.* Definite temporal information contains information describing a situation in which all time indications associated with some fact are absolute time indications. The temporal information is precisely known.
- *Indefinite temporal information* [24]. Indefinite temporal information contains information describing a situation in which the time indication associated with some fact has not been fully defined. E.g., consider an event that in fact occurred but it is not known exactly when.
- *Infinite temporal uncertain information* [51]. Infinite temporal uncertain information contains information describing a situation in which an infinite number of time indications are associated with some fact. This is usually found in recurrent events like meetings. E.g., consider meetings that take place every Wednesday at noon. Some systems (usually with different information providers) may dispute the occurrence and/or the duration of a fact.

3.2 Representation of Imperfect Information

As mentioned before, information systems may have to deal with time indications which contain vagueness. Even for some specific events or facts, the temporal indications may become imprecise. Therefore, a time point might be specified by means of a time interval of which the boundaries may not be precisely known. An example.

Example 2 Consider a speaker and a hearer. The speaker wants to make an appointment with the hearer. Now, consider the speaker saying:

'We will meet each other tomorrow around 10'

The hearer will now usually instinctively agree that the appointment will be in e.g., the time interval between 9.55 and 10.05 h.

The study of the semantics of 'around' in temporal [22] indications has shown that the size of the time interval associated with the imprecise specification of a time point depends on the distance with respect to the current time. E.g., consider now that the speaker is talking about something that happened *'during last week'*, then the hearer would consider a time interval of more or less 10 days.

Some proposals [13, 16, 54, 62] conclude that the best representation for incomplete temporal knowledge is therefore based on time intervals, even if they refer to a fact that happen at a time point. This means that, as Allen proposed in [1], the primitive units (the chronons) in a time domain, used in an information system should be intervals.

In order to represent and manage uncertain temporal information properly, several theoretical frameworks have been proposed:

- *Probability theory.* Probability theory [24, 46, 56] is usually employed when uncertainty concerning a time interval allows a probability to be associated to the time interval. The use of probability theory is very usual in logistics information systems. E.g., *'The package will arrive at its destination on Monday morning with a probability of 0.8'.*
- *Possibility theory.* Using possibility theory [28], a possibility degree is associated to the temporal fact or event. Possibility theory is widely used to model uncertainty and vagueness in time [23, 26, 30, 62]. Several works [65, 73] present fuzzy versions of the temporal relations proposed by Allen [1]. The aim of these works is generally to obtain a flexible way to compare uncertain, ill-known temporal intervals by means of temporal relationships. The study of imperfect temporal metadata is done in [8, 9]. In [77] a proposal to use in fuzzy databases temporal fuzzy linguistic terms is studied. Burney [11, 12] has studied recently the combination of fuzzy databases with temporal data.
- *Rough sets.* Rough set theory [66] has been used to represent uncertainty in time intervals. The two dimensional representation of time intervals and the temporal relationships between them has been studied in [70]. In [10] a rough set-based model for temporal databases is presented. The study of temporal relationships between rough time interval is studied in [3].

Fig. 3 Example for the Allen
relationship *'after'*. (1) The
event bounded within time
points A and B. (2) The crisp
version of the *'after'* operator.
(3) A fuzzy version of the after
operator. (4) Another event,
bounded within time points
$[C, D]$

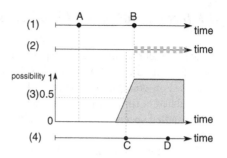

3.3 Imperfections in Temporal Relationships

As the existence of temporal relationships allows to compare temporal notions, many
approaches have been concerned with finding similar temporal relationships, able to
support imperfections in the temporal information which is described by temporal
notions or even by the temporal relationships themselves [26, 62, 65, 73]. These
approaches are often based on Allen's operators [1]. Example 3 presents a short
example concerning one of Allen's relationships.

Example 3 Consider an event which takes place between time points A and B. Thus,
the event comprises time interval $[A, B]$ (this is visualized in part (1) of Fig. 3). The
classical Allen relationship 'after' returns an interval $[B, \infty]$ as shown in part (2)
of Fig. 3. A fuzzified version of Allen's 'after' operator is illustrated in part (3) of
Fig. 3. The comparison between two time intervals results in a possibility degree in
the unit interval. The shape of the possibility distribution is shown in part (3) of
Fig. 3. Note that all the points strictly after the point B results in a possibility degree
of 1 whereas there is an area near the point B in which the possibility degree runs
smoothly between 0 and 1.

Consider now the interval given by $[C, D]$, illustrated in part (4) of Fig. 3. The
user wants to know if $[C, D]$ is after $[A, B]$. The crisp version of the 'after' operator
would return *'no'* as an answer. The fuzzy version for the same operator would return
'yes, with a possibility of 0.5'.

4 Basic Concepts and Issues in Temporal Databases

A temporal database can generally be seen as a database that manages some temporal aspects in its schema [5, 39]. In Sect. 4.1, some main concepts and properties
concerning temporal databases and their definitions are presented and explained. In
Sects. 4.2 and 4.3, some main issues of relational temporal databases are presented
and discussed. Finally, Sect. 4.4 presents an overview of some commercial temporal
database systems.

4.1 Basic Concepts and Properties

A database schema models some part of reality. As mentioned in the introduction, the part of reality a temporal database schema tries to model, contains some temporal aspects. For example, in this part of reality, some concepts or objects could have time-related or time-variant properties. The modelling of these temporal aspects has to be handled specifically in order for the database to maintain a consistent model of reality.

Thus, a temporal database will contain *temporal values*, i.e. values representing (indications of) time. Temporal values in a temporal database can be classified into the following types based on their interpretation and modelling purpose. The definitions and explanations of these types can be found in [36, 63] and more information can be found in [49, 63, 75].

Definition 14 Valid Time [36]
The *valid-time* (VT) of a fact is the time when the fact is true in the modeled reality.

Definition 15 Transaction Time [36]
A database fact is stored in a database at some point in time, and after it is stored, it is current until logically deleted. The *transaction-time* (TT) of a database fact is the time when the fact is current in the database and may be retrieved.

Definition 16 Decision Time [63]
Decision time (DT) denotes the time when an event was decided to happen.

Definition 17 User-defined Time [36]
User-defined time (UDT) is an uninterpreted attribute domain of date and time.

Valid times are usually provided by the user, whereas *transaction-times* are usually system-generated and -supplied [36]. Temporal values of type UDT are not given any extraordinary interpretation and have thus no extraordinary query language support [36].

A *temporal database* can now formally be defined as follows:

Definition 18 Temporal Database [36]
A *temporal database* supports some aspect of time, not counting user-defined time.

In a relational temporal database, temporal values will of course be in the tuples of the extensions of temporal relations:

Definition 19   Valid-time Relation [36]
A *valid-time relation* is a relation with exactly one system supported valid-time.

Definition 20   Transaction-time Relation [36]
A *transaction-time relation* is a relation with exactly one system supported transaction-time.

A *valid-time*, respectively *transaction-time relational database* is now defined as containing one or more valid-time, respectively transaction-time relations [36]. Next to this, *bitemporal* relational databases contain both valid-time and transaction-time [36] and tritemporal databases contain valid-time, transaction-time and decision-time [63].

A very extensive list of the most well-known temporal database models can be found in [82]. As it is of course necessary to define a consistent way to query the temporal data, there are several proposals concerned with query languages and query language adaptations for temporal databases like [64, 76].

In the rest of the chapter, the focus will be on concepts and issues concerning valid-time relations and aspects of valid-time relations. For this reason, the next two sections will present and discuss some main issues concerning temporal databases, specifically applied to or presented in the context of valid-time relations.

4.2   Primary Keys in Valid-Time Relation Design

Generally, when designing a relation based on a relational database model, a subset of the relation's attribute set is usually chosen as primary key. The values of a tuple for these attributes will then uniquely determine this tuple, hence no two distinct tuples may have the exact same values for every attribute in this primary key. Next to attributes unrelated to time, a valid-time relation schema will typically contain one or more attributes which model the valid-time aspects and behavior of the real objects and concepts modelled by the relation schema. In this work, these attributes are called *valid-time attributes*. In valid-time relation extensions, distinct tuples can exist containing the exact same values for every attribute except the valid-time attributes. These distinct tuples represent distinct versions of the same real object or concept, valid during different time periods. To allow the existence of such tuples when designing a valid-time relation using a relational database model, the most common solution is to include the valid-time attributes in the primary key.

Table 1 Example relation modelling the employees of a company

ID	Name	Birthday	Supervisor	Start	End
1	Peter	24/10/1985	3	2010	–
2	Maria	03/04/1984	3	2001	–
3	John	21/02/1964	–	1999	–
4	Sarah	29/11/1985	2	2005	2009

Values for the 'Birthday' attribute are visualized here in 'dd/mm/yyyy' format

The following example illustrates this primary key issue.

Example 4 Consider the example valid-time relation visualized in Table 1, which models when certain people worked as employees in a certain company and under whose supervision they worked during that time. The valid-time attributes 'Start' and 'End' describe the year when an employee started, respectively finished working for the company. For example, the last tuple visualized in Table 1 represents that the employee represented by this tuple started working for the company in 2005 and finished in 2009. The attributes 'Name', 'Birthday' and 'Supervisor' describe respectively the name, birthday date and unique identifier of the supervisor of an employee in the time during which he or she worked for the company. When correct, the date of an employee's birthday never changes and as such, the modelling of birthday dates has no effect on the database consistency. The 'Birthday' attribute thus describes UDT values. The attribute 'ID' describes employee identifiers. For each tuple, this identifier (a number) uniquely identifies the employee represented by the tuple.

Now consider {ID} being the primary key and consider the company wanting to hire Sarah again in 2010. This would be represented by another tuple in the relation, containing value 4 for attribute 'ID'. The existence of such a tuple is of course not allowed by the primary key, because it would mean the existence of two distinct tuples containing value 4 for attribute 'ID'. This problem can now be solved by defining a new primary key: {ID, Start, End}, which allows for the existence of distinct tuples with value 4 for attribute 'ID', as long as they have different values for attributes 'Start' or 'End'. The resulting relation is shown in Table 2.

Table 2 Example relation after including the valid-time attributes in the primary key and adding a tuple

ID	Name	Birthday	Supervisor	Start	End
1	Peter	24/10/1985	3	2010	–
2	Maria	03/04/1984	3	2001	–
3	John	21/02/1964	–	1999	–
4	Sarah	29/11/1985	2	2005	2009
4	Sarah	29/11/1985	2	2010	–

4.3 Consistency in Valid-Time Relation Content Modification

The solution presented in Sect. 4.2 concerns relation design and consists of including the valid-time attributes in the primary key. Unfortunately, implementing this solution as such allows for the existence of records whose values imply inconsistencies with respect to the modelling of reality.

Consider a valid-time relation of which the primary key can be partitioned into two sets of attributes. One set contains attributes totally unrelated to time, for which the values of a record allow to uniquely identify the object or concept represented by the record. The other set contains the valid-time attribute(s). Because the valid-time attribute(s) is(are) included in the primary key, the existence of distinct records with exactly the same values for all time-unrelated attributes and distinct values for at least one valid-time attribute is not prohibited. Thus, inserting such records into the relation is not prohibited either, even if the information represented by the values for the valid-time attributes shows clear inconsistencies. An example.

Example 5 Consider the example valid-time relation visualized in Table 3, which is based on the relation visualized in Table 1. The primary key is again {ID, Start, End}. The last record in the relation represents a person named 'Sarah' started working for the company in 2007 and finished in 2008, with supervisor 'John'. However, the fourth record represents the same person (the value for attribute 'ID' is the same) started working for the company in 2005 and finished in 2009, with supervisor 'Maria'. The intention is clear: Sarah worked in the company from 2005 to 2009, first for Maria, then for John, then again for Maria. It is of course possible for an employee to change supervisors, but it is of course impossible for a person to start working in the same company twice at different times, for different supervisors, without stopping to work for one in between, as it is impossible to stop working for a supervisor twice at different times, without working for another one in between. The valid-time information represented by the last record is clearly not consistent with the valid-time information represented by the fourth record, or vice versa.

The most usual approach to deal with this inconsistency problem is to adapt the DML used by the DBMS, as to enforce consistency towards time with respect to the modelled reality.

Example 6 Consider the problem presented in Example 5. The inconsistency arises when the last record in Table 3 is inserted. Because the record's values for the valid-

	ID	Name	Birthday	Supervisor	Start	End
Table 3 Example relation with records whose values for the valid-time attributes violate consistency	1	Peter	24/10/1985	3	2010	–
	2	Maria	03/04/1984	3	2001	–
	3	John	21/02/1964	–	1999	–
	4	Sarah	29/11/1985	2	2005	2009
	4	Sarah	29/11/1985	3	2007	2008

Table 4 Example relation
updated maintaining
consistency

ID	Name	Birthday	Supervisor	Start	End
1	Peter	24/10/1985	3	2010	–
2	Maria	03/04/1984	–	2001	–
3	John	21/02/1964	–	1999	2010
3	John	21/02/1964	–	2010	–
4	Sarah	29/11/1985	2	2005	2007
4	Sarah	29/11/1985	3	2007	2008
4	Sarah	29/11/1985	2	2008	2009

time attributes differ from those of the fourth record, the last record is accepted. The DML statement used was (the table is called 'Employees'):

```
INSERT INTO Employees VALUES
   (4, 'Sarah', '29/11/1985', 3, 2007, 2008);
```

The inconsistency problem can now be solved by replacing this statement with:

```
UPDATE Employees SET 'End' = '2007' WHERE
   (ID = 4) AND (Start = 2005) AND (End = 2009);
INSERT INTO Employees VALUES
   (4, 'Sarah', '29/11/1985', 3, 2007, 2008);
INSERT INTO Employees VALUES
   (4, 'Sarah', '29/11/1985', 2, 2008, 2009);
```

The resulting relation is visualized in Table 4.

4.4 Commercial Temporal Database Systems

Several commercial temporal DBMS exist. Table 5 gives an overview of some of the more well-known temporal DBMS and provides references for more information.

Oracle workspace manager [15] and TimeDB [79] are libraries for dealing with time in OracleDB. On another note, TimeDB and Postgree Temporal [69] are similar: both are simple implementations that implement a subset of the Allen operators and some operations for the creation and manipulation of temporal attributes (valid-time, transaction-time or both times are supported). Teradata [78] is mainly a business intelligence system designed for data mining. Secondo [25] is an extensible database system in which the core of the database may be replaced by a customized algebra. It is designed for non-standard applications and it supports both valid and transaction-times.

The most complete implementation is Workspace Manager.

Unfortunately, none of these systems take data imperfections into account, neither in data storage nor in querying.

Table 5 Commercial temporal database systems

Name	Time supported	Comments	Reference
Oracle Workspace Manager	VT and TT	Package for Oracle DB	[15]
TimeDB	VT and TT	Interface for Oracle DB	[79]
Postgree Temporal	VT	Package for Postgree SQL	[69]
Teradata	VT and TT	Used for data-mining	[78]
Secondo	VT and TT	Spatio-temporal database	[45]

5 Data Imperfections in Temporal Databases

Consider a logistics company which transports packages. At the moment a package leaves, the time when it will arrive at its destination may be estimated, but will typically not be known precisely. For such companies and in many other situations, information systems able to handle imperfection with respect to certain temporal aspects of the objects modelled by the system are necessary.

5.1 Data Imperfections in Temporal Databases

Data and information imperfections and techniques to represent them correctly in databases and queries are usually the focus of research in fuzzy databases. Proposals from this field may present an approach based somehow on fuzzy set theory [59] or possibility theory [29], although other theories support information imperfections too. Comparably, many proposals concerning the introduction of data imperfections or information imperfections in temporal databases present approaches based somehow on fuzzy set theory [5, 42, 43, 62] or possibility theory [18, 30, 68], although proposals based on other theories exist [21, 38, 70]. As possibility theory is usually seen as a theory of confidence, aimed at dealing with uncertainty, in some proposals, possibility theory is used specifically to handle uncertainty in temporal information. In fuzzy databases [41], uncertainty is usually expected to appear in the database content, whereas other types of imperfection, notably imprecision, are usually expected to appear in querying.

Concerning temporal databases, there are several approaches to handle uncertainty in temporal data stored in a database. Many of these approaches concern several different types of time notions (VT, TT or DT), but most of these approaches focus somehow on valid-time [42, 43].

In the following subsection, a novel approach to representing uncertainty concerning valid-time notions and a corresponding technique to query similar valid-time indications in a valid-time relation are proposed. The presented proposal is based on concepts introduced in [27] and on the framework proposed in [68].

5.2 Handling Uncertainty in a Valid-Time Relation

A valid-time indication usually takes the form of a time interval. Such a valid-time interval can be described (and stored in a valid-time relation record) using its boundaries (endpoints) or using one endpoint and the interval length. Usually, a valid-time interval is represented using its endpoints, which is also the approach adopted by the presented proposal.

Generally, the uncertainty concerning a set of values might be described by a possibility distribution on the powerset of which one of the elements can be the intended set [27]. This representation, however, introduces some issues in practice or in practical applications. Therefore, in the presented proposal, possibility theory is used to model uncertainty, but only uncertainty concerning the exact values of the start and end point of a valid-time interval is considered and the uncertainty in both the start point and the end point are modelled using possibility distributions.

In fact, to model the uncertainty related to a valid-time interval using possibility theory, the presented proposal introduces so-called ill-known time intervals, relying on the concept of ill-known sets [27].

5.2.1 Ill-known Time Intervals

To represent valid-time indications which might contain uncertainty, the presented proposal introduces the concept of ill-known valid-time intervals, which relies on the concept of ill-known sets [27]. To correctly explain the concept of ill-known sets, the concept of possibilistic variables should be introduced first. In the presented proposal, the definition of possibilistic variables of [68] is followed. In [68], a *possibilistic variable* is defined as follows:

> **Definition 21 Possibilistic variable** [68]
> A possibilistic variable X over a universe U is defined as a variable taking exactly one value in U, but for which this value is (partially) unknown. Its possibility distribution π_X on U models the available knowledge about the value that X takes: for each $u \in U$, $\pi_X(u)$ represents the possibility that X takes the value u. In the presented work, this possibility is interpreted as a measure of how plausible it is that X takes the value u, given (partial) knowledge about the value X takes.

Now, consider a set R, which contains single values (and not collections of values). When a possibilistic variable X_v is defined on such a set R, the unique value X_v takes, which is (partially) unknown, is called an *ill-known value* in this work [27].

When a possibilistic variable is defined on the powerset $\mathscr{P}(R)$ of some universe R, the unique value the variable takes will be a crisp set and its possibility distribution

Fig. 4 A closed ill-known
time interval $[X, Y]$, where
triangular possibility distributions describe the ill-known
values defining the start and
end points

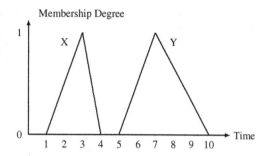

on the powerset $\mathscr{P}(R)$ will describe the possibility of each crisp subset of R to be the
value the variable takes. This value (a crisp set) the variable takes, which is (partially)
unknown, is now called an *ill-known set* [27].

Finally, consider a set R, which contains single values (and not collections of
values) and its powerset $\mathscr{P}(R)$. Now consider a subset $\mathscr{P}_I(R)$ of $\mathscr{P}(R)$ and let this
subset contain every element of $\mathscr{P}(R)$ that is an interval, but no other elements. When
a possibilistic variable X_i is defined on the subset $\mathscr{P}_I(R)$ of the powerset $\mathscr{P}(R)$
of some set R, the unique value X_i takes will be a crisp interval and the possibility
distribution π_{X_i} of X_i will be a possibility distribution on $\mathscr{P}_I(R)$. This π_{X_i} will
define the possibility of each value of $\mathscr{P}_I(R)$ (a value of $\mathscr{P}_I(R)$ is a crisp interval
which is a subset of R) being the value X_i takes. This exact value (a crisp interval)
the variable takes, which is (partially) unknown, is called an *ill-known interval* here.

The presented proposal will deal with ill-known time intervals. Ill-known time
intervals are ill-known intervals in a time domain. In the presented proposal, an ill-
known time interval will be defined and represented via its start and end point, which
will be ill-known values. The elements of the ill-known time interval are the values
between its start and end point, including the start and end points themselves.[1] It
should be clear that this approach to representing ill-known time intervals differs
from the approach based on a single possibility distribution on a set $\mathscr{P}_I(R)$ of a set
R. These approaches have a different behavior and can be used to describe different
ill-known time intervals. The correspondences, interactions and transitions between
these different representations of ill-known intervals and their interpretations are part
of the authors current research.

In the presented proposal, a closed ill-known time interval with start point defined
by possibilistic variable X and end point by possibilistic variable Y is noted $[X, Y]$.
Figure 4 shows a closed ill-known time interval.

Several authors work with concepts very similar to these ill-known time intervals
and some of them [43] propose transformations of the describing functions in order
to optimize the storage of such ill-known valid-time intervals, though recent research
might seem to indicate some minor issues with respect to some of these transformations [68]. A comparison between the transformations in [43] and the framework in

---

[1] The presented proposal only deals with closed ill-known time intervals. Dealing with halfopen or
open ill-known intervals is part of the current research of the authors.

Fig. 5 Transformation based
in the convex hull from the
two ill-known points X and Y

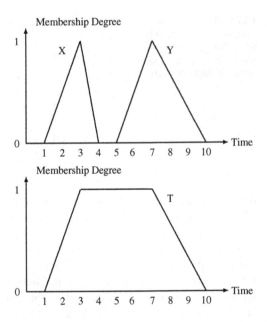

[68] is presented in [67]. Figure 5 illustrates a transformation based on the 'convex hull' approach [43].

In the presented proposal, ill-known valid-time intervals will be used to represent valid-time indications in a valid-time relation and to model the uncertainty these may contain.

To evaluate the temporal demands in queries issued by users to query a valid-time relation containing ill-known valid-time intervals, the presented proposal introduces a technique based on the concept of ill-known time constraints, which is based on the concept of ill-known constraints as presented in [68]. Both concepts are treated in Sect. 5.2.2.

Before ill-known time constraints can be introduced, another notion related to possibilistic variables shoud be paid attention to. In fact, a specific application of possibilistic variables is obtained when the set under consideration is the set of boolean values, denoted $\mathbb{B} = \{T, F\}$, where T denotes 'true' and F denotes 'false' [68]. Indeed, any boolean proposition p takes exactly one value in $\mathbb{B}$. If the knowledge about which value this proposition p takes, is given by a possibility distribution π_p, then proposition p can be seen as a possibilistic variable. In the presented proposal, the interest lies with the case where the proposition holds (denoted $p = T$) and the possibility and necessity that $p = T$ demand most attention. In the following sections, the following notations are used, based on previous notations:

$$\text{Possibility that } p = T: \quad Pos(p) = \pi_p(T) \tag{2}$$
$$\text{Necessity that } p = T: \quad Nec(p) = 1 - \pi_p(F) \tag{3}$$

5.2.2 Ill-Known Time Constraints

The presented proposal contains a technique for evaluating user queries used to query a valid-time relation in which the valid-time indications are represented by ill-known time intervals. Part of this query evaluation technique relies on the concept of ill-known time constraints, which is based on the concept of ill-known constraints as presented in [68]. These concepts are presented below. Following [68], an *ill-known constraint* is defined as follows.

Definition 22 Given a universe U, an *ill-known constraint* C is specified by means of a binary relation $R \subseteq U^2$ and a fixed, ill-known value defined by its possibilistic variable V on U, i.e.:

$$C \triangleq (V, R)$$

A set $A \subseteq U$ now satisfies this constraint C if and only if:

$$\forall a \in A : (V, a) \in R$$

An example of an ill-known constraint is given by:

$$C_< \triangleq (X, <)$$

Some set A then satisfies $C_<$ if

$$\forall a \in A : X < a$$

The satisfaction of a constraint $C \triangleq (V, R)$ by a set A is basically a Boolean matter (either the set satisfies the constraint or not) and can thus be seen as a boolean proposition, but due to the uncertainty inherent to the ill-known value V, it can be uncertain whether C is satisfied by A or not [68]. Based on the possibility distribution π_V of V, the possibility and necessity that A satisfies C can be found. This proposition can thus be seen as a possibilistic variable on $\mathbb{B}$. The required possibility and necessity are calculated using the following formulas [68].

$$Pos(A \text{ satisfies } C) = \min_{a \in A} \left(\sup_{(w,a) \in R} \pi_V(w)) \right) \tag{4}$$

$$Nec(A \text{ satisfies } C) = \min_{a \in A} \left(\inf_{(w,a) \notin R} 1 - \pi_V(w) \right) \tag{5}$$

Now, to calculate the possibility or necessity of a set A satisfying multiple constraints, the min t-norm operator is used to express a conjunction of constraints. For example:

$Pos((A \text{ satisfies } C_1) \text{ and } (A \text{ satisfies } C_2)) = \min_{a \in A}(Pos(A \text{ satisfies } C_1), Pos(A \text{ satisfies } C_2))$

$Nec((A \text{ satisfies } C_1) \text{ and } (A \text{ satisfies } C_2)) = \min_{a \in A}(Nec(A \text{ satisfies } C_1), Nec(A \text{ satisfies } C_2))$

Accordingly, the max t-conorm operator is used to express a disjunction of constraints. For example:

$Pos((A \text{ satisfies } C_1) \text{ or } (A \text{ satisfies } C_2)) = \max_{a \in A}(Pos(A \text{ satisfies } C_1), Pos(A \text{ satisfies } C_2))$

$Nec((A \text{ satisfies } C_1) \text{ or } (A \text{ satisfies } C_2)) = \max_{a \in A}(Nec(A \text{ satisfies } C_1), Nec(A \text{ satisfies } C_2))$

In the presented proposal, *ill-known time constraints* are considered, which are ill-known constraints of which the considered universe is a time domain.

In the next subsection, the core of the presented proposal is described.

5.2.3 Querying Valid-Time Relations containing Ill-known Valid-Time Intervals

One of the main purposes of the existence of (relational) databases is to allow information retrieval. The standard query language for (relational) databases is SQL [60], but several proposals to extend the SQL language for transaction-time databases [72], valid-time databases [40] and bitemporal databases [64] exist and some authors have studied how to support temporal querying in standard SQL [76].

As mentioned before, the presented proposal deals with querying a valid-time relation. In this subsection, the core of the presented proposal is described. First the particular structure of the relation is described, along with the nature and structure of its supposed contents. Next, the particular query structure is presented. Finally, the particular method for evaluating queries and for ranking the result records are presented.

Relation Structure

In the presented proposal, a valid-time relation is considered, in which every record contains just one valid-time indication. This valid-time indication is represented by a closed ill-known time interval, to allow uncertainty in the valid-time information. As explained above, the ill-known time intervals used here will be defined and represented via their start and end points, which are ill-known values in the valid-time domain.

Query Structure

In the presented proposal, a query consists of two separate constructs of user demands.

Definition 23 Query A query $\tilde{Q}$ is given by:

$$\tilde{Q} = \left(Q^{time}, Q \right) \tag{6}$$

Here, Q denotes the construct of all (possibly fuzzy) non-temporal user query demands. These comprise all constraints and demands unrelated to valid-time and thus unrelated to the valid-time indications in the queried relation. Q^{time} denotes the temporal demand specified by the user.

The presented query structure allows the user to specify a single temporal demand, denoted by Q^{time}.

Definition 24 Temporal demand A temporal demand Q^{time} is defined by:

$$Q^{time} = (I, AR) \tag{7}$$

Here, I denotes a crisp time interval (which can be specified in any way required) and AR denotes one of the Allen relations (cf. Sect. 2.3).

The interpretation of such a temporal demand $Q^{time} = (I, AR)$ is that, for a record with an ill-known valid-time interval J, the user demands that I AR J holds.

Table 6 Constructs of constraints related to their respective Allen relations, as used in the presented work

Allen relation	Construct of constraints
I before J	$C_1 \triangleq (<, X)$
I equal J	$(C_1 \triangleq (\geq, X)) \wedge \neg (C_2 \triangleq (\neq, X)) \wedge (C_3 \triangleq (\leq, Y)) \wedge \neg (C_4 \triangleq (\neq, Y))$
I meets J	$(C_1 \triangleq (\leq, X)) \wedge \neg (C_2 \triangleq (\neq, X))$
I overlaps J	$(C_1 \triangleq (<, Y)) \wedge \neg (C_2 \triangleq (\leq, X)) \wedge \neg (C_3 \triangleq (\geq, X))$
I during J	$((C_1 \triangleq (>, X)) \wedge (C_2 \triangleq (\leq, Y))) \vee ((C_3 \triangleq (\geq, X)) \wedge (C_4 \triangleq (<, Y)))$
I starts J	$(C_1 \triangleq (\geq, X)) \wedge \neg (C_2 \triangleq (\neq, X))$
I finishes J	$(C_1 \triangleq (\leq, Y)) \wedge \neg (C_2 \triangleq (\neq, Y))$

In this table, the ill-known time interval $J = [X, Y]$ in a record r has a start point described by possibilistic variable X and an end point described by possibilistic variable Y. The crisp time interval in the user's temporal demand is denoted I

Query Evaluation

Query satisfaction in a fuzzy relational database is usually a matter of degree. Typically, the evaluation of the query demands for a record results in a *satisfaction degree* s, which is typically in the unit interval, i.e. $s \in [0, 1]$. This satisfaction degree then models the extent to which the record satisfies the query demands. As such, a satisfaction degree of 0 indicates total dissatisfaction while a degree of 1 indicates total satisfaction and every level of satisfaction between total satisfaction and total dissatisfaction is indicated by a satisfaction degre between 0 and 1.

In the presented approach, for every record r, each part of a query $\tilde{Q} = (Q^{time}, Q)$ is evaluated independently:

- The user preferences expressed in the non-temporal part Q, are evaluated, resulting in a satisfaction degree denoted $e_Q(r)$. The presented approach accepts any valid, sound method of calculating this evaluation, as long as the method is well-founded and $e_Q(r) \in [0, 1]$.
- The evaluation of the temporal demand expressed in the temporal part, $Q^{time} = (I, AR)$, depends on AR. A specific construct of ill-known constraints (cf. Sect. 5.2.2) is considered, depending on the Allen relation denoted by AR. The exact construct of constraints is an instantiation based on the formulas which can be found in Table 6, for every possible value of AR. The form and capacity of these constraints are based on [68]. Then, using Eqs. (4) and (5), the exact formulas to calculate the possibility $Pos_{Q^{time}}(r)$ and the necessity $Nec_{Q^{time}}(r)$ that record r satisfies this construct of ill-known time constraints are derived from this construct of constraints. As mentioned, $Pos_{Q^{time}}(r)$ and $Nec_{Q^{time}}(r)$ denote the possibility, respectively the necessity that the considered record r satisfies the construct of constraints corresponding to the temporal demand Q^{time} and thus the possibility, respectively the necessity that r satisfies Q^{time}.

Aggregation and Ranking

In this subsection, the notations used in the previous subsection are followed. To be able to present the most appropriate results to the user most prominently, for every record r, an aggregation method is used to aggregate $\text{Pos}_{Qtime}(r)$ and $\text{Nec}_{Qtime}(r)$ into a temporal record rank $e_{Qtime}(r)$ and after this, a convex combination combining $e_{Qtime}(r)$ and $e_Q(r)$ will provide the final record rank $e_{final}(r)$.

To calculate $e_{Qtime}(r)$, an a simple and crude method is used:

$$e_{Qtime}(r) = \left(\frac{\text{Pos}_{Qtime}(r) + \text{Nec}_{Qtime}(r)}{2}\right) \tag{8}$$

This method aims to provide the result records with a natural ranking based on the users temporal constraint. $e_{Qtime}(r)$ will of course be a value in $[0, 1]$, as both $\text{Pos}_{Qtime}(r) \in [0, 1]$ and $\text{Nec}_{Qtime}(r) \in [0, 1]$. The purpose is that records which fit the users temporal demand better get a higher score than records fitting the temporal demand worse. Here, this aim is reached because the necessity degree $\text{Nec}_{Qtime}(r)$ cannot exceed 0 unless the possibility degree $\text{Pos}_{Qtime}(r)$ equals 1.

The final ranking $e_{final}(r)$ for a record r is now given by a convex combination of both temporal and non-temporal evaluation scores.

$$e_{final}(r) = \omega * e_Q(r) + (1 - \omega) * e_{Qtime} \tag{9}$$

A convex combination is used mainly for 2 reasons:

- The use of this convex combination allows a record to make up for a low temporal evaluation score with a high non-temporal evaluation score and vice versa.
- The exact value of ω can now be modified to ascribe more value to either the fulfillment of the user's temporal demands or the fulfillment of the user's non-temporal constraints.

In the next subsection, some main concepts and issues concerning bipolarity in the context of temporal databases are presented and discussed.

5.3 Bipolarity in Temporal Databases

Humans express their preferences using both positive and negative statements, where positive statements express what is desired or acceptable and negative statements express what is undesired or unacceptable [5]. This realization is interesting with regard to database querying, because sometimes a user does not exactly know his or her preferences or can't express them in only positive statements, but prefers to use negative statements to express what he or she dislikes or doesn't need. This introduces the need for *bipolar querying*, a technique to model both positive and negative user preferences in a database query. Sometimes positive and negative preferences are clearly symmetric, making it possible to derive one from the other. For example, a person may define the concept of 'tall' as '1.80 m or higher'. The negative would then be the opposite: not tall would be 'anything less than 1.80 m'. However, in some cases, positive preferences can not be directly obtained from negative preferences or vice versa. E.g., when a person prefers to buy a black motorbike, this does not necessarily mean the person would totally reject a very dark blue motorbike. This phenomenon is called *heterogeneous bipolarity* [33, 34].

The use of imprecise query preference formulation in bipolar querying is well discussed in existing literature [20, 34, 55]. In [55], desired and mandatory query conditions are used, instead of positive and negative preferences. However, the inverse of a mandatory preference expresses what should be rejected and this could be seen as negative information, whereas desired query conditions can be seen as positive preferences. However, the combination of bipolar querying and the use of imprecise query preferences in the context of temporal databases is not so well discussed in existing literature. A proposal for the bipolar querying of valid-time databases has been made by Billiet et al. [5]. The model presented there deals with a fuzzy valid-time specification based on [43].

Bipolarity can be handled using different concepts, such as intuitionistic fuzzy sets [2], interval valued fuzzy sets [83] Grattan-Guiness [44], Janh [47], Sambuc [71], [32] or twofold fuzzy sets [31].

From a theoretical point of view, bipolarity might be found either in the queries presented to a database system or in a database managed by a database system.

When bipolarity is found in queries, it is possible to distinguish between:

- Bipolarity inside query criteria: each individual query criterion may be specified using both positive and negative preferences. For example when querying a car database, the user can express that he or she wants a black car, but definitely not a red neither a blue one. Bipolarity resides here within the car color criterion.
- Bipolarity outside query criteria: the query is specified using a global positive and a global negative preference part. For example when querying a car database, the user can express that he or she wants a black car, but definitely not a car with a fuel consumption of 6 l or more.

Concerning bipolarity inside a database, it should be possible to specify both positive and negative real world object or concept aspects, even at record level. Nevertheless, not so much research exists concerning bipolarity in databases.

6 Conclusions and Further Research

In this chapter, some of the main concepts concerning information imperfections in temporal modelling and information imperfections in temporal modelling in information systems and the terminology corresponding with these concepts are introduced and explained and some of the main properties of and issues with these concepts are presented and discussed. An overview of some commercial temporal DBMS is briefly introduced. Finally, a novel technique for querying valid-time relations using imperfect query specifications is presented.

Further research work could follow several general directions. First of all, a theoretical model for dealing with uncertainty in both the database and the query at the same time could be researched and defined. Next, implementations including both DDL and DML could be proposed and constructed.

Acknowledgments Part of this research is supported by the grant BES-2009-013805 within the research project TIN2008-02066: *Fuzzy Temporal Information treatment in relational DBMS*.

References

1. Allen, J.F.: Maintaining knowledge about temporal intervals. Commun. ACM **26**, 832–843 (1983)
2. Atanassov, K.T.: Intuitionistic fuzzy sets. Fuzzy Sets Syst. **20**, 87–96 (1986)
3. Bassiri, A., Malek, M., Alesheikh, A., Amirian, P.: Temporal relationships between rough time intervals. In: Gervasi, O., Taniar, D., Murgante, B., Lagan, A., Mun, Y., Gavrilova, M. (eds.) Computational Science and Its Applications ICCSA 2009. Lecture Notes in Computer Science, vol. 5592, pp. 543–552. Springer, Berlin (2009). http://dx.doi.org.10.1007/978-3-642-02454-2_39
4. Benthem, J.F.K.A.V.: The logic of time: a model-theoretic investigation into the varieties of temporal ontology and temporal discourse. Reidel, Hingham (1982)
5. Billiet, C., Pons, J.E., Matthé, T., De Tré, G., Pons Capote, O.: Bipolar fuzzy querying of temporal databases. In: Lecture Notes in Artificial Intelligence, vol. 7022, pp. 60–71. Springer, Ghent (2011)
6. Böhlen, M., Busatto, R., Jensen, C.: Point-versus interval-based temporal data models. In: Proceedings of the 14th International Conference on Data Engineering, pp. 192–200 (1998)
7. Bolour, A., Anderson, T.L., Dekeyser, L.J., Wong, H.K.T.: The role of time in information processing: a survey. ACM SIGMOD Record **12**, 27–50 (1982)
8. Bordogna, G., et al.: Advanced database query systems. In: Flexible Querying of Imperfect Temporal Metadata in Spatial Data Infrastructures: Techniques, Applications and Technologies, p. 140. IGI Global (2011).10.4018/978-1-60960-475-2.ch006
9. Bordogna, G., Carrara, P., Pagani, M., Pepe, M., Rampini, A.: Managing imperfect temporal metadata in the catalog services of spatial data infrastructures compliant with inspire. In: Carvalho, J.P., Dubois, D., Kaymak, U., da Costa Sousa, J.M. (eds.) IFSA/EUSFLAT conference, pp. 915–920 (2009). http://dblp.uni-trier.de/db/conf/eusflat/eusflat2009.html#BordognaCPPR09
10. Burney, A., Mahmood, N., Abbas, Z.: Advances in fuzzy rough set theory for temporal databases. In: Proceedings of the 11th WSEAS International Conference on Artificial Intelligence, Knowledge Engineering and Data Bases, AIKED'12, pp. 237–242. World Scientific and Engineering Academy and Society (WSEAS), Stevens Point, Wisconsin, USA (2009). http://dl.acm.org/citation.cfm?id=2183067.2183107
11. Burney, A., Mahmood, N., Ahsan, K.: Tempr-pdm: a conceptual temporal relational model for managing patient data. In: Proceedings of the 9th WSEAS International Conference on Artificial Intelligence, Knowledge Engineering and Data Bases, AIKED'10, pp. 237–243. World Scientific and Engineering Academy and Society (WSEAS), Stevens Point, Wisconsin, USA (2010). http://dl.acm.org/citation.cfm?id=1808036.1808078
12. Burney, A., Mahmood, N., Jilani, T., Saleem, H.: Conceptual fuzzy temporal relational model (ftrm) for patient data. WSEAS Trans. Info. Sci. Appl. 7(5), 725–734 (2010). http://dl.acm.org/citation.cfm?id=1852534.1852546

13. Chountas, P., Petrounias, I.: Modelling and representation of uncertain temporal information. Requir. Eng. **5**, 144–156 (2000)
14. Clifford, J., Tansel, A.U.: On an algebra for historical relational databases: two views. SIGMOD Rec. **14**, 247–265 (1985)
15. Corp., O.: Oracle database 11g., Workspace manager overview
16. Van der Cruyssen, B., De Caluwe R. and De Tré, G.: A theoretical fuzzy time model based on granularities. EUFIT'97, pp. 1127–1131 (1997)
17. De Caluwe, R., Van der Cruyssen, B., De Tré, G., Devos, F., Maesfranckx, P.: Fuzzy Time Indications in Natural Languages Interfaces, pp. 163–185. Kluwer Academic Publishers, Norwell (1997)
18. De Caluwe, R., De Tré, G., Van Der Cruyssen, B., Devos, F., Maesfranckx, P.: Time management in fuzzy and uncertain object-oriented databases. In: Knowledge Management in Fuzzy Databases, vol. 39, pp. 67–88. Physica-Verlag, Heidelberg (2000)
19. De Tré, G., De Caluwe, R., Van der Cruyssen, B.: Dealing with time in fuzzy and uncertain object-oriented database models. EUFIT'97, pp. 1157–1161 (1997)
20. De Tré, G., Zadrozny, E.A.: Dealing with positive and negative query criteria in fuzzy database querying bipolar satisfaction degrees. In: Proceedings of 8th International Conference on FQAS, pp. 593–604. Springer, Berlin (2009)
21. Dekhtyar, A., Ross, R., Subrahmanian, V.S.: Probabilistic temporal databases, I: algebra. ACM Trans. Database Syst. **26**, 41–95 (2001)
22. Devos, F., Maesfranckx, P., De Tré, G.: Granularity in the interpretation of around in approximative lexical time indications. J. Quant. Linguist. **5**, 167–173 (1998)
23. Devos, F., Van Gyseghem, N., Vandenberghe, R., De Caluwe, R.: Modelling vague lexical time expressions by means of fuzzy set theory. J. Quant. Linguist. **1**(3), 189–194 (1994)
24. Dey, D., Sarkar, S.: A probabilistic relational model and algebra. ACM Trans. Database Syst. **21**(3), 339–369 (1996)
25. Dieker, S., Güting, R.H.: Plug and play with query algebras: Secondo-a generic dbms development environment. In: Proceedings of the 2000 International Symposium on Database Engineering & Applications, IDEAS '00, pp. 380–392. IEEE Computer Society, Washington, DC, USA (2000)
26. Dubois, D., HadjAli, A., Prade, H.: Fuzziness and uncertainty in temporal reasoning. J. Univ. Comput. Sci. **9**(9), 1168–1194 (2003)
27. Dubois, D., Prade, H.: Incomplete conjunctive information. Comput. Math. Appl. **15**, 797–810 (1988)
28. Dubois, D., Prade H.: Possibility Theory. Plenum Press, New York (1988)
29. Dubois, D., Prade, H.: Possibility Theory: An Approach to Computerized Processing of Uncertainty. Plenum Press, New York (1988)
30. Dubois, D., Prade, H.: Processing fuzzy temporal knowledge. IEEE Trans. Syst. Man Cybern. B Cybern. **19**, 729–744 (1989)
31. Dubois, D., Prade, H.: Bipolarity in flexible querying. In: Proceedings of the 5th International Conference on Flexible Query Answering Systems, FQAS '02, pp. 174–182. Springer, London (2002)
32. Dubois, D., Prade, H.: Interval-valued fuzzy sets, possibility theory and imprecise probability. In: Proceedings of International Conference in Fuzzy Logic and Technology, pp. 314–319 (2005)
33. Dubois, D., Prade, H.: Rough sets and current trends in computing. In: Lecture Notes in Computer Science. Bipolar Representations in Reasoning, Knowledge Extraction and Decision Processes, vol. 4259, pp. 15–26. Springer, Heidelberg (2006)
34. Dubois, D., Prade, H.: Handling bipolar queries in Fuzzy Information Processing. In: Handbook of Research on Fuzzy Information Processing in Databases, pp. 97–114. Information Science Reference, New York (2008)
35. Dutta, S.: An event based fuzzy temporal logic. In: Proceedings of the Eighteenth International Symposium on Multiple-Valued Logic, pp. 64–71 (1988)

36. Dyreson, C.: Grandi, F.e.a.: A consensus glossary of temporal database concepts. SIGMOD Rec. **23**, 52–64 (1994)
37. Dyreson, C., Snodgrass, R.: Temporal granularity and indeterminacy: two sides of the same coin. Technical report tr 94–06, Computer Science Department, University of Arizona, Tucson, USA (1994).
38. Dyreson, C.E., Snodgrass, R.T.: Supporting valid-time indeterminacy. ACM Trans. Database Syst. **23**, 1–57 (1998)
39. Etzion, O., Jajodia, S., Sripada, S.: Temporal Databases: Research and Practice. Lecture Notes in Computer Science. Springer, Berlin (1998)
40. Gadia., S.K.: A seamless generic extension of sql for querying temporal data. TR 92-02, Iowa State Univerity, Department of Computer Science (1992)
41. Galindo, J.: Fuzzy Databases: Modeling, Design, and Implementation. IGI Publishing, Hershey (2006)
42. Galindo, J., Medina, J.M.: Ftsql2: Fuzzy time in relational databases. In: EUSFLAT Conference '01, pp. 47–50 (2001)
43. Garrido, C., Marin, N., Pons, O.: Fuzzy intervals to represent fuzzy valid time in a temporal relational database. Int. J. Uncertainty Fuzziness Knowl. Based Syst. **17**, 173–192 (2009)
44. Grattan-Guinness, I.: Fuzzy membership mapped onto intervals and many-valued quantities. Math. Logic Quart. **22**(1), 149–160 (1976)
45. Güting, R.H., Schneider, M.: Moving objects databases. http://dna.fernuni-hagen.de/Lehre-offen/Kurse/1675/KE1.pdf
46. Haddawy, P.: Believing change and changing belief. IEEE Trans. Syst. Man Cybern. Spec. Issue on High. Ord. Uncertainty 26 (1996)
47. Jahn, K.U.: Intervall-wertige mengen. Mathematische Nachrichten **68**(1), 115–132 (1975)
48. Jensen, C., Dyreson, C., Böhlen, M., Clifford, J., Elmasri, R., Gadia, S., Grandi, F., et al.: The consensus glossary of temporal database concepts—february 1998 version. In: Lecture Notes in Computer Science, pp. 367–405 (1998)
49. Jensen, C.S., Mark, L., Roussopoulos, N.: Incremental implementation model for relational databases with transaction time. IEEE Trans. Knowl. Data Eng. **3**, 461–473 (1991)
50. Jensen, C.S., Snodgrass, R.T., Soo, M.D.: The tsql2 data model. In: The TSQL2 Temporal Query, Language, pp. 153–238 (1995)
51. Kabanza, F., m. Stevenne, J., Wolper, P.: Handling infinite temporal data. J. Comput. Syst. Sci. 392–403 (1990)
52. Klein, W.: Time in language. Routledge, London (1994)
53. Klopprogge, M.R., Lockemann, P.C.: Modelling information preserving databases: consequences of the concept of time. In: Proceedings of the 9th International Conference on Very Large Data Bases, pp. 399–416. Morgan Kaufmann Publishers Inc., San Francisco (1983)
54. Knight, B., Ma, J.: Time representation: a taxonomy of temporal models. Artif. Intell. Rev. **7**, 401–419 (1993)
55. Lacroix, M., Lavency, P.: Preferences; putting more knowledge into queries. In: Proceedings of the 13th International Conference on Very Large Data Bases, VLDB '87, pp. 217–225. Morgan Kaufmann Publishers Inc., San Francisco (1987)
56. Lakshmanan, L.V.S., Leone, N., Ross, R., Subrahmanian, V.S.: Probview: a flexible probabilistic database system. ACM Trans. Database Syst. **22**(3), 419–469 (1997)
57. Lin, H., Jensen, C.S., Böhlen, M.H.: Efficient conversion between temporal granularities. TR 19, The University of Arizona (1997)
58. Lorentzos, N.A.: A formal extension of the relational model for the representation of generic intervals. Ph.D. thesis, Birkbeck College, University of London (1988)
59. Zadeh, L.: Fuzzy sets. Inf. Control **8**, 338–353 (1965)
60. Melton, J., Simon, A.R.: Understanding the new SQL: a complete guide. Morgan Kaufmann Publishers Inc., San Francisco (1993)
61. Mitra, D., et al.: A possibilistic interval constraint problem: fuzzy temporal reasoning. In: Proceedings of the Third IEEE Conference on Fuzzy Systems: IEEE World Congress on Computational Intelligence, vol. 2, pp. 1434–1439 (1994)

62. Nagypál, G., Motik, B.: A fuzzy model for representing uncertain, subjective, and vague temporal knowledge in ontologies. In: On The Move to Meaningful Internet Systems 2003: CoopIS, DOA, and ODBASE, LNCS, vol. 2888, pp. 906–923. Springer, Heidelberg (2003)
63. Nascimento, M.A., Eich, M.H.: Decision time in temporal databases. In: Proceedings of the Second International Workshop on Temporal Representation and Reasoning, pp. 157–162 (1995)
64. Navathe, S., Ahmed, R.: Tsql: a language interface for history databases. In: Workshop on Temporal Databases-TDB, pp. 109–122 (1987)
65. Ohlbach, H.J.: Relations between fuzzy time intervals. In: International Symposium on Temporal Representation and Reasoning, pp. 44–51 (2004)
66. Pawlak, Z., Grymala-Busse, J., Slowinski, R., Ziarko, W.: Rough sets. Commun. ACM **38**(6), 88–95 (1995)
67. Pons, J., Bronselaer, A., Pons, O., de Tre, G.: Possibilistic evaluation of fuzzy temporal intervals. In: Sainz, G., Alcalá, J. (eds.) Actas del XVI Congreso Espa nol sobre Tecnologías y Lógica Fuzzy. Valladolid, Spain (2012)
68. Bronselaer, A., Pons, J.E., De Tré, G., Pons, O.: Possibilistic evaluation of sets. Int, J. Unc. Fuzz. Knowl. Based Syst. 21(3), 325–346 (2013)
69. Postgree: Temporal postgreesql (2011). http://pgfoundry.org/projects/temporal/
70. Qiang, Y., Asmussen, K., Delafontaine, M., De Tré, G., Stichelbaut, B., De Maeyer, P., Van de Weghe, N.: Visualising rough time intervals in a two-dimensional space. In: 2009 IFSA World Congress / EUSFLAT Conference, Proceedings (2001)
71. Sambuc, R.: Fonctions ϕ-floues. application l'aide au diagnostic en pathologie thyroidienne. Ph.D. thesis, Univ. Marseille, France (1975)
72. Sarda, N.L.: Extensions to sql for historical databases. IEEE Trans. Knowl. Data Eng. **2**, 220–230 (1990)
73. Schockaert, S., De Cock, M., Kerre, E.: Fuzzifying allen's temporal interval relations. IEEE Trans. Fuzzy Syst. **16**(2), 517–533 (2008)
74. Shackle, G.: Decision, order and time in human affairs. Cambridge University Press, Cambridge (1961)
75. Snodgrass, R.: The temporal query language tquel. In: Proceedings of the 3rd ACM SIGACT-SIGMOD Symposium on Principles of Database Systems, PODS '84, pp. 204–213. ACM, New York (1984)
76. Snodgrass, R.T., Jensen, C.S., Jensen, C.S., Jensen, C.S., Steiner, A., Böhlen, M.H., Busatto, R., Gregersen, H.: Transitioning temporal support in tsql2 to sql3. In: Etzion, O., Jajodia, S., Sripada S. (eds.) Temporal Databases: Research and Practice, Lecture Notes in Computer Science, vol. 1399. Springer, Berlin (1998)
77. Soysangwarn, S., Chittayasothorn, S.: Toward fuzzy temporal databases with temporal fuzzy linguistic terms. In: Second International Conference on the Applications of Digital Information and Web Technologies, 2009. ICADIWT '09, pp. 8–13 (2009).10.1109/ICADIWT.2009. 5273853
78. Teradata: Teradata Temporal (2011). http://www.teradata.com/database/teradata-temporal/
79. TimeDB: A temporal relational dbms (2011) http://www.timeconsult.com/Software/Software. html
80. Virant, J., Zimic, N.: Attention to time in fuzzy logic. Fuzzy Sets Syst. **82**(1), 39–49 (1996)
81. Wang, X.S., Jajodia, S., Subrahmanian, V.S.: Temporal modules: an approach toward federated temporal databases. In: Information Systems, pp. 227–236 (1993)
82. Wu, Y., Jajodia, S., Wang, X.: Temporal database bibliography update. In: O. Etzion, S. Jajodia, S. Sripada (eds.) Temporal Databases: Research and Practice, Lecture Notes in Computer Science, vol. 1399, pp. 338–366. Springer, Berlin (1998)
83. Zadeh, L.A.: The concept of a linguistic variable and its application to approximate reasoning—I. Inf. Sci. **8**(3), 199–249 (1975)

Part IV
Flexible Queries Over Nonstandard Data

Chapter 10
A Unifying Model of Flexible Queries with Distinct Semantics of Search Term Weights

Gloria Bordogna, Gabriella Pasi and Giuseppe Psaila

Abstract When querying documents archives there is often the need to specify importance weights of the search terms that define flexible selection conditions on documents representation. Several interpretations of the semantics of these weights have been proposed within distinct information retrieval models. In this contribution we define a unifying model of information retrieval based on a vector p-norm, were importance weights with distinct semantics can be specified in flexible queries.

1 Introduction

One important aspect of flexible queries in databases and information retrieval systems is the possibility of expressing both flexible selection conditions and soft aggregations of the selection conditions themselves. When aggregating soft selection conditions which admit degrees of satisfaction, a trade off of these satisfaction degrees must be computed. This issue occurs in flexible queries to both databases and documents archives managed by Information Retrieval Systems (IRSs). In this last context, search terms define soft selection conditions on the documents representation which generally consists of pairs (term - index term weight) where the weight indicates the significance degree of the term in representing the document's seman-

G. Bordogna (✉)
CNR IDPA, via Pasubio 5, I-24044 Dalmine, BG, Italy
e-mail: gloria.bordogna@idpa.cnr.it

G. Pasi
Dipartimento Informatica Sistemistica e Comunicazione, University of Milano Bicocca,
Viale Sarca, 336, 20126 Milano, MI, Italy
e-mail: pasi@disco.unimib.it

G. Psaila
Department of Engineering, University of Bergamo, via Marconi 5, I-24044 Dalmine, BG, Italy
e-mail: psaila@unibg.it

O. Pivert and S. Zadrożny (eds.), *Flexible Approaches in Data, Information*
and Knowledge Management, Studies in Computational Intelligence 497,
DOI: 10.1007/978-3-319-00954-4_10, © Springer International Publishing Switzerland 2014

tics and is generally computed based on the frequency of the term in the document and in the whole archive [21]. A search term is thus interpreted as a soft condition on the index term's significance degree: the greater the significance degree of a search term in a document, the greater the query satisfaction by that document.

When aggregating the soft conditions in a query one can consider them as all mandatory, i.e., necessary, thus not allowing any compensation among their satisfaction; or one may consider them as all replaceable, i.e., the satisfaction of just one of them is considered as a sufficient condition to retrieve a document. Furthermore, they can be regarded as a mixture of mandatory/replaceable search terms, or even as optional, desirable to satisfy [2].

Within fuzzy set theory these distinct semantics of soft aggregations have been modeled by distinct aggregation operators. While mandatory conditions are aggregated by t-norms (conjunctive query), fully replaceable conditions are aggregated by t-conorms (disjunctive query) [20]. Concerning partially mandatory/replaceable conditions, several approaches have been proposed based on either the Ordered Weighted Averaging operators (OWA) [2, 6, 14], or the Conjunctive Disjunctive Function [12], or the vector p-norm operator [3, 22]. Finally, mandatory and optional conditions can be aggregated by bipolar aggregation operators (non monotonic conjunctive query) [2, 10, 19, 25].

Further flexibility of soft aggregations in queries to document archives can be provided by associating distinct importance weights with the search terms.

As outlined in [1], the introduction of query weights raises the problem of their semantic interpretation: in the vast literature on information retrieval models, query weights have been intended as either relative appraisal of search terms, which tells to what extent a term is more important than another in determining the global satisfaction degree of the query, or as ideal, desired (or undesired) significance degrees of the search terms in documents, or still as minimum acceptance levels of satisfaction of the conditions, i.e., thresholds on the term significance in document representations.

Nevertheless, a IR model that unifies in a common framework the evaluation of flexible queries with soft aggregations of weighed terms, and distinct semantics of importance weights has not been defined yet.

A first attempt to define a unifying IR model was proposed in [4], by making it possible to interpret the importance weights with distinct semantics in conjunctive queries.

In this contribution, first we recap this model; then we analyze some approaches for modeling the soft aggregations of flexible conditions with unequal importance weights in information retrieval systems and databases, by discussing their characteristics and semantics of the importance weights. Finally, we define a unifying model, based on the vector p-norm that allows modeling several semantics of the importance weights of search terms in both (partially) conjunctive and (partially) disjunctive queries.

2 Modeling Distinct Semantics of Query Weights in Conjunctive Queries Based on the Division of Fuzzy Relations

Information Retrieval Systems are aimed at both handling huge document repositories, and retrieving those documents which correspond to user needs expressed in a query. Documents generally consist of texts which are indexed to represent their contents (index terms), and queries are based on the specification of terms used as selection conditions for identifying topics of interest [1]. In most commercial IRSs conjunctive queries can be specified by users to locate the documents which are related to the specified search terms.

Most IR models represent the significance of the index terms in describing the documents' semantics by a numeric index term weight, and some of them allow specifying the importance of terms in queries. Since these models are based on different formalisms their comparison is not easy. Their common basis is the introduction of weights both in document representation and in the query language.

In [4] it has been shown that a unique formalization of the retrieval activity of IR models can be based on the notion of the division of fuzzy relations that is general enough to model distinct semantics of search terms weights. The only limitation of the above proposal is related to the aggregation of the weighted terms that are assumed to be ANDed, i.e., necessary in order to retrieve a document.

Hereafter we recap how the distinct semantics of query term weights can be modelled by calling upon the notion of division of fuzzy relations.

In the framework of the relational data model, a universe is modeled by a set of relations (in a mathematical sense, i.e., a relation R_i is a subset of the Cartesian product of some domains), which can be manipulated with the help of relational algebra operators (set operations, selection, projection).

Among these operations, the division of the relation R(A,X) by the relation S(A), denoted by $R[A \div A]S$, where A is a set of attributes common to R and S, aims at determining the X-values connected in R with all the A-values appearing in S [5]. This operation can be defined in the following ways:

$$x \in [A \div A]S \Leftrightarrow \forall a \in S, (x,a) \in R \tag{1}$$

$$x \in [A \div A]S \Leftrightarrow a \in S \Rightarrow (x,a) \in R \tag{2}$$

$$x \in [A \div A]S \Leftrightarrow S \subseteq \Gamma(x) \text{ where } \Gamma(x) = \{a|(x, a) \in R\} \tag{3}$$

$$R[A \div A]S = \cap_{a \in S} \Gamma^{-1}(a) \text{ where } \Gamma^{-1} \text{ is the inverse function of } \Gamma \tag{4}$$

where formulas (1)–(4) are equivalent.

Let us consider a document d described as a set of terms $d = \{t_1, \ldots, t_m\}$, with $t_i \in T$, where T is the set of all the n index terms. Moreover, let us restrict to the case in which a query q looks for those documents indexed by all the search terms $P = \{t_1, \ldots, t_k\}$. Notice that P can contain keywords freely chosen by the user and thus not necessarily contained in T.

Table 1 Relations representing an archive and a query

Archive		Expected-terms
Doc.	Index term	
d_1	t_1	t_1
d_1	t_4	t_2
d_2	t_1	t_3
d_2	t_2	
d_2	t_3	
d_3	t_2	

The set of documents of the archive may be represented as an un-normalized relation (UR) where a tuple has the form: $<d, t_1, \ldots, t_m>$, or as a normalized relation (NR) where the information stored in the previous tuple is split through m tuples: $<d, t_1>, \ldots, <d, t_m>$.

The search terms in P appearing in the query may be seen as a unary relation (P) and the query may be answered as the division of NR by P.

In the example in Table 1, the result of the division

ARCHIVE [index term $\div$ search term] EXPECTED-TERMS

returns the document d_2, which corresponds to the only document containing the three desired search terms $\{t_1, t_2, t_3\}$.

Now, let us represent a document by a fuzzy set of index terms:

$R(d) = \{\mu_{t_1}(d)/t_1, \ldots, \mu_{t_m}(d)/t_{m>}\}$, in which $\mu_{t_i}(d) \in [0,1]$ is the weight associated with the index term t_i to represent its significance in document d.

Analogously, to draw a parallel with the fuzzy database context $\mu_{t_i}(d)$ can be considered as the degree of satisfaction of a flexible condition imposed on the value of attribute t_i of a fuzzy relation $R(d)$, where d is a tuple of a database.

A query q asking for those documents indexed by a set of expected (weighted) terms can also be represented by a fuzzy set of terms $S(q) = \{i_1/e_1, \ldots, w_n/e_n\}$ in which $i_i \in [0,1]$ is the weight associated with the term e_i to express its desired importance. Analogously, in a database, i_i can be the importance weight of the flexible selection condition imposed on the values of attribute t_i.

A document can then be viewed as a fuzzy relation $R: D \times T \rightarrow [0,1]$ in which D is the set of documents in the archive and T the set of the index terms, such that with each pair (d,t) a membership degree is associated $\mu_R(d,t) = \mu_t(d) \in [0, 1]$. A query can be represented as a unary relation $S: T \rightarrow [0,1]$, with membership degrees $\mu_S(t)$ corresponding to the importance weights $i_t \in [0, 1]$ of the search terms (see Tables 2 and 3).

Then, the answer to a flexible query q may be defined as the generalization of the one exemplified above, namely the division of two fuzzy relations R and S.

In this case, the result of the division is defined as a fuzzy set, i.e., a fuzzy relation $R[T \div T]S$, and a natural extension stems from (3) where the usual set inclusion operator is changed into a grade of inclusion g defined as:

$$\mu_{R[T\div T]S}(d) = g(S[T] \subseteq \Gamma(d)) \tag{5}$$

$\Gamma(d)$ being a fuzzy set defined as:

$$\Gamma(d) = \{\mu_R(d,t)/t | \mu_R(d,t)/(d,t) \in R, \text{ and } d \in D, t \in T\}. \tag{6}$$

The membership degree $\mu_R(d,t)$ is generally computed based on the schema term-frequency* inverse-document-frequency [21]. Then, the semantics of the division depends on both the choice of the inclusion degree g and on the intended meaning of the weights $\mu_R(d,t)$ and $\mu_S(t)$ in relations R and S respectively [7, 8].

A view of the fuzzy inclusion consists in defining the degree of inclusion $g(S[T] \subseteq \Gamma(d))$ by using a fuzzy implication:

$$S \subseteq \Gamma(d) \Leftrightarrow \forall t((t \in S) \Rightarrow t \in \Gamma(d)))$$
$$\Leftrightarrow \forall t(\mu_S(t) \to \mu_\Gamma(t)),$$
$$\Leftrightarrow \forall t(\mu_S(t) \to \mu_R(d,t)),$$

and then, we obtain the following:

$$g(S[T] \subseteq \Gamma(d)) = \min_{t \in S}(\mu_S(t) \to \mu_R(d,t)) \tag{7}$$

Two slightly different interpretations may be distinguished depending on the nature of the interaction of the degrees in the two relations. In the first case, the degree $\mu_S(t)$ is seen as a ***threshold*** and the complete satisfaction requires that this threshold is attained by $\mu_R(d,t)$ for each value t of S. When the threshold is not reached, a penalty is applied and several choices are possible: 0, $\mu_R(d,t)$, $\mu_R(d,t)/\mu_S(t)$ or also $(1 - \mu_S(t)+\mu_R(d,t))$: these cases are modeled by the Rescher-Gaines (RG), Gödel (Gd), Goguen (Gg) and Lukasiewicz (L) implications respectively:

$$\mu_{R[T \div T]S}(d) = \min_{t \in S}\mu_S(t) \to_{RG} \mu_R(d, t) \tag{8}$$

where $x \to_{RG} y = 1$ if $x \le y$, 0 otherwise

$$\mu_{R[T \div T]S}(d) = \min_{t \in S}\mu_S(t) \to_{Gd} \mu_R(d, t) \tag{9}$$

where $x \to_{Gd} y = 1$ if $x \le y$, y otherwise

$$\mu_{R[T \div T]S}(d) = \min_{t \in S}\mu_S(t) \to_{Gg} \mu_R(d, t) \tag{10}$$

where $x \to_{Gg} y = 1$ if $x \le y$, y/x otherwise

$$\mu_{R[T \div T]S}(d) = \min_{t \in S}\mu_S(t) \to_L \mu_R(d, t) \tag{11}$$

where $x \to_L y = 1$ if $x \le y$, $1 + y - x$ otherwise.

With all definitions from (8) to (11) full satisfaction 1 is obtained only when y is greater equal x, otherwise a penalty to full satisfaction is applied. With the RG definition (8) we have a Boolean result that is a full penalty is applied and the result is zero. With the Gd definition (9) the penalty is equal to y. With Gg definition (10) the penalty y/x is applied. With L definition (11) the penalty is equal to the difference between x and y.

In the second interpretation, $\mu_S(t)$ defines the **relative importance** of value x (and then the degree $\mu_R(d,t)$ is modulated). In the logical framework imposed by an implication, the underlying notion is the one of a guaranteed satisfaction when this importance is under 1: when $\mu_S(t) < 1$ the requirement is fully important, and it can be forgotten to some extent. The complete satisfaction requires that $\mu_R(d,t)$ equals 1 for each value x of S whatever its importance and $\mu_{R[T \div T]S}(d) = 0$ only if for at least one t in S, both $\mu_S(t) = 1$ (the requirement has the maximum level of importance) and $\mu_R(d,t) = 0$ (the document does not fulfill the requirement at all). This behavior is modeled by using the Dienes implication ($x \rightarrow_D y = \max(1 - x, y)$) in defining the division:

$$\mu_{R[T \div T]S}(d) = \min_{t \in S} \mu_S(t) \rightarrow_D \mu_R(d, t)$$
$$= \min_{t \in S} \max(1 - \mu_S(t), \mu_R(d, t)) \qquad (12)$$

where S is a normalized fuzzy relation ($\exists u, \mu_S(u) = 1$), so as to have an appropriate scaling of the levels of importance.

This approach is logical and conjunctive and an "absorption effect" occurs: the division operator only retains the smallest degree of implication between S and R. The S-grades, i.e., the query importance weights, can express either a **threshold** or a **relative importance**. If we assume that the degree $\mu_R(d,t)$ of a term t in a document d (hereafter, for sake of simplicity, we name $\mu_R(d,t) = \mu_t(d)$) refers to the relevance of d with respect to t, the weight $\mu_S(t)$ tied to the search term t stands for a minimal relevance with the threshold interpretation while it represents the importance of t with the second interpretation. Consequently, the solutions suggested before for the division of fuzzy relations may be an interesting basis for plausible interpretations of document retrieval.

Example. Let us consider the archive represented by the fuzzy relation in Table 2 and the queries q and q' represented by the fuzzy relations in Table 3.

Depending on the chosen semantics, the result of the queries q and q' are given in Table 4. We can see that with the relative semantics modeled by Dienes definition neither document gets full relevance. With the threshold semantics, a distinct penalty is applied to the full relevance of d_1 due to the under satisfaction of the threshold 0.5 imposed on the index term weight of t_4 by q'.

	t_1	t_2	t_3	t_4
d_1	1	0.9	1	0.2
d_2	0.7	0.6	0.3	0.8

Table 2 Relation representing an archive of documents as a fuzzy relation R

Table 3 Each row is a fuzzy relation S representing a query

	t_1	t_2	t_3	t_4
q	1	0.4	0	0.6
q'	0.6	0.6	0.3	0.5

Table 4 Result of the queries of Table 3 referred to the archive of Table 2

	Query weight	Semantics	d_1	d_2
q	Relative importance	Dienes (11)	0.2	0.6
		Gödel (8)	0.1	1
q'	Threshold	Goguen (9)	0.33	1
		Lukasiewicz (10)	0.7	1

3 Soft Aggregations of Conditions with Unequal Importance

In the previous modeling framework we did not consider the distinct semantics of the aggregation of query terms, but we assumed that they were ANDed, i.e., search terms are considered mandatory selection conditions.

Nevertheless, when submitting queries to search engines an implicit relaxation of their aggregation is assumed so that either a cascading of the conditions is applied where the importance of the search terms decreases from the first term to the last one listed in the query, or the Pareto principle holds: this means that the quantity of the important terms (conditions) that are met by a document must positively increase the overall relevance of the document [9, 16, 17].

In fuzzy Information Retrieval Systems and databases, distinct models of soft aggregations of partially mandatory/replaceable conditions having unequal importance weights have been proposed. A well known approach is based on the use of OWA operators that hereafter we synthesize.

3.1 Soft Aggregations Based on the OWA Operator

In order to make the expression of the semantics of the aggregation of terms easier one can use relative monotone non decreasing linguistic quantifier Q, such as *most*, defined by a fuzzy set $\mu_Q: [0,1] \rightarrow [0,1]$ [2, 15].

When aggregating n terms (or satisfaction degrees of the flexible selection conditions in databases) with distinct importance weights $[i_1, \ldots, i_n]$ the weighting vector $W = [w_1, \ldots, w_n]$ associated with the OWA_Q operator can be obtained as follows:

$$w_i = \mu_Q\left(\frac{1}{e}\sum_{k=1}^{i} e_k\right) - \mu_Q\left(\frac{1}{e}\sum_{k=0}^{i-1} e_k\right) \quad e = \sum_{k=1}^{n} e_k = \sum_{k=1}^{n} i_k \quad (13)$$

where e_k is the importance weight of the k-th greatest significant term. The subtraction in formula (13) allows computing w_i as the increment of the relevance degree due to having the i-th query term with not null index term weight in the representation of the documents. Further this increment is proportional to the importance of the i-th query term.

The OWA_Q operator is then defined as in [23].

Given a document d and $\mu_{t1}(d), \ldots, \mu_{tn}(d) \in [0,1]$, its degrees of satisfaction of the query conditions $t_1, \ldots, t_n$, the OWA_Q operator is defined as follows:

$$OWA_Q(\mu_{t1}(d), \ldots, \mu_{tn}(d)) = \sum_{i=1}^{N} w_i \, Sup_i(\mu_{t1}(d), \ldots, \mu_{tn}(d)) \qquad (14)$$

in which $Sup_i(\mu_{t1}(d), \ldots, \mu_{tn}(d))$ denotes the i-th highest of its arguments.

Notice that by applying definition (13) the importance weights affect the weighting vector W of the OWA_Q operator, so that its semantics is different from the original definition provided by μ_Q. For example, if one specifies as aggregation *"all"* (with $\mu_{all}(1) = 1, \mu_{all}(x) = 0, \forall x \in [0, 1]$), having all terms with equal importance would produce an OWA_{all} selecting the smallest significance degree, i.e., the OWA_{all} behaves like a t-norm in accordance with μ_{all}. On the contrary, if only the most significant term is important while the others have zero importance the OWA_{all} operator would select the highest significance degree as the satisfaction degree, in this case behaving like a t-conorm, then completely in contrast with μ_{all}.

A user should be careful in specifying the linguistic quantifier in conjunction with soft section conditions having distinct importance, in order not to express contradictory requirements in the query [24]. Further, while the semantic of *"most"* makes sense when aggregating a large number of terms, it is unintuitive when aggregating only two or three terms as it commonly happens when querying search engines over the Internet.

3.2 Soft Aggregations Based on the Generalized Conjunction Disjunction Function

A second approach, originally proposed for evaluating soft aggregations of criteria in decision making is based on the Generalized Conjunction Disjunction Function (GCD) [11, 12]. GCD enables a continuous transition from the full conjunction to the full disjunction, using a parameter p that specifies a desired level of conjunction or disjunction of the soft aggregation.

The Generalized Conjunction/Disjunction function (GCD) has been defined in [12] as the weighted power mean. Given $[\mu_{t1}(d), \ldots, \mu_{tn}(d)]$ in [0,1] the satisfaction degrees of n selection conditions imposed by the search terms $t_1, \ldots, t_n$ on a document d, with importance weights $[i_1, \ldots, i_n]$, with $i_i \in [0,1]$ and $\Sigma_{k=1,\ldots,n} \, i_k = 1$, the GCD_p aggregation function is defined as follows:

$$GCD_p(\mu_{t1}(d), \ldots, \mu_{tn}(d)) = \left(\sum_{k=1}^{n} i_k \times (\mu_{tk}(d))^p \right)^{1/p} \quad -\infty \le p \le +\infty \wedge p \ne 0$$

(15)

The exponent p is used to set the logic properties of the aggregation function. By varying the value of p the GCD_p function can model distinct basic aggregations:

- *Simultaneity* aggregator: full conjunction, i.e., AND aggregation, is obtained with $p = -\infty$; partial conjunction, i.e., AND-OR aggregation, models situations in which *most* terms are mandatory, thus should be satisfied simultaneously, and is obtained with $-\infty < p < 1$.
- *Replaceability* aggregator: full disjunction, i.e., OR aggregation, is obtained with $p = +\infty$; partial disjunction, i.e., OR-AND aggregation, is obtained with $1 < p < +\infty$. With this kind of aggregators one wants to model the situation in which *a single* or *a few* terms are sufficient, thus the presence of *a single* term, or of *a few* terms, can be replaced by the presence of another one, or a few other ones.
- *Neutrality* aggregator: vector product is obtained with $p = 1$. This aggregator is exactly in the middle between AND and OR. By choosing it, one wants a balance between simultaneity and replaceability of the terms, i.e., a weighted average.

The semantics of a GCD_p function can be captured by computing its fundamental properties of *Andness* for (partial) conjunctions, and *Orness* for (partial) disjunction aggregations. The *Andness*(GCD_p) and *Orness*(GCD_p) of a GCD_p function measure the similarity between the semantics of GCD_p function and the semantics of the full conjunction and full disjunction respectively. There are various ways to define *Andness* and *Orness* (see [13]), hereafter we only need to know that they take values in [0,1] and that they are complementary and satisfy the following:

Andness(GCD_p) = 1-*Orness*(GCD_p)
Andness($GCD_{-\infty}$) = 1
Orness($GCD_{+\infty}$) = 1,
Andness(GCD_1) = *Orness*(GCD_1) = 0.5.

Therefore, a GCD_p function has a mixture of conjunctive and disjunctive properties. In the case of partial conjunction, conjunctive properties predominate: *Andness*($GCD_{-\infty<p<1}$) > 0.5, viceversa in the case of partial disjunction: *Orness*($GCD_{1<p<+\infty}$) > 0.5. From the above definitions it follows that the semantic of the aggregation is specified by selecting the parameter p.

Nevertheless, the precision of humans in specifying the desired level of andness/orness is limited, since a practical way is to allow them to select the proper GCD function from a table whose entries report the parameter settings for discrete increments of Andness [11]. Further, when associating importance weights to search terms one might encounter inconsistency between a low weight and a high andness of the aggregation operator, as in the case of the OWA. In fact, the low importance weight is interpreted as a low relative importance, thus based on this the condition should be disregarded. However, a high andness of the aggregation means the requirement for high simultaneity, which indirectly means that all terms must be present and consequently they are very important. So, a low weight and a high andness may sometimes represent a contradiction, and should be avoided.

Thus modeling soft aggregation of search terms based on the GCD function is not very practical in flexible queries to information retrieval systems at least from a user point of view, due to the difficulty in understanding the semantics of the formulations.

3.3 Soft Aggregations Based on the p-Norm Operator

A third model of soft aggregation of search terms in IR and of soft conditions in fuzzy databases was proposed in [3, 22] based on the vector p-norm. This approach has a geometric interpretation, since conjunctive and disjunctive queries are regarded as distinct vectors in a metric space, where each dimension corresponds to an index term. The satisfaction of a query is then based on a p-norm distance of the documents (tuples) from the query vectors, and the contribution of each dimension (term or condition) to the distance is penalized or rewarded based on the importance weight of the term (condition). Hereafter, we recap this model.

In [3, 22] query terms (conditions) with distinct importance weighs are aggregated by means of a *vector p-norm* operator.

Given a vector $V = [v_1, \ldots, v_n]$ with $v_k \in [0, 1]$, and $k = 1, \ldots n$, a *p-norm* operator is defined as the *Minkowski* distance:

$$||V||_p \text{normalized} = \left(\frac{1}{n}\right)^{\frac{1}{p}} ||V||_p = \left[\frac{\sum_{k=1}^{n} v_k^p}{n}\right]^{\frac{1}{p}} \text{ with } 1 \le p \le \infty. \quad (16)$$

By varying p, distinct metrics are obtained: for example, with $p = 1$ definition (16) is the *Hamming* distance, while with $p = 2$ it is the *Euclidean* distance.

Each document d (or tuple t of a database) is represented by a vector of coordinates $[\mu_{t1}(d), \ldots, \mu_{t_n}(d)]$ in an n-dimensional space, where each axis corresponds to an index term t, and the coordinates of a vector are the significance degrees of the term in the document (or the satisfaction degrees of the soft condition by the tuple). Notice that a vector of this space can represent multiple documents (tuples), those that contain the terms with the same significance degrees (satisfy to the same degrees the n conditions).

A **Conjunctive query** identifies an ideal vector with coordinates $[1, \ldots, 1]$ since it is best satisfied by documents as far as possible *similar* to $[1, \ldots, 1]$, i.e., documents that contain with full significance all query terms. Further, it is partially satisfied by documents that correspond to vectors close to $[1, \ldots, 1]$, and their overall query satisfaction degrees increases as their similarity to $[1, \ldots, 1]$ increases. On the other side, a **disjunctive query** demands that the documents to be retrieved satisfy, at least a little, one condition. Then, a disjunctive query identifies the undesired vector $[0, \ldots, 0]$ and the relevance of documents to a query increases with their **distance** from $[0, \ldots, 0]$ corresponding to no search terms in the document representation.

When associating unequal importance weights $[i_1, \ldots, i_n]$, with $i_k \in [0,1]$, $\forall k$, the impact of satisfying (or not satisfying) a condition imposed by a search term decreases as the importance of the condition decreases.

This is the ***relative importance*** semantics of the query weights.

A conjunctive query of n ANDed search terms $t_1, \ldots t_n$, with importance weights $[i_1, \ldots, i_n]$ is evaluated by computing for each document d a similarity measure with respect to vector $[1, \ldots 1]$ ($sim(d, q^p_{AND})$) as defined in formula (17) below. On the other side a disjunctive query of n ORed search terms $t_1, \ldots t_n$, with importance weights $[i_1, \ldots, i_n]$ is evaluated by computing for each document d a distance measure with respect to vector $[0, \ldots, 0]$ ($dist(d, q^p_{OR})$) as defined in formula (17):

$$sim(d, q^p_{AND}) \equiv 1 - \left[\frac{\sum_{k=1}^n i_k^p \times (1 - \mu t_k(d))^p}{\sum_{k=1^n} i_k^p} \right]^{\frac{1}{p}}$$

$$dist(d, q_{OR^p}) \equiv \left[\frac{\sum_{k=1}^n i_k^p \times (\mu t_k(d))^p}{\sum_{k=1}^n i_k^p} \right]^{\frac{1}{p}} \qquad (17)$$

When $p = 1$, $sim(d, q^1_{AND}) = dist(d, q^1_{OR})$, then the meaning of AND and OR is no more modeled but both conjunctive and disjunctive queries reduce to the weighted mean.

When $p = \infty$ and all the conditions have equal importance, we obtain the fuzzy modeling of full mandatory and full replaceable conditions, respectively [22]: $sim(d, q^\infty_{AND}) = min(\mu_{t1}(d), \ldots, \mu_{tn}(d))$; $dist(d, q^\infty_{OR}) = max(\mu_{t1}(d), \ldots, \mu_{tn}(d))$.

So, by increasing p above 1 one can model by definitions (17) more and more mandatory q_{AND} and replaceable q_{OR} semantics of search terms, respectively.

In [3] it has been shown that when $p = \infty$ and the search terms have unequal importance weights, definitions (17) reduce to the importance weighted transformation functions defined for ANDed and ORed aggregations respectively, that model the relative importance semantics of $[i_1, \ldots, i_n]$ [3, 18].

The p-norm model provides an intuitive geometric interpretation of the query semantics. Documents that get the same overall query satisfaction lie on equidistant lines (hyper surfaces in an n dimensional metric space) having centroid vector $[1, \ldots, 1]$ for (partially) conjunctive queries and $[0, \ldots, 0]$ for (partially) disjunctive queries.

With the Euclidean distance metrics, $p = 2$, these hyper surfaces are hyper spheres centered in $[1, \ldots, 1]$ and $[0, \ldots, 0]$ respectively.

In Fig. 1 an example of two documents d_1 and d_2 in a bi-dimensional space defined by two index terms t_1 and t_2 is shown. The dotted circle centered in $[1,1]$ represents the equidistance line around the ideal document $[1,1]$. This means that without importance weights both documents d_1 and d_2 lie on the same equidistance line and thus have the same relevance to a conjunctive query t_1 AND t_2.

The specification of distinct importance weights of the terms, affects the shape of the hyper surfaces: the greater the importance of a search term (an axis), the smaller

Fig. 1 Equidistant lines
(*grey ellipses*) from the query
[1,1] having unequal relative
importance weights $i_2 > i_1$
and $p = 2$

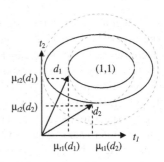

(greater) the projection of the equidistant hyper surfaces in case of conjunctive query
(in case of disjunctive query) on the associated axis.

As depicted in Fig. 1 the concentric dotted circles around the ideal vector [1,1]
in case of equal importance ($i_1 = i_2$), become ellipses (continuous lines) when
having unequal important terms with $i_2 > i_1$. A greater distance from [1,1] on the
most important axis t_2 (y axis in Fig. 1) is valuated as a smaller distance on the less
important one t_1, (x-axis in Fig. 1). When having equal importance weights, the two
documents d_1 and d_2 lie on the same circle and then get the same rank, while with
distinct relative importance weights $i_2 > i_1$ document d_1 gets a higher rank than d_2
since it lies on the ellipses closer to [1,1]. This geometric representation is effective
and simple to understand both in presence of few and many terms.

4 Generalized p-Norm Aggregation with Distinct Interpretations of Query Importance Weights

A generalization of the previous definitions in (17) in order to allow other interpretations of the semantics of importance weights of soft conditions in queries to
databases has been formulated in [3]. Hereafter, we apply this model to the case of
information retrieval.

Partially conjunctive and disjunctive queries are evaluated as follows:

$$sim(d,q_{AND}^p) \equiv 1 - \left[\frac{\sum_{k=1}^{n} a_k^p \times |b_k - F_k(\mu_{t_k}(d)|^p}{\sum_{k=1}^{n} a_k^p} \right]^{\frac{1}{p}}$$

$$dist(d,q_{OR^p}) \equiv \left[\frac{\sum_{k=1}^{n} a_k^p \times |F_k(\mu_{t_k}(d)) - c_k|^p}{\sum_{k=1}^{n} a_k^p} \right]^{\frac{1}{p}} \qquad (18)$$

where $|z|$ denotes the absolute value of z.

$[a_1, \ldots, a_n], [b_1, \ldots, b_n,], [c_1, \ldots, c_n]$ are parameter vectors in the n dimensional space with values defined in [0,1], and $F_k : [0, 1] \to [0, 1] \; \forall k = 1, \ldots, n$ are non decreasing functions defining soft constraints on the document significance degrees of the search terms .

By varying the values of the parameters and by selecting distinct F_k functions, distinct semantics of the importance weights can be obtained.

The relative importance semantics of the query weights is obtained by setting:

$[a_1, \ldots, a_n] = [i_1, \ldots, i_n]$ equal to the importance weights of search terms,

$[b_1, \ldots, b_n] = [1, \ldots, 1]$,

$[c_1, \ldots, c_n] = [0, \ldots, 0]$,

and $F_k(x) = x \; \forall k = 1, \ldots, n$.

In this case it can be noticed that definitions (18) reduce to definitions (17).

4.1 Importance Weights as Thresholds on the Significance Degrees of Terms

A useful feature of flexible queries is the possibility to reduce the number of retrieved documents based on the specification of distinct minimum acceptable significance degrees of the search terms. With this semantics, a flexible query delimits the interesting portion of the n-dimensional space for retrieving the documents.

A conjunctive and a disjunctive query of n ANDed and ORed terms with minimum acceptance levels $[i_1, \ldots, i_n] \in [0, 1]$ can be evaluated by applying definitions (18) with the following setting of the parameters:

$[a_1, \ldots, a_n] = [1, \ldots, 1]$,

$[b_1, \ldots, b_n] = [1, \ldots, 1]$,

$[c_1, \ldots, c_n] = [0, \ldots, 0]$,

and $F_k(x) = x$ if $x \geq i_k \quad \wedge \quad F_k(x) = 0$ if $x < i_k$, for $k = 1, \ldots, n$.

It can be seen that, based on (18), we obtain the following definitions:

$$sim(d, q_{AND^p}) \equiv 1 - \left[\frac{\sum_{k=1}^{n} \begin{cases} (1 - \mu_{t_k}(d))^p & for\, \mu_{t_k}(d) \geq i_k \\ 1 & for\, \mu_{t_k}(d) < i_k \end{cases}}{n} \right]^{\frac{1}{p}}$$

$$dist(d, q_{OR^p}) \equiv \left[\frac{\sum_{k=1}^{n} \begin{cases} (\mu_{t_k}(d))^p & for\, \mu_{t_k}(d) \geq i_k \\ 0 & for\, \mu_{t_k}(d) < i_k \end{cases}}{n} \right]^{\frac{1}{p}} \tag{19}$$

Fig. 2 The acceptance
levels (threshold importance
weights) $i_1 > i_2$ constrain
the interesting portion of the
space ($p=2$) to satisfy $x > i_1$
and $y > i_2$

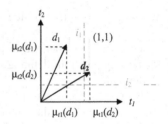

Function F_k imposes a full penalty when the acceptance level i_k is not reached by μ_{tk} (d). In fact, $F_k(\mu_{tk}(d)) = 0$ if μ_t (d) $< i_k$. In this case μ_{tk} (d) does not contribute to the global query satisfaction degree.

Based on definition (19) the n crisp acceptance levels $[i_1, \ldots, i_n]$ delimit the subspace of interest by defining a crisp region with projections $[i_1, \ldots, i_n]$ on the axis that includes the point $[1,\ldots,1]$ and that does not include the point $[0,\ldots,0]$.

To clarify the semantics of the minimum acceptance levels, let us look at Fig. 2 that depicts the same two documents of Fig. 1 equidistant to [1,1] in the Euclidean bi-dimensional metric space. It can be seen that the distinct minimum acceptance levels $i_1 > i_2$ delimit the interesting portion of the space so that while both the index term weights $\mu_{t1}(d_2)$ and $\mu_{t2}(d_2)$ contribute to compute the ranking of d_2 (since they are both above the thresholds $\mu_{t1}(d_2) > i_1$ and $\mu_{t2}(d_2) > i_2$) only $\mu_{t2}(d_1)$ contributes to the ranking of d_1, since the index term weight of t_1 does not reach the threshold ($\mu_{t1}(d_1) < i_1$). Thus d_2 is ranked first than d_1.

Tolerant acceptance levels (with a tolerance degree $0 < \delta < i_k \forall k = 1, \ldots, n$) can be defined so as to obtain a region with a broad boundary by setting the following F_k functions:

$$F_k(\mu_{tk}(d)) = \mu_{tk}(d) \text{ if } \mu_{tk}(d) \geq i_k$$
$$F_k(\mu_{tk}(d)) = 0 \text{ if } \mu_{tk}(d) < i_k - \delta \qquad\qquad (20)$$
$$F_k(\mu_{tk}(d)) = \delta \text{ if } i_k - \delta \ < \mu_{tk}(d) < i_k$$

This way, the n thresholds $[i_1, \ldots, i_n]$ delimit the core of the interesting region of the subspace but still a contribution equal to δ is provided by $\mu_{tk}(d)$ when it is below the threshold i_k but above $i - \delta$. So the documents vectors belonging to the broad boundary region still contribute to the satisfaction of the query to an extent δ.

Figure 3 depicts the situation in which unequal tolerant acceptance levels are imposed on the index term weights. Differently than with the crisp acceptance levels in Fig. 2, now d_1 lies on the broad boundary defined by i_1, and thus $\mu_{t1}(d_1)$ contributes to the ranking of d_1 to the degree δ.

Fig. 3 The tolerant acceptance levels i_1 and i_2 define
soft constraints on the index
term weights so that the interesting portion of the space has
a broad boundary ($p = 2$)

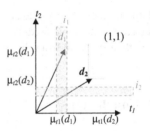

4.2 Importance Weights as Ideal Satisfaction Degrees

Another possible interpretation of importance weights in flexible queries is to consider them as desired (undesired) significance degrees of the index terms. The more
(the less) the $\mu_t(d)$ is close to i the greater (the lower) should be its contribution to
the global satisfaction degree of the query. This semantics may be useful to model
the Rocchio feedback algorithm in the p-norm vector space [21].

Let us assume that a graphic user interface for visualizing the documents as points
in the n-dimensional space exists by which a user can explore the documents retrieved
by an initial query. One might project documents on a sub-space by selecting the
interesting axes and might click on the points representing the documents to see
their contents. Now, let us suppose that the user discovers that one of these documents with coordinates $[x_1, \ldots, x_n]$ is very relevant (not at all relevant) to his/her
needs. These coordinates can be considered as those of an ideal document (desired
or undesired) to retrieve. To reorder the documents based on the user positive (negative) feedback we consider a conjunctive q_{AND} query (disjunctive q_{OR} query) in
which the index terms have as importance weights the coordinates of the ideal point
$[i_1, \ldots, i_n] = [x_1, \ldots, x_n]$, and re-rank the documents by applying formula (18) by
setting the following:
$[a_1, \ldots, a_n] = [1, \ldots, 1]$,
$[b_1, \ldots, b_n] = [c_1, \ldots, c_n] = [i_1, \ldots, i_n]$,
and $F_k(x) = x \forall k = 1, \ldots, n$.

With these settings, definitions (18) reduce to the following:

$$sim(d, q_{ANDp}) \equiv 1 - \left[\frac{\sum_{k=1}^{n} |i_k - \mu_{t_k}(d)|^p}{n} \right]^{\frac{1}{p}} \qquad dist(d, q_{ORp}) \equiv \left[\frac{\sum_{k=1}^{n} |\mu_{t_k}(d) - i_k|^p}{n} \right]^{\frac{1}{p}}$$

$$(21)$$

The equidistant hyper surfaces are centered in the ideal vector $[i_1, \ldots, i_n]$. In the case
of positive feedback one reorders the documents with respect to $sim(d, q_{and}^p)$, i.e.,

their closeness to the ideal document $[i_1, \ldots, i_n]$: the closest vectors are ranked first. In the case of negative feedback the documents are ranked with respect to $dist(d, q_{or}^p)$, i.e., their distance from the undesired document $[i_1, \ldots, i_n]$.

5 Conclusions

This contribution proposes a unifying information retrieval model for evaluating flexible queries to documents archives, in which both soft aggregations of the search terms, and distinct semantics of importance of the search terms can be specified in queries. This approach generalizes the extended Boolean Information Retrieval models defined in the literature to manage query importance weights [1, 3]. Specifically, we have outlined the drawbacks of the common approaches. Further, we have proposed a generalization of the p-norm model defined in [22]. Besides the relative importance semantics of search terms, as in the original p-norm model [22], the proposal in this contribution can model also both ideal (desired or undesired) significance degrees of search terms and minimum (crisp or broad) acceptance levels of the significance degrees. The geometric representation provided by the p-norm model, where the documents identify vectors, and the query defines hyper surfaces of equi-ranked documents supports an intuitive understanding of the flexible query semantics.

References

1. Bordogna, G., Pasi, G.: Modeling vagueness in information retrieval. Lectures on Information Retrieval Lecture Notes in Computer Science. Springer, Berlin 1980/2001, pp. 207–241 (2001)
2. Bordogna, G., Pasi, G.: Linguistic aggregation operators of selection criteria in fuzzy information retrieval. Int. J. Intell. Syst. **10**, 233–248 (1995)
3. Bordogna, G., Psaila, G.: (2009) Soft Aggregation in Flexible Databases Querying based on the Vector p-norm. Int. J. Uncertainty Fuzziness Knowl. Based Syst. (World Scientific Publishing Company) **17**(1), 25–40 (2009)
4. Bordogna, G., Bosc, P., Pasi, G.: Extended Boolean information retrieval in terms of fuzzy inclusion. In: Pons, O., Vila, M.A., Kacprzyk, J. (eds.) Knowledge Management in Fuzzy Databases. Studies in Fuzziness and Soft Computing series, vol. 39, pp. 234–247. Physica-Verlag, New York. ISBN: 3790812552 (2000)
5. Bosc, P., Dubois, D., Pivert, O., Prade H.: Fuzzy division for regular relational databases. In: Proceedings of 4th International IEEE Conference on Fuzzy Systems (FUZZ-IEEE/IFES'95), pp. 729–734, Yokohama (Japan), March 1995
6. Bosc, P., Lietard, L.: Quantified statements and some interpretations for the OWA operators. In: Yager, R., Kacprzyk, J. (eds.) The Ordered Weighted Averaging Operators: Theory and Applications, pp. 241–257. Kluwer, Boston (1997)
7. Bosc, P.: Some views of the division of fuzzy relations. In: Proceedings of 5th International Workshop on Current Issues on Fuzzy Technologies (CIFT'95), pp. 14–22, Trento (Italy), June 1995
8. Bosc, P., Pivert, O.: Some approaches for relational databases flexible querying. J. Intell. Inf. Syst. **1**, 323–354 (1992)

9. Chomicki, J.: Preference formulas in relational queries. ACM Trans. Database Syst. **28**(4), 427–466 (2003)

10. Dubois, D., Prade, H.: Bipolarity in flexible querying. In: Andreasen, T., Motro, A., Christiansen, H., Larsen, H.L. (eds.) Flexible Query Answering Systems, FQAS 2002. Lecture Notes in Computer Science, vol. 2522, pp. 174–182. Springer, Berlin (2002)

11. Dujmović, J.J., Fang, W.Y.: An empirical analysis of assessment errors for weights and andness in LSP criteria. In: Torra, V., Narukawa, Y. (eds.) Modeling Decisions for Artificial Intelligence. LNAI, vol. 3131, pp. 139–150. Springer, Berlin (2004)

12. Dujmović, J.J., Larsen, H.L.: Generalized conjunction/disjunction. Int. J. Approximate Reasoning **46**(3), 423–446 (2007)

13. Dujmović, J.J.: Nine forms of andness/orness. In: Kovalerchuk, B. (ed.) Proceedings of the Second IASTED International Conference on Computational Intelligence, pp. 276–281. ISBN Hardcopy: 0-88986-602-3/CD: 0-88986-603-1 (2006)

14. Kacprzyk, J., Zadrozny, S.: Implementation of OWA operators in fuzzy querying for Microsoft Access. In: Yager, R.R., Kacprzyk, J. (eds.) The Ordered Weighted Averaging Operators: Theory and Applications, pp. 293–306. Kluwer, Boston (1997)

15. Kacprzyk, J., Ziolkowski, A.: Database queries with fuzzy linguistic quantifiers. IEEE Trans. Syst. Man Cybern. **16**, 474–479 (1986)

16. Kießling, W., Kostler, G.: Preference SQL—design, implementation, experiences. In: Proceedings of the 28th VLDB Conference, Hong Kong (2002)

17. Lacroix M., Lavency P.: Preferences: putting more knowledge into queries. In: Proceedings of 13th Conference VLDB, pp. 217–225, Brighton, GB (1987)

18. Larsen, H.L.: Importance weighted OWA aggregation of multicriteria queries. In: Larsen, H.L. (ed.) Proceedings of NAFIPS 1999, pp. 740–744, New York (1999)

19. Lietard, L., Rocacher, D., Tbahriti, S.E.: Towards an extended SQLf: bipolar query language with preferences. Int. J. Appl. Math. Comput. Sci. **4**(1), 58–63 (2008)

20. Petry, F.E.: Fuzzy Databases. Kluwer Academic Pub., Boston (1996)

21. Salton, G., McGill, M.J.: Introduction to Modern Information Retrieval. McGraw-Hill, New York, NY (1983)

22. Salton, G., Fox, E., Wu, H.: Extended Boolean information retrieval. Commun. ACM **26**(12), 1022–1036 (1983)

23. Yager, R.R.: On ordered weighted averaging aggregation operators in multi-criteria decision making. IEEE Trans. Syst. Man Cybern. **18**, 183–190 (1988)

24. Zadrozny, S., Kacprzyk, J.: Issues in the practical use of the OWA operators in fuzzy querying. J. Intell. Inf. Syst. **33**, 307–325 (2009)

25. Zadrozny, S., Kacprzyk, J.: Bipolar queries: an approach and its various interpretations. In: Proceedings of IFSA-EUSFLAT 2009, pp. 1288–1293. ISBN: 978-989-95079-6-8 (2009)

Chapter 11
Social Network Database Querying Based on Computing with Words

Ronald R. Yager

Abstract Fuzzy relationships and their role in modeling weighted social relational networks are discussed. We describe how the idea of computing with words can provide a bridge between a network analyst's linguistic description of social network concepts and the formal model of the network. We then turn to some examples of taking an analyst's network concepts and formally representing them in terms of network properties. We first do this for the concept of clique and then for the idea of node importance. Finally we introduce the idea of vector–valued nodes and begin developing a technology of social network database theory.

1 Introduction

Social relational networks have rapidly become an important technology in our digital based information intense world [1–8]. Among the notable examples of social network sites are Facebook and LinkedIn. In addition to providing an ability for enabling people from all over the world to connect with each other each they provide a vast source of information about the individual participants in the network, the so called nodes. Each of these participants can be viewed as a kind of database containing information about themselves. We see here that a social network can be viewed as kind of network of databases. This leads us to understand that the development of appropriate technologies to manage social networks requires a combination of the use of network technologies, graph theory and database technologies [9]. Furthermore the development of intelligent social network management requires the extension of these two technologies by the introduction of ideas from soft and granular computing, computational intelligence [8]. In addition to the intelligent extension of these two technologies with soft computing an important task on the road map

R. R. Yager(✉)
Machine Intelligence Institute, Iona College, New Rochelle, NY 10801, USA
e-mail: yager@panix.com

O. Pivert and S. Zadrożny (eds.), *Flexible Approaches in Data, Information and Knowledge Management*, Studies in Computational Intelligence 497, DOI: 10.1007/978-3-319-00954-4_11, © Springer International Publishing Switzerland 2014

of developing intelligent social networks is the combining of network theory with database theory. Since many advances have been made toward the development of intelligent database technologies, especially by contributors to this volume, here we shall take some steps in the other two tasks, the intelligent extension of social network theory and the combining of social networks with databases. We note that one area that has received a considerable amount of research is the mining of social networks [4, 5]. We emphasize that our interest here is not on this problem but we shall be more interested in the issue of querying the social network with its nodes seen as individual databases. We shall refer to this as SONDAB-Q as an acronym for **SO**cial Network **DA**ta**B**ase **Q**uerying.

The current social network technology can be extended and enriched to help in modeling these newly emerging applications by introducing ideas from fuzzy sets and related granular computing technologies [10–16]. We can provide this extension and enrichment in two ways. The first is with the introduction of fuzzy graphs representing the networks [17–19]. This allows a generalization of the types of connection between nodes in a network from simply connected or not to weighted or fuzzy connections. Here the idea of strength of connection becomes important. The second and perhaps more interesting extension is the use of Zadeh's fuzzy set based paradigm of computing with words [10–12] to provide a bridge between a human network analyst's linguistic description of social network concepts and the formal model of the network. Fundamental to this capability is the realization that both formal network models and the paradigm of computing with words are built upon set based technologies. More specifically, the formal representation of a social network is in terms of a mathematical set object called a relationship [1] and computing with words uses a set object, fuzzy subsets, to formally represent the semantics of linguistic terms. This common underlying set based technology allows us to take human concepts and formally represent them in terms of network properties. This in turn allows an analyst to estimate the truth or falsity of observations about a network as well helps in the mining of social relation networks. In an attempt to help the reader get an understanding of the technology useful in this approach we provide few examples of how we would model some social network concepts.

Another useful idea we discuss is vector-valued nodes. Here we associate with each node a vector whose components are the attribute values of the node. Combining this with the machinery of Zadeh's computing with words we are then able to intelligently query the network with questions that involve both attributes and connections. We see this as the basis of an emerging discipline of social network database querying, SONDAB-Q

2 Fuzzy Graphs

Since a social network can be formally viewed as graph whose nodes represent the members of the social network we begin by introducing some ideas from fuzzy graphs. Here we first describe the idea of a fuzzy relationship. The concept of a fuzzy relationship plays a fundamental role in modeling a type of weighted graph called a fuzzy graph [17–19]. Let X be a set of elements. A fuzzy relationship on X is a mapping $R : X \times X \to [0, 1]$ where $R(x, y)$ indicates the degree of relationship between x and y. We note that we can view a fuzzy relationship as a fuzzy subset on $X \times X$. This allows us to use much of the formalism of fuzzy sets. For example we can say that $R_1 \subseteq R_2$ if $R_1(x, y) \leq R_2(x, y)$ for all (x, y). We note that $R_1 \subseteq R_2$ means that R_1 is a subset of R_2. We shall also use the terminology $R_1 \leq R_2$.

Some notable properties that can be associated with fuzzy relationships are

(1) **Reflexivity**: $R(x, x) = 1$ for all x
(2) **Symmetry**: $R(x, y) = R(y, x)$
(3) **Transitivity**: $R(x, z) \geq \text{Max}_y[R(x, y) \land R(y, z)]$

An important operation on fuzzy relations is composition. Assume R_1 and R_2 are two relations on X. The composition $R = R_1 \blacklozenge R_2$ is also a relationship on X such that

$$R(x, z) = \text{Max}_y[R_1(x, y) \land R_2(y, z)]$$

The composition operation can be shown to be associative

$$(R_1 \blacklozenge R_2) \blacklozenge R_3 = R_1 \blacklozenge (R_2 \blacklozenge R_3)$$

The associativity property allows us to use the notation $R^k = R \blacklozenge R \blacklozenge \ldots \blacklozenge R$ for the composition of R with itself k times. In addition we shall define R^0 to be such that $R^0(x, y) = 0$ for all x and y.

If R is reflexive then $R^{k_2} \supseteq R^{k_1}$ for $k_2 > k_1$. On the other hand if R is transitive, it can be shown that $R^{k_2} \subseteq R^{k_1}$ if $k_2 > k_1$. From this we see that if R is reflexive and transitive then $R^{k_2} = R^{k_1}$ for all k_1 and $k_2 \neq 0$.

We shall now define various types of fuzzy graphs. Let X be a set of elements, which we shall refer to using graph terminology as nodes or vertices. We shall further assume R is a reflexive fuzzy relationship on X. The pair $<X, R>$ can be seen as defining a fuzzy or weighted graph in which $R(x, y)$ is the weight associated with the arc $x \to y$, (x, y). More generally if F is a fuzzy subset of X we can also define a fuzzy graph as $<X, F, R>$. Here in addition to having degrees of connection we also have a degree to which each of the nodes belongs to the network. In this case we let R.F be a relationship on X defined such that $R.F(x, y) = R(x, y) \land F(x) \land F(y)$ and say that R.F is the relationship R on F. We note here that $R.F(x, x) = F(x)$. If $F = X$ then $R.F(x, y) = R(x, y)$. It can be shown here that $R.F^{k_1} \subseteq R.F^{k_2}$ if $k_2 > k_1$. We note that if $F = X$ then $<X, F, R> = <X, R>$.

If R is symmetric we shall say $<X, F, R>$ is an undirected fuzzy graph. We note that if R is symmetric then R.F is also symmetric. If R is not symmetric we shall refer to $<X, F, R>$ as a directed graph and we refer to a pair (x, y) as an arc. Here the weight on the arc (x, y) and arc (y, x) may be different. In the case of where R symmetric we shall refer to the pair (x, y) as an edge. Since we shall primarily be concerned with undirected graphs, we shall simply use the unmodified term graph or network to refer to this case where R is symmetric.

At times, especially when working with undirected graphs, we shall find it convenient to consider the space U which consists of all unordered pairs of distinct elements which we denote as $\{x, y\}$. In this case we can refer to R_U as the reflection of R on U. In this $R_U(\{x, y\}) = R(x, y) = R(y, x)$. We note that at a formal level we can also view U as consisting of all subsets of X consisting of two elements.

Assume $G = <X, F, R>$ is a fuzzy graph. A path ρ in G is a sequence of distinct nodes $x_0 x_1 \cdots x_n$. The number of links in the path is n. The strength of the path is defined as

$$ST(\rho) = \underset{i=1 \text{ to } n}{\text{Min}} [R.F(x_{i-1}, x_i)].$$

If $F = X$ then $ST(\rho) = \underset{i=1 \text{ to } n}{\text{Min}} [R(x_{i-1}, x_i)]$.

Two nodes for which there is a path ρ with $ST(\rho) > 0$ between them are called connected. We call ρ a cycle if $n \geq 2$ and $x_0 = x_n$

Consider the graph $G = <X, R>$ let us now consider R^k. We can show that $R^k(x, y)$ is the strength of the strongest path from x to y containing at most k links. We see that if X has n nodes then R^{n-1} provides the strongest connection between two nodes using any number of links. If $R^k(x, y) \neq 0$ we can say x and y are connected at least of order k.

We note that if $G = <X, F, R>$ we can make statements similar to the above about R.F.

Assume $G = <X, R>$ is a fuzzy graph. Let $\rho = x_0 x_1 \cdots x_n$ be a path in X. A concept introduced by Rosenfeld [17] is the length of the path. He defined

$$L(\rho) = \sum_{i=1}^{n} \frac{1}{R(x_{i-1}, x_i)}$$

Clearly $L(\rho) \geq n$. We note that if there exists one $R(x_{i-1}, x_i) = 0$ then $L(\rho) = \infty$ and $St(\rho) = 0$. We note that if R is crisp and $St(\rho) \neq 0$ then $L(\rho) = n$. Using this idea we can define the distance between two nodes x and y in the graph G as

$$\delta(x, y) = \underset{\text{all paths } x \text{ to } y}{\text{Min}} [L(\rho)]$$

It is the length of the shortest path from x to y. It can be shown that δ is a metric [17], $\delta(x, x) = 0$, $\delta(x, y) = \delta(y, x)$ and $\delta(x, z) \leq \delta(x, y) + \delta(y, z)$. Using the terminology common in social network analysis [1, 2] we can refer to the path ρ such that $L(\rho) = \delta(x, y)$ as the geodesic between x and y and refer to $\delta(x, y)$ as the geodesic distance.

While there appears to be some inverse connection between strength of a path and its length as for example in the case where $ST(\rho) = 0$ implies $L(\rho) = \infty$ this is not a strict correlation. Consider for example the two paths ρ_1 and ρ_2 shown in Fig. 1. We see that $ST(\rho_1) = 0.75 > Str(\rho_2) = 0.5$. On the other hand $L(\rho_1) = \frac{4}{3} + \frac{4}{3} + \frac{4}{3} = 4 \geq L(\rho_2) = 1$

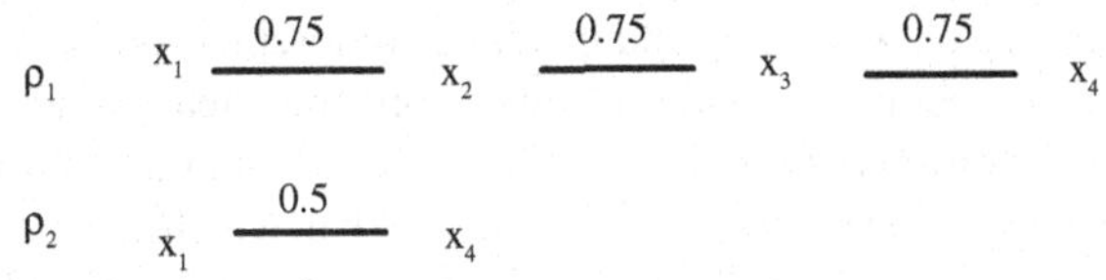

Fig. 1 Two paths

$$\rho_1 \quad x_1 \underline{\quad 0.75 \quad} x_2 \underline{\quad 0.75 \quad} x_3 \underline{\quad 0.75 \quad} x_4$$

$$\rho_2 \quad x_1 \underline{\quad 0.5 \quad} x_4$$

Let $<X, R>$ be a weighted graph. At times it can be useful to view this from the level set point of view [20]. This will allow us to make use of the representation theorem [20] to extend operations to these fuzzy relationships. We recall that if R is a fuzzy relationship on $X \times X$ then $R_\alpha = \{(x, y)/R(x, y) \geq \alpha\}$ is called the α level set of R. We see that each R_α is a crisp relationship on X.

We note that if R^k is the k composition of R then $R_\alpha^k = \{(x, y)/R^k(x, y) \geq \alpha\}$. It can also be shown that [21]

$$R_\alpha^k = R_\alpha \blacklozenge R_\alpha \quad \blacklozenge \quad \blacklozenge R_\alpha,$$

the k composition of R_α is also R_α^k.

The representation theorem allows us to represent fuzzy relationship in terms of the collection of its level sets. This can be used to extend operations that are well defined on crisp sets to be defined on fuzzy sets. Using this we can express $R^k = \overset{1}{\underset{\alpha=0}{\smile}} \{\frac{\alpha}{R_\alpha^k}\}$, here we are using the standard fuzzy set notation where $\frac{\alpha}{R_\alpha^k}$ indicates that α is the membership grade of R_α^k .

Undirected fuzzy graphs, which are also transitive, provide a very interesting class of graph. In these graphs if x is related to y and y is related to z then x is related to z. In social networks transitivity captures the property "friend of a friend is a friend".

Many of the concepts introduced in the preceding are valid for both directed and undirected graphs. A fundamental difference is the following. Assume $x_0 x_1 \cdots x_n$ is a sequence of points that constitute a path. In an undirected graph its transpose $x_n x_{n-1} \cdots x_0$ is a path having the same strength and length. In a directed graph we can't say anything about the transpose sequence. Whenever possible in an undirected graph we shall refer a path as between x_0 and x_n while in a directed graph we shall refer to the path from x_0 to x_n.

3 Computing with Words

Our goal here is to extend our capabilities for analyzing social relational networks by associating with these networks concepts with which human beings view social network relationships in such a way that they are comprehensible to both humans and machines. On one hand human beings predominantly use linguistic terms in which to communicate, reason and understand the world. Machines on the other hand require much more formal symbols. One of the most useful approaches to providing a bridge between man and machine comprehension is the general framework provided by granular computing [8] and more specifically Zadeh's fuzzy set based paradigm of computing with words. This technology allows for a high level of man-machine cooperation by providing a framework in which concepts can be modeled in a manner amenable to both. The potential for applying fuzzy set based technologies within the domain of social network analysis is particularly promising given that the computer modeling of these networks is in terms of mathematical relationships which as we already noted are equivalent to fuzzy sets.

In the following we introduce some ideas from the fuzzy set based approach to computing with words. Let U be some attribute that takes its value in the space Y. An example of such an attribute is age, which, for human beings takes its value in the set $Y = \{0, \ldots, 100\}$. A fundamental concept in computing with words is the idea of linguistic value [22]. A linguistic value is some word used to express the value of U. In the case of age, some examples of linguistic values are "old", "young", "about 30", A linguistic value can be seen as a granule, it is a collection of values from the space Y. As we noted it is with the aid of linguistic values that human beings best understand and reason about their environment.

By a vocabulary we shall mean a collection of commonly understood words that are used to express the linguistic values associated with an attribute. These are the words that people use to communicate with each other. They are also the words people use to reason with about the attribute. Fuzzy sets provide a useful tool for formalizing the idea of a vocabulary in a way that allows machine computation and understanding. If W is a word in the vocabulary associated with the variable U we can express W as a fuzzy subset W of the domain of U. Here then for any element $y \in Y$ the membership grade of y in W, W(y), indicates the compatibility of the value y with the linguistic value W. Thus the fuzzy subset W provides a machine comprehensible interpretation of the word W.

We are now in a position to bridge the gap in man-machine communication with respect to the analysis of social relational network by allowing the human to build a vocabulary of linguistic terms associated with an attribute and then provide a representation for these terms by fuzzy sets (see Fig. 2). Thus here now we have a communal vocabulary coherent to both the human and the machine.

What must be emphasized here is that the choice of the vocabulary as well as the associated meaning of the words in terms of fuzzy sets is completely in the hands of the human partner. This greatly simplifies this task. The vocabulary that will be used is imposed, we are giving the computer meaning. Particularly noteworthy is the fact that learning algorithms need not necessarily be required, the computer is told these are the terms I will be using and this is what they mean in your language, fuzzy

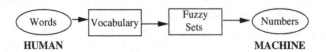

Fig. 2 Paradigm of man-machine understanding

Fig. 3 Fuzzy subset of term *strong*

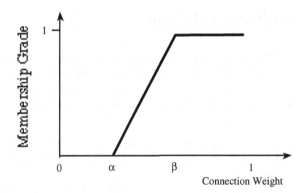

sets. This is not to say that the construction of the communal vocabulary is not an important and complex task to which future research must be dedicated so that it is thoughtfully and appropriately done but only to reflect the fact that it is within our power to impose what we have decided. This situation allows us in the following to assume the availability of a communal vocabulary associated with attributes of interest.

In analyzing weighted relational networks there are a number of attributes about which it will be useful to have a vocabulary of commonly accepted terms. One such attribute is strength of connection. This is an attribute whose domain is the unit interval, $I = [0, 1]$. Terms like *strong, weak, none* would be part of a vocabulary associated with this attribute. In this case we would define the word *strong* as a fuzzy subset S of $[0, 1]$ such that for any $y \in [0, 1]$ the value $S(y)$ would indicate the degree to which y satisfies the working definition of the concept *strong* connection. We would assume that S would be such that $S(0) = 0$, $S(1) = 1$ and S is monotonic, $S(y_1) \geq S(y_2)$ if $y_1 \geq y_2$. A prototypical example of the definition of the term *strong* would be the piecewise linear fuzzy subset shown in Fig. 3.

Another attribute for which it will be useful to have a communal vocabulary is the number of links in a path, path length. Some words in such a vocabulary would be *short* and *long*. In the case of this attribute we would provide a semantics for the words of the vocabulary, in terms of fuzzy subsets of the domain $H = \{0, 1, \ldots, n\}$ when n is the number of vertices in the network.

Concepts, in addition to actual physical objects, can provide attributes of interest. One such concept that we shall find useful are proportions. Here U is an attribute that takes its value in the set $I = [0, 1]$ where $r \in [0, 1]$ is a proportion. Examples of linguistic values that could be part of a communal vocabulary associated with this attribute are *many, most* and *about half*. As noted by Zadeh these terms provide a generalization of the quantifiers "all" and "none" that are used in logic. We can refer to these as linguistic quantifiers.

4 Clusters and Cliques

An important idea in classical graph theory is the concept of a cluster; here we want to extend this to weighted graphs. Let us first review the ideas from crisp graph theory. Let $<X, R>$ be a graph where R is a crisp relationship. We are implicitly assuming our graph is undirected so R is symmetric. One approach is to call a subset C of X a cluster of order k if

(a) For all node pairs x and y in C we have $R^k(x, y) = 1$
(b) For all nodes $z \notin C$ there is some $x \in C$ such that $R^k(x, z) = 0$.

Note: When $k = 1$ we call C a clique.
Note: The order k of the cluster is reflected in the term R^k.

In [17] Rosenfeld suggested extending these ideas to fuzzy graphs. Here we let $<X, R>$ be a fuzzy graph where R is a symmetric fuzzy relationship. A crisp subset $C \subset X$ is called a fuzzy cluster of order k if

$$\text{Min}_{x,y \in C}[R^k(x, y)] > \text{Sup}_{z \notin C}[\text{Inf}_{w \in C} R^k(w, z)].$$

In the following we suggest an alternative, more human meaningful, definition of a clique and then using the paradigm of computing with words we can provide a procedure for evaluating how well a subset of nodes satisfies our definition.

Let A be a subset of elements from X. We shall define A to be a clique if

C1: All elements in A are connected by a **short strong** path.
C2: No element not in A is connected to an element in A by a **strong** path.

Here then we have two criteria that need to be satisfied for a subset $A \subset X$ to be considered as a clique, C_1 and C_2. If we let $C_1(A) \in [0, 1]$ be the degree to which it is true that A satisfies C_1 and $C_2(A) \in [0.1]$ be the degree to which it is true that A satisfies C_2 then $C(A) = \text{Min}[C_1(A), C_2(A)]$ is the degree to which it is true that A is a clique.

We must now formulate the two criteria in terms of features from the network $<X, R>$. Here we shall make use of the communal vocabularies for the attributes strength of connection and path length. We shall assume the availability of the word **strong** for strength of connection and **short** for length of path in the vocabulary, that is we have expressions for the meaning, semantics, of these words as fuzzy subsets.

We first focus on C_1. Here we will make use of the term "strong connection" which we will represent as a fuzzy subset S of the unit interval I. In addition we need use the linguistic term "short path" which we represent as a fuzzy subset SH of the space N.

Assume x_i and x_j are two nodes in the proposed clique A. Here $R^k(x_i, x_j)$ indicates the strength of connection of x_i and x_j for a path of at most k links. For any $R^k(x_i, x_j)$ the value $S(R^k(x_i, x_j))$ is the degree to which $R^k(x_i, x_j)$ is a strong connection. We recall that $R^k(x_i, x_j)$ is monotonic in k, that is if $k_2 > k_1$ then $R^{k_2}(x_i, x_j) \geq R^{k_1}(x_i, x_j)$. In addition as we noted earlier S, strong connection, is also monotonic

Fig. 4 Fuzzy set representation of short path

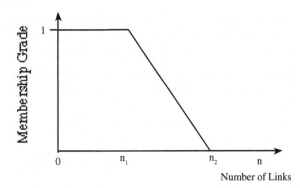

in strength, $S(a) \geq S(b)$ if $a > b$. From this we can conclude that for $k_2 > k_1$ then $S(R^{k_2}(x_i, x_j)) \geq S(R^{k_1}(x_i, x_j))$.

The concept short path with respect to the number of links it contains can be defined as a fuzzy subset SH of $N = \{1, 2, \ldots, n\}$. Here we can observe that generically SH should have the following properties; $SH(1) = 1$, $SH(n) = 0$ and $SH(k_1) \geq SH(k_2)$ if $k_1 < k_2$. Here it is monotonic decreasing in k, increasing k leads to decrease the satisfaction. A prototypical example of this concept is the piecewise linear fuzzy subset as shown in Fig. 4.

Using these ideas we can determine the degree to which there exists a short-strong connection between x_i and x_j. We define this using a form of the Sugeno integral [23] as

$$C_1(x_i, x_j) = \max_{k=1 \text{ to } n} [SH(k) \wedge S(R^k(x_i, x_j))]$$

In the above $\wedge$ is the Min operator. We see that as k increases $SH(k)$ tends to get smaller while $S(R^k(x_i, x_j))$ tends to get bigger.

We now can use this to determine the degree to which all elements in A are connected by a short-strong path,

$$C_1(A) = \min_{\substack{x_i, x_j \in A \\ x_i \neq x_j}} [C_1(x_i, x_j)].$$

It is very important to observe the marriage of different types of set objects used in making up the definition of $C_1(x_i, x_j)$. We first see that we have used $R^k(x_i, x_j)$, which is essentially the set–based definition of the network. In addition we have used the fuzzy sets SH and S, which are the fuzzy set definitions of the words short and strong. Here we have taken advantage of the fact that both the basic representation of a graph (network) and the meaning of words can be formulated using a set-based formulation to mix language and network representation. In [8] we referred to this approach as PISNA, the **P**aradigm for **I**ntelligent **S**ocial **N**etwork **A**nalysis.

We now consider the second criteria, C_2, no element not in A has a strong connection with an element in A. In the following we shall let $x \in A$ and $z \notin A$. For these elements $R^n(x, z)$ is the strength of the strongest path between x and z of any length. Here then $S(R^n(x, z))$ is the degree to which there exists a strong path between x and z. If we calculate $\underset{x \in A, z \notin A}{\text{Max}} [S(R^n(x, z))]$ we obtain the degree to which there exists a strong path between an element in A and an element not in A. Using this can calculate $C_2(A)$ as

$$C_2(A) = 1 - \underset{x \in A, z \notin A}{\text{Max}} [S(R^n(x, z))]$$

We now can use these formulations to determine the degree to which a subset A is a clique.

In the following we will make some observations regarding the preceding approach to defining the concept of clique.

In the preceding we defined the concept of short path in an absolute way as a fuzzy subset SH of the set of number of elements in a graph. It is possible to express short in a more universal way as a subset of the proportion of the number of elements in the graph. Thus here we can define "short path" as a fuzzy subset SH_p defined on the unit interval where for any $r \in [0, 1]$ the value $SH_p(r)$ indicates the degree to which the proportion r of elements in the network constitute a "short path." Thus here we are defining "short path" as a proportion of number of vertices in the network. This allows us to have a universal definition of the concept of "short path" independent of the number of nodes in the network. In this case where we have SH_p we calculate

$$C_1(x_i, x_j) = \underset{k=1 \text{ to } n}{\text{Max}} \left[SH_p \left(\frac{k}{n} \right) \wedge S(R^k(x_i, x_j)) \right]$$

We note that SH_p will have the same form as SH, $SH_p(r)$ decreases as r increases.

We also note that it is possible to somewhat relax the second criteria. Instead of having

C_2 = no element, not in A, is connected to an element in A by a strong path.

We can say that

C_2 = no element not in A is connected to an element in A by a strong path that is not long.

Here we need obtain the word **Long** from our communal vocabulary. A typical example of a fuzzy subset representing such a definition is shown in Fig. 5.

If L is the fuzzy subset defined on N corresponding to Long then not Long is a fuzzy subset N.L defined such that $N.L(k) = 1 - L(k)$. Using this we obtain

$$\underset{k=1 \text{ to } n}{\text{Max}} [N.L(k) \wedge S(R^k(x, z))]$$

as the degree to which there is a strong and not long path between x and z. Finally from this we obtain

$$C_2(A) = 1 - \underset{x \in A, z \notin A}{\text{Max}} \left(\underset{k=1 \text{ to } n}{\text{Max}} \left[N.L(k) \wedge S \left(R^k(x, z) \right) \right] \right)$$

Fig. 5 Fuzzy set representing term Long

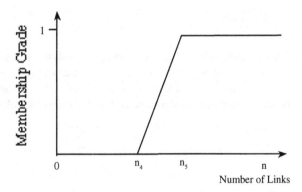

We now consider the situation in what we allow the set A to be a fuzzy set. Thus we want to determine if A is a fuzzy clique. We shall consider the same two original conditions C_1 and C_2 as defining a clique. We shall use $\tilde{A}$ to indicate our fuzzy Clique.

Here we first look at C_1. In this case we still get for any two nodes x and y $\in$ X.

$$C_1(x, y) = \underset{k=1 \text{ to } n}{\text{Max}} [SH(k) \wedge S(R^k(x, y))]$$

as the degree to which there is a short and strong path between the nodes x and y. Using this we obtain

$$C_1(\tilde{A}) = \underset{x \neq y \in X}{\text{Min}} [(1 - \tilde{A}(x)) \vee (1 - \tilde{A}(y)) \vee C_1(x, y))]$$

We note that in the case where $\tilde{A}$ is crisp this reduces to our original definition of C_1. For if x or y are in $\tilde{A}$ then the disjunction reduces to for $C_1(x, y)$ while if either x or y is not in $\tilde{A}$ then the argument becomes 1.

In this case of C_2, for a fuzzy clique we get

$$C_2(\tilde{A}) = 1 - \underset{\substack{x, z \in X \\ x \neq z}}{\text{Max}} [\tilde{A}(x) \wedge (1 - \tilde{A}(z)) \wedge S(R^n(x, z))]$$

We note that if $\tilde{A}$ is crisp this reduces to our previous definition. For if $x \in \tilde{A}$ and if $z \notin \tilde{A}$ then it becomes the original format. If either $x \notin \tilde{A}$ or $z \in \tilde{A}$ then $\tilde{A}(x) \wedge (1 - \tilde{A}(z)) \wedge S(R^n(x, z)) = 0$ and it doesn't effect the calculation of $C_2(\tilde{A})$.

5 Centrality

An important concept in social network analysis is centrality [1, 2]. The centrality of a node is closely related to its importance in the network. Assume $<X, R>$ is a relational network where R is a crisp relation. The measure of centrality of node x_i is the number of nodes connected to it by at most k links. In this case

$$C^k(x_i) = \sum_{\substack{j=1 \\ j \neq i}}^{n} R^k(x_i, x_j)$$

is the measure of the centrality of node x_i.

If we have a network where R is a weighted graph a straightforward way to extend this is to calculate $C^k(x_i)$ as in the above. One problem that can arise here is that a large number of weak connections, small values of $R^k(x_i, x_j)$, can add up to appear as a strong connection. Here we can suggest some other alternative methods for obtaining this measure of centrality.

One method is to use the level set representation and obtain a fuzzy set representation for the centrality. Here we can express

$$C_\alpha^k(x_i) = \sum_{\substack{j=1 \\ j \neq i}}^{n} R_\alpha^k(x_i, x_j)$$

Thus here $C_\alpha^k(x_i)$ is the number of nodes connected to x_i with strength of at least α using at most k links. Using this we can define

$$\tilde{C}^k(x_i) = \bigvee_{\alpha \in [0,1]} \left\{ \frac{\alpha}{C_\alpha^k(x_i)} \right\},$$

here we are using the standard fuzzy notation indicating that α is the membership grade of $C_\alpha^k(x_i)$.

In this case $\tilde{C}^k(x_i)$ is a fuzzy number. We should note that since for $\alpha_2 > \alpha_1$ we have $R_{\alpha_2}^k \subseteq R_{\alpha_1}^k$ then $C_{\alpha_1}^k C_{\alpha_2}^k$.

Let us consider another way to extend the concept of centrality to the case of a weighted graph. An alternative and perhaps more appropriate definition of centrality would be the "number of **strong** connections using at most k links." Here we shall define the concept "strong" as a fuzzy subset of unit interval. For example see Fig. 3. Using this definition we can obtain

$$\tilde{C}^k(x_i) = \sum_{\substack{j=1 \\ j \neq i}}^{n} \text{Strong } R^k(x_i, x_j)$$

Thus here we transform the scores via the concept strong.

6 Social Network Databases

In the following we shall consider a weighted network $<X, R>$ where each of the nodes has an associated vector of attribute (feature) values. In these types of networks each of the node objects have various attributes, properties and features. This structure can be viewed as the combination of a network and database.

In these networks we have a collection of q attributes $U_1, \ldots, U_q$. In the case where the nodes are people, examples of attributes could be nationality, age or income. Each of the attributes takes a value from a domain Y_i. In this situation, each node has an associated q vector whose ith component is the value of the ith attribute for that node.

We shall use the notation $U_i(x_j)$ to indicate the variable corresponding to the attribute U_i in the case of node x_j. For example with U_i being the attribute age then $U_i(x_j)$ would indicate the variable age of x_j. If U_i is the attribute country of birth then Y_i would be a list of countries and $U_i(x_j)$ would be the variable corresponding the country of birth of x_j. We shall let v_{ij} indicate the value of the variable $U_i(x_j)$, thus $U_i(x_j) = v_{ij}$. Thus in this case any node in our network has an associated vector V_j whose i component v_{ij} corresponds to the value of attribute U_i for node x_j. We should observe the above network could in some ways be viewed as a kind of database.

In the following we shall begin to describe techniques that can be used to analyze, investigate and question networks with vector-valued nodes. Here we shall be using flexible/fuzzy-querying techniques [9, 24].

In the following we shall assume that country of residence is one of the attributes, we shall denote this as U_1. Thus $U_1(x_j)$ is the variable denoting the country of residence of x_j. In this case the domain Y_1 associated with U_1 is the set of countries. In addition, a communal vocabulary associated with this attribute would consist of terms such as "Middle East", "North America", "South America" and "Southeast Asia". Other terms such as "mountainous country", "Spanish speaking", and "Oil producing" can also be part of the vocabulary. Each of the terms in the vocabulary would be defined in terms of subsets of Y_1. Some of these terms can be defined using crisp subsets while others will require fuzzy subsets.

In addition we shall assume age is another of the attributes associated with the network nodes. We shall denote this as U_2 with its domain Y_2 being the set of non-negative integers. Here we shall also assume the availability of mutually understandable vocabulary of commonly used linguistic terms to describe a person's age. These terms will be defined in terms of subsets of Y_2.

Assume x_j is some node in our network. We can ask "To what extent is x_j strongly connected to a person residing in South America?" In the following we shall let SA indicate the subset of Y_1 corresponding to South America. Using this we can obtain as the answer to our question

$$\underset{i, i \neq j}{\text{Max}}[SA(U_1(X_i)) \wedge R^n(x_i, x_j)]$$

More specifically we can ask "To what extent is x_j strongly connected to a young person residing in South America?" In this case with U_2 being the attribute age and Young being the subset corresponding concept young person we get as the answer

$$\underset{i,i\neq j}{\mathrm{Max}}[SA(U_1(X_i)) \wedge \mathrm{Young}(U_2(X_i)) \wedge R^n(x_i, x_j)]$$

We note this value is also the truth of the question "Does x_j have a connection to a Young South American".

We note that if we want to find out "does x_j have a *strong* connection to a Young South American" then we obtain the truth of this as

$$\underset{i,i\neq j}{\mathrm{Max}}[SA(U_1(X_i)) \wedge \mathrm{Young}(U_2(X_i)) \wedge \mathrm{Strong}(R^n(x_i, x_j)))]$$

Here we have replaced the predicate $R^n(x_i, x_j)$ with Strong $(R^n(x_i, x_j))$

A related question is the following. Let B be some crisp subset of X. We now ask what is the strongest connection between an element in B and a Young South American not in B. The answer is then obtained from the following

$$\underset{X_i \in B}{\mathrm{Max}} \left[\underset{x_j}{\mathrm{Max}}[SA(U_1(X_i)) \wedge \mathrm{Young}(U_2(X_j)) \wedge R^n(x_i, x_j) \wedge \overline{B}(X_j)] \right]$$

If in the above we are interested in only direct connections rather than any connection we replace R^n by R.

We now consider the question: Do all people residing in South America have a strong connection with each other? We shall denote the truth of this question Tr[Q] We calculate this truth-value as

$$\mathrm{Tr}(Q) = \underset{x_i, x_j \in X}{\mathbf{Min}} [(1 - SA(U_1(x_i)) \vee (1 - SA(U_1(x_j)) \vee \mathrm{Strong}(R^n(x_i, x_j))]$$

Let us look at this for the special case where SA is a crisp set. We first see that in the case of a pair (x_i, x_j) in which at least one of the elements do not reside in South America then either $SA(U_1(x_i)) = 0 \text{ or } SA(U_1(x_j)) = 0$ and therefore

$$(1 - SA(U_1(x_i)) \vee (1 - SA(U_1(x_j)) \vee \mathrm{Strong}(R^n(x_i, x_j)) = 1.$$

This case will not be the min. For the case in which both x_i and x_j reside in South America then $SA(U_1(x_i)) = SA(U_1(x_j)) = 1$ and hence

$$(1 - SA(U_1(x_i)) \vee (1 - SA(U_1(x_j)) \vee \mathrm{Strong}(R^n(x_i, x_j)) = \mathrm{Strong}(R^n(x_i, x_j))$$

From this we get as expected $\mathrm{Tr}(Q) = \underset{x_i x_j \in SA}{\mathrm{Min}} [\mathrm{Strong}(R^n(x_i, x_j))]$

Now we shall consider the slightly more complicated question of whether "*most* of the people residing in *western* countries have *strong* connections with each other?"

We shall here assume the term *most* is available in our common vocabulary where it is defined as a fuzzy subset over the unit interval. In addition we shall assume that the concept western country, is a concept that is defined by the fuzzy subset W over the domain Y_1. In this case for each $x_i \in X$ we have $W(U_1(x_i))$ indicates the degree to which it is true that x_i is from a western country. In the following we shall set P to be the set of all unordered pairs of distinct elements from X. P is the set of all the subsets of X consisting of two elements. We see that the number of elements in P, $n_P = \frac{(n)(n-1)}{2}$ where n is the number of elements in X. We shall denote an element $\{x_i, x_j\}$ in P as p_k. Here then k goes from 1 to n_P.

For each pair $p_k = \{x_i, x_j\}$ we obtain two values. The first $V_k = Min(W(U_1(x_i)), W(U_1(x_j)))$ indicates the degree to which p_k consists of pair of elements both from a western country. The second value is $S_k = Strong(R^n(x_i, x_j))$, indicates the degree to which there is a strong connection between the pair $\{x_i, x_j\}$. We shall use the technology of OWA operators to answer our question [25]. We proceed to obtain the answer as follows:

(1) Order the S_k and let ind(j) be the index of the *j*th largest of the S_k. Thus here $S_{ind(j)}$ is the *j*th largest of the S_k and $V_{ind(j)}$ is its associated V value.
(2) We next calculate

$$R = \sum_{j=1}^{n_P} V_{ind}(j)$$

(3) We next obtain a set of weight w_j for $j = 1$ to n_P where

$$w_j = Most(\frac{R_j}{R}) - Most(\frac{R_j - 1}{R})$$

here $R_j = \sum_{i=1}^{j} V_{ind(i)}$
(4) We finally calculate the truth of the question as

$$Tr(q) = \sum_{j=1}^{n_P} w_j S_{ind(j)}$$

An interesting special case of the preceding occurs if the subset W, Western Country, is a crisp set. In this case $V_k = Min(W(U_1(x_i)), W(U_1(x_j)))$ is a binary value, either one or zero. It is one if both x_i and x_j are from western countries and zero of either is not from a western country.

Another question we can ask is whether the young people form a clique. Since the young people provide a fuzzy subset over the space X and we have previously indicated a process for determining whether a fuzzy set is a clique we can answer this question.

7 Conclusion

We discussed the idea of fuzzy relationships and their role in modeling weighted social relational networks. The paradigm of computing with words was introduced and the role that fuzzy sets play in representing linguistic concepts was described. We discussed how these technologies can provide a bridge between a network analyst's linguistic description of social network concepts and the formal model of the network. Some examples of taking an analyst's network concepts and formally representing them in terms of network properties were provided. We applied this to the concept of clique and then to the idea of node centrality. Finally we introduced the idea of vector–valued nodes and began developing a technology of social network database theory. Clearly this newly introduced idea of social network database theory will provide many applications and will need a more formal mathematical framework, a task for future research.

References

1. Wasserman, S., Faust, K.: Social Network Analysis: Methods and Applications. Cambridge University Press, New York (1994)
2. Scott, J.: Social Network Analysis. SAGE Publishers, Los Angeles (2000)
3. Newman, M.: Networks: An Introduction. Oxford University Press, New York (2010)
4. Russell, M.A.: Mining the Social Web: Analyzing Data from Facebook, Twitter, LinkedIn, and Other Social Media Sites. O'Reilly Media, Schasptopol, CA (2011)
5. Aggarwal, C.C.: Social Network Data Analytics. Springer, New York (2011)
6. Prell, C.: Social Network Analysis: History, Theory and Methodology. Sage Publisher, London (2012)
7. Kadushin, C.: Understanding Social Networks: Theories, Concepts and Findings. Oxford University, New York (2012)
8. Yager, R.R.: Intelligent social network analysis using granular computing. Int. J. Intell. Syst. **23**, 1196–1219 (2008)
9. Pivert, O., Bosc, P.: Fuzzy Preference Queries to Relational Databases. World Scientific, Singapore (2012)
10. Zadeh, L.A.: Fuzzy logic = computing with words. IEEE Trans. Fuzzy Syst. **4**, 103–111 (1996)
11. Zadeh, L.A.: From computing with numbers to computing with words—from manipulation of measurements to manipulations of perceptions. IEEE Trans. Circuits Syst. **45**, 105–119 (1999)
12. Zadeh, L.A.: Outline of a computational theory of perceptions based on computing with words. In: Sinha, N.K., Gupta, M.M. (eds.) Soft Computing and Intelligent Systems, pp. 3–22. Academic Press, Boston (1999)
13. Zadeh, L.A.: Generalized theory of uncertainty (GTU)-principal concepts and ideas. Comput. Stat. Data Anal. **51**, 15–46 (2006)
14. Zadeh, L.A.: Fuzzy logic. In: Meyers, A.R. (ed.) Encyclopedia of Complexity and Systems Science. Springer, Heidelberg (2009)
15. Bargiela, A., Pedrycz, W.: Granular Computing: An Introduction. Kluwer Academic Publishers, Amsterdam (2003)
16. Yager, R.R.: Human behavioral modeling using fuzzy and Dempster-Shafer theory. In: Liu, H., Salerno, J.J., Young, M.J. (eds.) Social Computing, Behavioral Modeling and Prediction, pp. 89–99. Springer, Berlin (2008)

17. Rosenfeld, A.: Fuzzy graphs. In: Zadeh, L.A., Fu, K.S., Tanaka, K., Shimura, M. (eds.) Fuzzy Sets and their Applications to Cognitive and Decision Processes, pp. 77–97. Academic Press, New York (1975)
18. Delgado, M., Verdegay, J.L., Vila, M.A.: On the valuation and optimization problems in fuzzy graphs (a general approach and some particuar cases). ORSA J. Comput. **2**, 74–83 (1990)
19. Koczy, L.T.: Fuzzy graph in the evaluation and optimization of networks. Fuzzy Sets Syst. **46**, 307–319 (1992)
20. Yager, R.R.: Level sets and the extension principle for interval valued fuzzy sets and its application to uncertainty. Inf. Sci. **178**, 3565–3576 (2008)
21. Zadeh, L.A.: Similarity relations and fuzzy orderings. Inf. Sci. **3**, 177–200 (1971)
22. Zadeh, L.: The concept of a linguistic variable and its application to approximate reasoning: Part 1. Inf. Sci. **8**, 199–249 (1975)
23. Murofushi, T., Sugeno, M.: Fuzzy measures and fuzzy integrals. In: Grabisch, M., Murofushi, T., Sugeno, M. (eds.) Fuzzy Measures and Integrals, pp. 3–41. Physica-Verlag, Heidelberg (2000)
24. Zadrozny, S., de Tré, G., de Caluwe, R., Kacprzyk, J.: An overview of fuzzy approaches to database querying. In: Galindo, J. (ed.) Handbook of Research on Fuzzy Information Processing in Databases, vol. 1, pp. 34–54. Information Science Reference, Hershey, PA (2008)
25. Yager, R.R.: Quantifier guided aggregation using OWA operators. Int. J. Intell. Syst. **11**, 49–73 (1996)

Part V
Fuzzy Knowledge Discovery and Exploitation

Chapter 12
Fuzzy Cardinalities as a Basis
to Cooperative Answering

Grégory Smits, Olivier Pivert and Allel Hadjali

Abstract Cooperative approaches to relational database querying help users retrieve the tuples that are the most relevant with respect to their information needs. In this chapter we propose a unified framework that relies on a fuzzy cardinality-based summary of the database. We show how this summary can be efficiently used to explain failing queries or to revise queries returning a plethoric answer set.

1 Introduction

The paradigm of cooperative answering is originated from the works in the context of natural-language question-answering done by Kaplan [20] in the end of the seventies. One of the aims of such works is to avoid natural-language query systems to produce "there is zero result" when a query fails. Cooperative intelligent systems should rather correct any false presupposition of the user, anticipate follow-up queries and provide information not explicitly requested by the user.

Cooperative responses to a query are indirect responses that are more helpful to the user than direct, literal responses would be. Interest in cooperative responses in the database field arises in the middle of the eighties [13, 14, 17, 25, 31]. In this context, cooperative answering represents intensional, qualified or approximate answers. They may explain the failure of a query to produce results, build queries that are related to the original one and re-submit them for an evaluation. Most cooperative

G. Smits (✉)
IRISA-IUT, Lannion, France
e-mail: gregory.smits@univ-rennes1.fr

O. Pivert · A. Hadjali
IRISA-ENSSAT, Lannion, France
e-mail: pivert@enssat.fr

A. Hadjali
e-mail: hadjali@enssat.fr

O. Pivert and S. Zadrożny (eds.), *Flexible Approaches in Data, Information and Knowledge Management*, Studies in Computational Intelligence 497, DOI: 10.1007/978-3-319-00954-4_12, © Springer International Publishing Switzerland 2014

techniques proposed in the literature deal with the empty answer problem in a crisp query setting.

In this chapter, we consider fuzzy queries which express preferences modeled using fuzzy set membership functions (that describe the preference profiles of the user on each attribute domain involved in the query). We address two problematic situations users can be faced with when querying relational databases: their query returns (i) an empty set of answers or, (ii) a plethoric answer set. We propose a uniform solution to these two symmetrical problematic situations. This solution relies on the precomputation of a summary of the queried database according to a predefined shared vocabulary. This summary provides information about the distribution of the data over the definition domains of the different target attributes. This summarization strategy efficiently computes fuzzy cardinalities using a single scan of the database.

Recall that with respect to Boolean queries, fuzzy queries reduce the risk of obtaining an empty set of answers since the use of a finer discrimination scale—[0, 1] instead of {0, 1}—increases the chance for an element to be considered some-what satisfactory. Nevertheless, the situation may occur where none of the elements of the target database satisfies the query even to a low degree.

In the context of fuzzy queries, beside the empty answer set (EAS) problem, another situation deserves attention: that where the answer set is not empty but only contains elements which satisfy to a *low degree* the preferences specified in the user query. We show in this chapter that a generic—and very efficient—type of approach that leverages fuzzy cardinalities may be employed to provide explanations for both types of situations (empty or unsatisfactory answer set). Minimal failing subqueries [24] constitute useful explanations about the conflicts in a failing query. These explanations may (i) help the user revise or reformulate his/her initial query or (ii) be used to set up an automatic and targeted relaxation strategy.

As for the symmetrical problem, i.e. the plethoric answer set (PAS) problem, it has been intensively addressed by the information retrieval community and two main approaches have been proposed for Boolean queries. The first one, that may be called data-oriented, aims at ranking the answers in order to return the best k ones to the user. However, this strategy is often faced with the difficulty of comparing and distinguishing among tuples that satisfy the initial query. In this data-oriented type of approach, we can also mention works which aim at summarizing the answer set to a query [36].

The second type of approach may be called query-oriented as it performs a mod-ification of the initial query in order to make it more selective. For instance, a strat-egy consists in strengthening the specified predicates (as an example, a predicate $A \in [a_1, a_2]$ becomes $A \in [a_1 + \gamma, a_2 - \gamma]$) [6]. However, for some predicates, this strengthening (if applied in an interated way) can lead to a deep modification of the meaning of the initial predicate. Another type of approach advocates the use of user-defined preferences on attributes which are not involved in the initial query [3, 12, 21]. Such a subjective knowledge can then be used to select the most preferred items among the initial answer set. Still another category of query-oriented approaches [26, 27] aims at automatically completing the initial query with additional predicates to make it more demanding. Our work belongs to this last family of approaches but its specificity concerns the way additional predicates are selected.

Indeed, we consider that the predicates added to the query must respect two properties: (i) they must reduce the size of the initial answer set, (ii) they must modify the semantic scope of the initial query as little as possible. Based on a predefined vocabulary materialized by fuzzy partitions that linguistically describes the attribute domains, we propose to identify the predicates which are the most correlated to the initial query. Moreover, we consider that the queries involve a user-specified quantitative threshold k corresponding to the approximate number of expected results (the best ones). To assist the user through the reduction of a plethoric answer set to a subset containing approximately k results, we again propose to make use of precomputed fuzzy cardinalities that constitute useful knowledge about the data distributions.

The remainder of the chapter is structured as follows. Section 2 provides a concise reminder about fuzzy sets and fuzzy queries. In Sect. 3, we present the context of our work and especially the fuzzy cardinality-based summarization process. Sections 4 and 5 respectively deal with the two symmetrical problems, i.e. the PAS problem and the explanation of failing queries. We address these issues using a uniform framework based on the notion of fuzzy cardinalities. Experimental results are presented and analyzed in Sect. 6. Section 7 discusses related work, whereas Sect. 8 recalls the main contributions and outlines perspectives for future work.

2 Preliminaries

2.1 Fuzzy Sets

Fuzzy set theory was introduced by Zadeh [22] for modeling classes or sets whose boundaries are not clear-cut. For such objects, the transition between full membership and full mismatch is gradual rather than crisp. Typical examples of such fuzzy classes are those described using adjectives of the natural language, such as *young, cheap, fast*, etc. Formally, a fuzzy set F on a referential U is characterized by a membership function $\mu_F : U \to [0, 1]$ where $\mu_F(u)$ denotes the grade of membership of u in F. In particular, $\mu_F(u) = 1$ reflects full membership of u in F, while $\mu_F(u) = 0$ expresses absolute non-membership. When $0 < \mu_F(u) < 1$, one speaks of partial membership.

Two crisp sets are of particular interest when defining a fuzzy set F:

- the core $C(F) = \{u \in U \mid \mu_F(u) = 1\}$, which gathers the *prototypes* of F,
- the support $S(F) = \{u \in U \mid \mu_F(u) > 0\}$.

The notion of an α-cut encompasses both these concepts. The α-cut (resp. strict α-cut) F_α (resp. $F_{\overline{\alpha}}$) of a fuzzy set F is defined as the set of elements from the referential which have a degree of membership to F at least equal to (resp. strictly greater than) α:

Fig. 1 Trapezoidal membership function

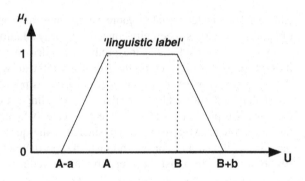

$$F_\alpha = \{u \in U \mid \mu_F(u) \geq \alpha\}$$

$$F_{\overline{\alpha}} = \{u \in U \mid \mu_F(u) > \alpha\}.$$

Straightforwardly, one has: $C(F) = F_1$ and $S(F) = F_{\overline{0}}$.

In practice, the membership function associated with F is often of a trapezoidal shape. Then, F is expressed by the quadruplet (A, B, a, b) where $C(F) = [A, B]$ and $S(F) = [A - a, B + b]$, see Fig. 1.

Let F and G be two fuzzy sets on the universe U, we say that $F \subseteq G$ iff $\mu_F(u) \leq \mu_G(u)$, $\forall u \in U$. The complement of F, denoted by F^c, is defined by $\mu_{F^c}(u) = 1 - \mu_F(u)$. Furthermore, $F \cap G$ (resp. $F \cup G$) is defined the following way: $\mu_{F \cap G}(u) = min(\mu_F(u), \mu_G(u))$ (resp. $\mu_{F \cup G}(u) = max(\mu_F(u), \mu_G(u))$).

As usual, the logical counterparts of the theoretical set operators $\cap$, $\cup$ and the complementation operator correspond respectively to the conjunction $\wedge$, disjunction $\vee$ and negation $\neg$. See [16] for more details.

2.2 Fuzzy Queries and SQLf

Fuzzy sets are convenient tools to model vague criteria and user's preferences. The underlying fuzzy set theory offers a large panoply of connectives to aggregate these preferences following different semantics. Fuzzy sets are used to model and represent common sense properties like 'recent', 'low', 'very cheap', 'large', ..., that correspond to familiar and easily understandable notions for end users. Moreover, in accordance with the imprecise nature of the concepts they represent, the fuzzy sets behind these properties introduce some graduality when checking the satisfaction of the items wrt. the user's preferences. This gradual satisfaction provides the necessary information to rank order the items that somewhat satisfy the user's requirements.

The language called SQLf described in [8, 29] extends SQL so as to support fuzzy queries. The general principle consists in introducing gradual predicates wherever it makes sense. The three clauses *select, from* and *where* of the base block of SQL are kept in SQLf and the *from* clause remains unchanged. The principal differences affect mainly two aspects :

- the calibration of the result since it is made with discriminated elements, which can be achieved through a number of desired answers (k), a minimal level of satisfaction (α), or both, and
- the nature of the authorized conditions as mentioned previously.

Therefore, the base block is expressed as:
select [**distinct**] [$k \mid \alpha \mid k, \alpha$] *attributes*
from *relations*
where *fuzzy-condition*
where *fuzzy-condition* may involve both Boolean and fuzzy predicates. This expression is interpreted as:

- the fuzzy selection of the Cartesian product of the relations appearing in the *from* clause,
- a projection over the attributes of the *select* clause (duplicates are kept by default, and if *distinct* is specified the maximal degree is attached to the representative in the result),
- the calibration of the result (top k elements and/or those whose score is over the user-specified threshold α_u).

The operations from relational algebra—on which SQLf is based—are extended to fuzzy relations by considering fuzzy relations as fuzzy sets on the one hand and by introducing gradual predicates in the appropriate operations (selections and joins especially) on the other hand. The definitions of these extended relational operators can be found in [4]. As an illustration, we give the definitions of the fuzzy selection and join operators hereafter, where r and s denote two fuzzy relations defined on the sets of attributes X and Y.

- $\mu_{select(r, cond)}(t) = \top(\mu_r(t), \mu_{cond}(t))$ where *cond* is a fuzzy predicate and $\top$ is a triangular norm (most usually, *min* is used),
- $\mu_{join(r, s, A \theta B)}(tu) = \top(\mu_r(t), \mu_s(u), \mu_\theta(t.A, u.B))$ where A (resp. B) is a subset of X (resp. Y), A and B are defined over the same domains, and θ is a binary relational operator (possibly fuzzy).

A typical example of a fuzzy query is: "retrieve the recent and low-mileage cars", where *recent* and *low-mileage* are gradual predicates represented by means of fuzzy sets as illustrated in Fig. 2.

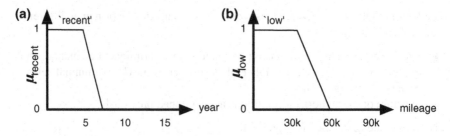

Fig. 2 Predicates: **a** recent and **b** low-mileage

3 Fuzzy-Cardinality-Based Database Summaries

3.1 A Fuzzy-Partition-Based Predefined Vocabulary

Fuzzy sets constitute an interesting framework for extracting knowledge on data that can be easily comprehensible by humans. Indeed, associated with a membership function and a linguistic label, a fuzzy set is a convenient way to formalize a gradual property. As noted in some previous works, especially in [27], such prior knowledge can be used to represent what the authors call a "macro expression of the database". Contrary to the approach presented in [27] where this knowledge is computed by means of a fuzzy classification process, it is, in our approach, defined *a priori* by means of a partition in the sense of Ruspini [33] of each attribute domain. These partitions form a predefined and shared vocabulary and it is assumed that the fuzzy sets involved in users' flexible queries are taken from this vocabulary.

Let R be a relation containing w tuples $\{t_1, t_2, \ldots, t_w\}$ defined on a set Z of q categorical or numerical attributes $\{Z_1, Z_2, \ldots, Z_q\}$. A shared predefined vocabulary on R is defined by means of fuzzy partitions of the q domains. A partition $\mathscr{P}_i$ associated with the domain of attribute Z_i is composed of m_i fuzzy predicates $\{P_{i,1}, P_{i,2}, \ldots, P_{i,m_i}\}$, such that for all Z_i and for all $t \in R$:

$$\sum_{j=1}^{m_i} \mu_{P_{ij}}(t) = 1.$$

As mentioned above, we consider Ruspini partitions for numerical attributes (Fig. 3), i.e., fuzzy partitions composed of fuzzy sets, where a set, say P_i, can only overlap with its predecessor P_{i-1} or/and its successor P_{i+1} (when they exist). For categorical attributes, we simply impose that for each tuple the sum of the satisfaction degrees on all elements of a partition is equal to 1. These partitions are specified by an expert during the database design step and represent "common sense partitions" of the domains. Each $\mathscr{P}_i$ is associated with a set of linguistic labels $\{L_{i,1}^p, L_{i,2}^p, \ldots, L_{i,m_i}^p\}$, each of them corresponding to an adjective which gives the meaning of the fuzzy

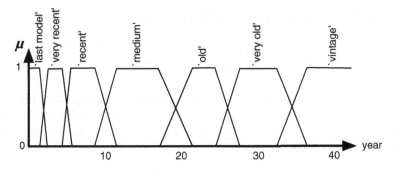

Fig. 3 A partition over the domain of attribute *year*

Table 1 A partition over the domain of attribute *make*

	dodge	jeep	...	honda	...	nissan	renault	peugeot	dacia	...	ARO	oltcit	...	vw	Lamborghini	Skoda	...
'American'	1	1	...	0	...	0	0	0	0	...	0	0	...	0	0	0	
'Asian'	0	0	...	1	...	0.6	0	0	0	...	0	0	...	0	0	0	...
...	...	...	...	...	...	...	...	...	...	...	...	...	...	...	...	...	...
'French'	0	0	...	0	...	0.4	1	1	0.4	...	0	0	...	0	0	0	...
'East-european'	0	0	...	0	...	0	0	0	0.6	...	1	1	...	0	0	0	...
'German'	0	0	...	0	...	0	0	0	0	...	0	0	...	1	0.5	0.6	...
...	...	...	...	...	...		...		...	...	...	...	...	...		...	...

predicate. A query Q to this relation R is a conjunction of fuzzy predicates chosen among the predefined ones which form the partitions.

As an example, let us consider a database containing ads about second hand cars and a view named *secondHandCars* of schema (*id*, *model*, *description*, *year*, *mileage*, *price*, *make*, *length*, *height*, *nbseats*, *consumption*, *acceleration*, *co2emission*) as the result of a join-query over the database. A common sense partition and labelling of the domain of attribute *year* is illustrated in Fig. 3. Table 1 shows a possible common sense partition and labelling of the domain of the categorical attribute *make*.

3.2 About Fuzzy Cardinalities and Their Computation

Hereafter, we describe a technique aimed at building fuzzy database summaries that can be helpful in a cooperative answering perspective.

In the context of flexible querying, fuzzy cardinalities appear to be a convenient formalism to represent how many tuples from a relation satisfy a fuzzy predicate to various degrees. We assume in the following that these various membership degrees are defined by a finite scale $1 = \sigma_1 > \sigma_2 > \cdots > \sigma_f > 0$. Such fuzzy cardinalities can be incrementally computed and maintained for each linguistic label and for the diverse conjunctive combinations of these labels. Fuzzy cardinalities are represented by means of a possibility distribution [15] like

$$F_{P^a} = 1/0 + \cdots 1/(n-1) + 1/n + \lambda_1/(n+1) + \cdots +$$
$$\lambda_k/(n+k) + 0/(n+k+1) + \cdots,$$

where $1 > \lambda_1 \geq \cdots \geq \lambda_k > \lambda_{k+1} = 0$ for a predicate P^a. This expression represents a cardinality that possibly equals at least n to degree 1 and possibly equals at least $(n+k)$ to degree λ_k. In this chapter, without loss of information, we use a more compact representation:

$$F_{P^a} = \sigma_1/c_1 + \sigma_2/c_2 + \cdots + \sigma_f/c_f,$$

where $c_i, i = 1..f$ is the number of tuples in the concerned relation that are P^a with a degree at least equal to σ_i. For the computation of cardinalities concerning a conjunction of q fuzzy predicates, like $F_{P^a \wedge P^b \wedge \cdots \wedge P^q}$, one takes into account the minimal satisfaction degree obtained by each tuple t for the concerned predicates, $min(\mu_{P^a}(t), \mu_{P^b}(t), \ldots, \mu_{P^q}(t))$.

As illustrated by Algorithm 1, the computation of the fuzzy cardinalities relies on a single scan of the database but for each tuple, one has to compute its satisfaction degree regarding every possible conjunction of the fuzzy predicates involved in the query. The number of all possible conjunctions to consider is equal to 2^q where q is the number of predicates in the query, but the computation has a linear data complexity and the process remains tractable as soon as q is reasonably small, which is the case in practice (in general, $q \leq 10$). Indeed, even though databases are getting larger and larger, the number of predicates involved in users queries remains stable around half a dozen of predicates. Section 6 illustrates this observation in a concrete applicative context.

The computation of fuzzy cardinalities relies on two steps. First a lattice is generated to store the fuzzy cardinalities according to all the possible conjunctions of predicates. Figure 4 shows the lattice generated for a set of three predicates $\{P^a, P^b, P^c\}$.

Then, for each tuple t from the concerned database $\mathcal{D}$, one computes its performance vector $\langle \mu_{P^a}(t), \mu_{P^b}(t), \ldots, \mu_{P^q}(t) \rangle$ that stores the satisfaction degrees of t

Fig. 4 Lattice of possible conjunctions for a set of three predicates $\{P^a, P^b, P^c\}$

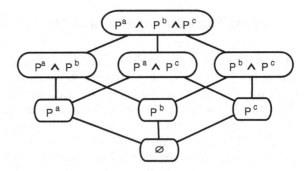

wrt. the atomic predicates $\{P^a, P^b, \ldots, P^q\}$ involved in the query. Using a depth or breadth first exploration of the lattice, one updates the fuzzy cardinalities according to the performance vector of the tuple. Thus, according to the currently analyzed predicate or conjunction of predicates P and a tuple t, if t satisfies P with a degree greater or equal than α_u then the function *updateCardinality* simply increments the stored fuzzy cardinality of P for each $\alpha \geq \alpha_u$. A t-norm, here the minimum, is used by the function *computeSatisDegree* to compute the satisfaction degree of a tuple according to a conjunction of predicates. This satisfaction degree is directly computed on the performance vector associated with the currently analyzed tuple (line 1.6). During the exploration of the lattice, for a tuple t, a path is discarded as soon as t does not satisfy at all the query, say Q, composed of the current conjunction of predicates since

$$\mu_Q(t) = 0 \Rightarrow \forall Q' \text{ such that } pred(Q) \subseteq pred(Q'), \mu_{Q'}(t) = 0$$

where $pred(Q)$ denotes the set of predicates involved in Q.

Input: a failing query $Q = P_a \wedge \ldots \wedge P_n$; a scale of degrees
$A = \alpha_f < \ldots < \alpha_2 < (\alpha_1 = 1)$; a user-defined qualitative
threshold α_u;
Output: L a lattice of fuzzy cardinalitites for Q;

1.1 **begin**
1.2 $\mathscr{R} \leftarrow execute(P_a \vee \ldots \vee P_n)$;
1.3 $L \leftarrow generateLattice(\{P_a, \ldots, P_n\})$;
1.4 $//L$ points to the entry node ($\emptyset$) of the lattice
1.5   **foreach** $t \in \mathscr{R}$ **do**
1.6 $\langle \mu_{Pa}(t), \mu_{Pb}(t), \ldots, \mu_{Pq}(t) \rangle \leftarrow computePerfVector(t, \{P_a, \ldots, P_n\})$;
1.7 $updateLattice(L, \langle \mu_{Pa}(t), \mu_{Pb}(t), \ldots, \mu_{Pq}(t) \rangle)$;
1.8   **end**
1.9 **end**

Algorithm 1: Fuzzy Cardinalities Computation

Two strategies can be envisaged to compute the fuzzy cardinalities: dynamically or *a priori*. The dynamic computation of fuzzy cardinalities allows for the use of user defined fuzzy predicates inside queries. However, to perform an efficient dynamic computation of the fuzzy cardinalities, it would be necessary to modify the optimizer of the DBMS so as to integrate this process in their execution plan.

Input: a lattice of fuzzy cardinalities L; a performance vector
$V = \langle \mu_{Pa}(t), \mu_{Pb}(t), ..., \mu_{Pq}(t) \rangle$; a scale of degrees
$A = \alpha_f < ... < \alpha_2 < (\alpha_1 = 1)$; a user-defined qualitative
threshold α_u;

2.1 **begin**
2.2 $N \leftarrow Parent(L)$;
2.3   **foreach** $node \in N$ **do**
2.4   let P be the predicate associated with the node N;
2.5 $\mu_{P(t)} = computeSatisDegree(V, P)$;
2.6   **if** $\mu_{P(t)} \geq \alpha_u$ **then**
2.7 $updateCardinality(N, \mu_{P(t)})$;
2.8 $updateLattice(N, V, A, \alpha_u)$;
2.9   **end**
2.10   **end**
2.11 **end**

Algorithm 2: Recursive function *updateLattice*

In the following, we assume that a shared vocabulary is *a priori* defined by means of fuzzy partitions over the domain of each searchable attribute. Thus, fuzzy cardinalities can be pre-computed for each possible conjunction of predicates taken from the shared vocabulary. More precisely, one computes the fuzzy cardinalities for all the possible conjunctions of predicates containing no more than one predicate of each attribute partition. Indeed, we consider that it does not make sense to explore conjunctions of predicates from the same attribute partition like '*year is young and year is old*'. Fuzzy cardinalities associated with conjunctions which are somewhat satisfied by at least one tuple are stored in a dedicated table of the database. This table can be easily maintained as the fuzzy cardinalities can be updated incrementally [5].

An index computed on the string representation of each conjunction makes it possible to efficiently access the different fuzzy cardinalities.

3.3 A Semantic Correlation Measure

In this subsection, we introduce a measure aimed at assessing the extent to which two fuzzy predicates are semantically correlated. This measure will be used in the approach presented in Sect. 5 as a basis to the augmentation of a query leading to a plethoric answer set.

Given two predicates P^a and P^b, an association rule denoted by $P^a \Rightarrow P^b$ expresses that tuples which are P^a are also P^b (P^a and P^b can be replaced by any conjunction of predicates). As suggested in [9], the confidence of such an association may be quantified by means of a scalar or by a fuzzy (relative) cardinality. The first representation (as a scalar) is used in our approach as it appears more convenient and easier to interpret. Thus, the confidence of an association rule $P^a \Rightarrow P^b$, denoted by $conf(P^a \Rightarrow P^b)$, is computed as follows:

$$conf(P^a \Rightarrow P^b) = \frac{\Gamma_{P^a \wedge P^b}}{\Gamma_{P^b}}. \tag{1}$$

Here, $\Gamma_{P^a \wedge P^b}$ and Γ_{P^a} correspond to scalar cardinalities, which are computed as the weighted sum of the elements belonging to the associated fuzzy cardinalities. For example, the scalar version of $\Gamma_{recent} = 1/6 + 0.6/7 + 0.2/8$ is $\Gamma_{recent} = 1 \times 6 + 0.6 \times (7 - 6) + 0.2 \times (8 - 7) = 6.8$.

To quantify the semantic link between a query Q and a predicate P, one computes a correlation degree denoted by $\mu_{cor}(P, \ Q)$, as:

$$\mu_{cor}(P, \ Q) = \top(conf(Q \Rightarrow P), \ conf(P \Rightarrow Q)) \tag{2}$$

where $\top$ stands for a t-norm and the minimum is used in our experimentation (Sect. 6.3). One can easily check that this correlation degree is both reflexive $\mu_{cor}(Q, \ Q) = 1$ and symmetric $\mu_{cor}(P, \ Q) = \mu_{cor}(Q, \ P)$.

4 Query Failure Explanation

In this section, we show how fuzzy cardinalities can be used inside a cooperative system to explain failing queries. We still assume that the queries are composed of fuzzy predicates chosen in a predefined vocabulary. Let us first recall the formal definition of the EAS problem.

Definition 1 Let Q be a fuzzy query and $\Sigma_Q = \{t \in D \mid \mu_Q(t) > 0\}$ the set of answers to Q against a given database D. We say that Q results in the EAS problem if $\Sigma_Q = \emptyset$.

4.1 About Minimal Failing and Unsatisfactory Subqueries

An empty set of answers associated with a fuzzy query $Q = P_1 \wedge P_2 \wedge \cdots \wedge P_n$ is necessarily due to an empty support (w.r.t. the current state of the database) for at least one of the *subqueries* of Q. The notion of an unsatisfactory set of answers generalizes this problem by considering an empty α-cut of Q where α is a user-defined qualitative threshold. As explained in Sect. 2.1, the support and the core of a fuzzy set are particular cases of α-cuts where α is respectively equal to 0^+ and 1. In the rest of the chapter we only use the notion of an empty α-cut to refer to failing queries as well as unsatisfactory ones.

Thus, an extreme case of a failing query corresponds to an empty 1-cut for Q only. The opposite extreme is when one or several predicates P_i have an empty 0^+-cut.

Between these two situations, it is of interest to detect the subqueries composed of more than one predicate and less than n predicates, which have an empty 0^+-cut. From an empty to an unsatisfactory set of answers, the problem defined above just has to be slightly revisited, where the condition of an empty 0^+-cut is transposed to α-cuts, where α is taken from a predefined scale of membership degrees $\mathcal{S} : 1 = \alpha_1 > \alpha_2 > \cdots > \alpha_f = 0^+$.

Definition 2 Let us consider a query $Q = P_1 \wedge P_2 \wedge \cdots \wedge P_n$, and let S and S' be two subsets of predicates such that $S' \subset S \subseteq \{P_1, P_2, \ldots, P_n\}$. A conjunction of elements from S (resp. S') is a *subquery* (resp. *strict subquery*) of Q.

If one wants to explain why the result of the initial query is empty (resp. unsatisfactory), and/or weaken the query by identifying the subqueries whose α-cut is empty, one must naturally require that such subqueries be minimal: a subquery Q' of a query Q constitutes a minimal explanation if the considered α-cut is empty and if no (strict) subquery of Q' has an empty α-cut. This corresponds to a generalization of the concept of a *Minimal Failing Subquery* (MFS) [18].

Let us denote by Σ_Q^α the set of answers to the α-cut of a query Q against a given database D: $\Sigma_Q^\alpha = \{t \in D \mid \mu_Q(t) \geq \alpha\}$.

Definition 3 A *Minimal Failing Subquery* of a query $Q = P_1 \wedge P_2 \wedge \cdots \wedge P_n$ for a given α is any subquery Q' of Q such that $\Sigma_{Q'}^\alpha = \emptyset$ and for all strict subquery Q'' of Q', $\Sigma_{Q''}^\alpha \neq \emptyset$.

When faced with an empty set of answers for a user-defined threshold α, the explanation process that we propose in this chapter generates layered MFSs for different satisfaction degrees $\alpha_i \in \mathcal{S}$, $\alpha_i \in [\alpha, 1]$.

Obviously, due to the monotonicity of inclusion of α-cuts, one has $\Sigma_Q^{\alpha_i} \subseteq \Sigma_Q^{\alpha_j}$ if $\alpha_i \geq \alpha_j$. Therefore, a query Q that fails for a given α_j also fails for higher satisfaction degrees $\alpha_i > \alpha_j$. However, this property is not satisfied by minimal failing subqueries. Indeed, a subquery Q' can be an MFS of Q for a given α_j without being minimal for higher satisfaction degrees $\alpha_i > \alpha_j$ as a strict subquery of Q', say Q'', may fail for α_i and not for α_j.

During the layered MFS detection step (Sect. 4.2), when a subquery Q' of an initial failing or unsatisfactory query Q is detected for a degree α_j, one has to check for each higher level $\alpha_i > \alpha_j$ if Q' is also minimal at the level α_i before considering Q' as an MFS for this level.

4.2 Cardinality-Based MFS Detection

Using the precomputed fuzzy cardinalities, one can detect the MFSs for different empty α-cuts of Q, starting from a user-defined qualitative threshold up to the highest satisfaction degree 1.

In the manner of Apriori [1], Algorithm 3 starts with atomic predicates and the first α_i-cut of interest, the one corresponding to the user-defined qualitative threshold α_u. To determine if an atomic predicate P_a is a failing subquery of Q, one just has to check the associated precomputed fuzzy cardinality. If no tuple satisfies P_a at least with the degree α_i then P_a, as an atomic predicate, is by definition an MFS of Q and is also an MFS for $\alpha_j > \alpha_i$. Then, for the second round of the loop (line 3.7 of Algorithm 3), conjunctions containing two non failing predicates are generated and for each of them (line 3.11) one checks the fuzzy cardinalities so as to determine if it is an MFS. If one of these conjunctions, say $P_a \wedge P_b \wedge P_c$, is an MFS for a degree α_i one tries to propagate it to higher satisfaction degrees (see Algorithm 4 where $isMFS(L, MFS_{\alpha_j}(Q))$ returns $true$ if $L \in MFS_{\alpha_j}(Q)$, $false$ otherwise). As the MFS property is not monotonic with respect to α-cuts, one checks with Algorithm 4 for each $\alpha_j > \alpha_i$ if a subquery of $P_a \wedge P_b \wedge P_c$ corresponds to a previously detected MFS for degree α_j; if it is not the case $P_a \wedge P_b \wedge P_c$ is stored as an MFS of Q for α_j. Obviously, an atomic failing query is an MFS for all α-cuts. Then, the algorithm goes back to the main loop (line 3.7) and conjunctions containing three predicates are generated for each considered satisfaction degree (line 3.8) taking care that these conjunctions do not contain an already identified MFS. This recursive process goes on until candidate conjunctions cannot be generated anymore.

The complexity of this algorithm is obviously exponential in the number of predicates involved in the failing query to explain, where the worst case corresponds to a single MFS Q for the maximal satisfaction degree of 1. In this case, the *foreach* loop (line 3.11) makes 2^n iterations where n is the number of predicates in Q. For a complete gradual explanation from $\alpha = 0^+$ to 1, the 2^n iterations are repeated f times, where f is the number of considered satisfaction degrees in $\mathscr{S} : 1 = \alpha_1 > \alpha_2 > \cdots > \alpha_f = 0^+$. Thus, the final complexity in the worst case is $f \times 2^n \in \theta(2^n)$. As we said previously, this is not a problem in practice as the number of predicates specified by a user is rather low (≤ 10) in most applicative contexts. Therefore, this process remains tractable as we will show experimentally in Sect. 6.

Once the MFSs have been detected, it is possible to inform the user about the conflicts in his/her query, which should help him/her revise the selection condition of the failing query.

5 Plethoric Answer Set Reduction

In this section, we address the problem symmetrical to that studied in Sect. 4: the Plethoric Answer Sets (PAS) problem. Let Q be a fuzzy query and Σ_Q (denoted also by $\Sigma_Q^{0^+}$) its set of answers against the database D. One can write $\Sigma_Q^{0^+} = \{\mu_1/t_1, \mu_2/t_2, \ldots, \mu_n/t_n\}$ where t_i is a tuple of D and μ_i its satisfaction degree w.r.t. Q. Assume that a user provides a number k of desired answers along with the query Q.

Input: a failing query $Q = P_1 \wedge \ldots \wedge P_n$;
      a scale of degrees $A = 0 < \alpha_f < \ldots < \alpha_2 < (\alpha_1 = 1)$;
      a user-defined qualitative threshold α_u;
Output: $MFS(Q)$ ordered sets of MFS's of Q, one set for each α-cut of Q.

3.1	**begin**
3.2	**foreach** $\alpha_i \in A \mid \alpha_i \geq \alpha_u$ **do**
3.3	$MFS_{\alpha_i}(Q) \leftarrow \emptyset$; $E_{\alpha_i} \leftarrow \{P_1, \ldots, P_n\}$;
3.4	$Cand_{\alpha_i} \leftarrow E_{\alpha_i}$;
3.5	**end**
3.6	$nbPred \leftarrow 1$;
3.7	**while** $Cand_{\alpha_1} \neq \emptyset$ **do**
3.8	**foreach** $\alpha_i \in A \mid \alpha_i \geq \alpha_u$ **do**
3.9	// generation of the candidates of size nbPred
3.10	$Cand_{\alpha_i} \leftarrow \{M$ composed of $nbPred$ predicates present in E_{α_i} such that $\forall M' \subset M, M' \notin MFS_{\alpha_i}(Q)\}$;
3.11	**foreach** L in $Cand_{\alpha_i}$ **do**
3.12	**if** $card(L_{\alpha_i}) = 0$ **then**
3.13	$MFS_{\alpha_i}(Q) \leftarrow MFS_{\alpha_i}(Q) \cup \{L\}$;
3.14	//E_L contains the atomic predicates that compose L
3.15	$E_{\alpha_i} \leftarrow E_{\alpha_i} - E_L$;
3.16	//Propagate L to higher satisfaction degrees
3.17	// $E = \cup_i E_{\alpha_i}$ and $MFS = \cup_i MFS_{\alpha_i}(Q)$
3.18	$propagate(\alpha_i, A, L, MFS, E)$;
3.19	**end**
3.20	**end**
3.21	**end**
3.22	$nbPred \rightarrow nbPred + 1$;
3.23	**end**
3.24	**end**

Algorithm 3: Gradual MFS computation

Definition 4 We say that a PAS problem occurs for Q if $\left| \Sigma_Q^{\mu_{max}(Q)} \right| \gg k$.

Where $\mu_{max}(Q) = sup_{t_i \in \Sigma_Q^{0^+}} \mu_i$ and $\Sigma_Q^{\mu_{max}(Q)} = \{t_i \in \Sigma_Q^{0^+} \mid \mu_i = \mu_{max}(Q)\}$.

This definition means that the set of answers $\Sigma_Q^{0^+}$ contains a large number of answers (with a maximal satisfaction degree) w.r.t. the number k of desired answers. The general idea of our solution is to augment a user query Q with predefined predicates which are semantically correlated with those present in Q, in order to reduce the initial answer set and get an answer subset whose cardinality is as close to k (the user-specified quantitative threshold) as possible.

Input: a satisfaction degree: α_i; a scale of degrees: A; detected MFS for α_i: L;
  a reference to the array of layered MFS: MFS; a reference to the array
  of predicates used for the generation of candidates: E;

4.1 procedure *propagate*(α_i, A, L, MFS, E) **begin**

4.2 | **foreach** $\alpha_j \in A \mid \alpha_j \geq \alpha_i$ **do**

4.3 | | **if** *isAtomic(L)* or *isMFS$(L, MFS_{\alpha_j}(Q))$* **then**

4.4 | | | $MFS_{\alpha_j}(Q) \leftarrow MFS_{\alpha_j}(Q) \cup \{L\}$;

4.5 | | | $E_{\alpha_j} \leftarrow E_{\alpha_j} - E_L$;

4.6 | | **else**

4.7 | | | break;

4.8 | | **end**

4.9 | **end**

4.10 | **end**

4.11 end

Algorithm 4: Procedure that propagates an MFS to higher satisfaction degrees

5.1 Correlation-Based Ranking

In the approach we propose, the new conjuncts to be added to the initial query are chosen among a set of possible predicates pertaining to the attributes of the schema of the database queried (see Sect. 3.1). This choice is mainly made according to their correlation with the initial query. A user query Q is composed of n (≥ 1) specified fuzzy predicates, denoted by $P^{s_1}, P^{s_2}, \ldots, P^{s_n}$, which come from the predefined vocabulary associated with the database (Sect. 3.1). The first step of the query augmentation process is to identify the predefined predicates most correlated to the initial query Q.

The notion of correlation introduced in Sect. 3.3 is used to qualify and quantify the extent to which two fuzzy sets (one associated with a predefined predicate $P^P_{i,j}$, the other associated with the initial query Q) are somewhat "semantically" linked.

Using the fuzzy-cardinality-based measure of correlation (cf. Formula 2), we can identify the predefined predicates most correlated to an under-specified query Q. In practice, we only consider the η most correlated predicates to a query, where η is a technical parameter which has been set to 5 in our experimentation. This limitation is motivated by the fact that an augmentation process involving more than η iterations, i.e., the addition of more than η predicates, could lead to important modifications of the scope of the initial query. Those η predicates most correlated to Q are denoted by $P^{c^1}_Q, P^{c^2}_Q, \ldots, P^{c^\eta}_Q$.

5.2 Reduction-Based Reranking

The second step of the query augmentation process aims at reranking the η predicates most correlated to the query according to their "reduction capability". It is assumed that the user specifies a value for the parameter k which defines the number of answers he/she expects. Let $F_{Q \wedge P_Q^{c^r}}$, $r = 1..\eta$, be the fuzzy cardinality of the answer set when Q is augmented with $P_Q^{c^r}$. $P_Q^{c^r}$ is all the more interesting for augmenting Q as $Q \wedge P_Q^{c^r}$ contains a σ_i-cut ($\sigma_i \in \mathscr{S}$ and $\sigma_i \geq \alpha_u$) with a cardinality c_i close to k and σ_i close to 1. To quantify how interesting $P_Q^{c^r}$ is, we compute for each σ_i-cut of $F_{Q \wedge P_Q^{c^r}}$ a "strengthening degree" which represents a compromise between its membership degree σ_i and its associated cardinality c_i. The global degree assigned to $F_{Q \wedge P_Q^{c^r}}$, denoted by $\mu_{stren}(F_{Q \wedge P_Q^{c^r}})$, is the maximum of its strengthening degrees over the different σ_i-cuts:

$$\mu_{stren}(F_{Q \wedge P_Q^{c^r}}) = sup_{1 \leq i \leq f} \top \left(1 - \frac{|c_i - k|}{max(k, |\Sigma_Q| - k)}, \sigma_i\right)$$

where $\top$ stands for a t-norm and the minimum is used in our experimentation. This reranking of the predicates the most correlated to Q can be carried out using the fuzzy cardinalities associated with each conjunction $Q \wedge P_Q^{c^r}$, $r = 1..\eta$.

Example 1 To illustrate this reranking strategy, let us consider a user query Q resulting in a PAS problem ($|\Sigma_Q^*| = 123$ and $|\Sigma_Q| = 412$), where k has been set to 50. As an example, let us consider the following candidates $P_Q^{c^1}, P_Q^{c^2}, P_Q^{c^3}, P_Q^{c^4}, P_Q^{c^5}$ and the respective fuzzy cardinalities:

- $F_{Q \wedge P_Q^{c^1}} = \{1/72 + 0.8/74 + 0.6/91 + 0.4/92 + 0.2/121\}, \mu_{stren}(F_{Q \wedge P_Q^{c^1}}) \simeq 0,94$
- $F_{Q \wedge P_Q^{c^2}} = \{1/89 + 0.8/101 + 0.6/135 + 0.4/165 + 0.2/169\}, \mu_{stren}(F_{Q \wedge P_Q^{c^2}}) \simeq 0,9$
- $F_{Q \wedge P_Q^{c^3}} = \{1/24 + 0.8/32 + 0.6/39 + 0.4/50 + 0.2/101\}, \mu_{stren}(F_{Q \wedge P_Q^{c^3}}) \simeq 0,93$
- $F_{Q \wedge P_Q^{c^4}} = \{1/37 + 0.8/51 + 0.6/80 + 0.4/94 + 0.2/221\}, \mu_{stren}(F_{Q \wedge P_Q^{c^4}}) \simeq 0,96$
- $F_{Q \wedge P_Q^{c^5}} = \{1/54 + 0.8/61 + 0.6/88 + 0.4/129 + 0.2/137\}, \mu_{stren}(F_{Q \wedge P_Q^{c^5}}) \simeq 0,99.$

According to the problem definition ($k = 50$) and the fuzzy cardinalities above, the following ranking is suggested to the user: 1) $P_Q^{c^5}$, 2) $P_Q^{c^4}$, 3) $P_Q^{c^1}$, 4) $P_Q^{c^3}$, 5) $P_Q^{c^2}$. Of course, to make this ranking more intelligible to the user, the candidates are proposed with their associated linguistic labels (cf. the concrete example about used cars given in Sect. 6.3).◇

5.3 Query Augmentation Process

Precomputed Knowledge

As the predicates specified by the user and those that we propose to add to the initial query are chosen among the predefined vocabulary, one can precompute some useful knowledge that will make the augmentation process faster. We propose to compute and maintain precomputed knowledge which is stored in two tables. The first one contains the precomputed fuzzy cardinalities introduced in Sect. 3.2, whereas the second one stores the correlation degrees between sets of predefined predicates (corresponding to the initial query) and any other predefined predicate. Using these correlation degrees, one can also determine and store, for each conjunction of predefined predicates, the most correlated predefined atomic predicates ranked in decreasing order of their correlation degrees. Both tables have to be updated after each (batch of) modification(s) performed on the data but these updates imply a simple incremental computation.

Interactive Augmentation Mechanism

The query augmentation process consists of the following steps. One first checks the table of fuzzy cardinalities in order to determine whether the user is faced with a PAS problem according to the value he/she has assigned to k. If so, one retrieves—still in constant time—up to η candidates that are then reranked according to k and presented to the user. Finally, as it is illustrated in Sect. 6.3, the user can decide to process the initial query, to process one of the suggested augmented queries, or to ask for another augmentation iteration of one of the augmented queries.

6 Experimentation

6.1 Context

The fuzzy-cardinality-based summarization process as well as the cooperative approaches described in Sects. 4 and 5 have been tested on a concrete database containing ads about second hand cars. This database is composed of a single relation named *secondHandCars* and contains 46,089 tuples with the following schema: {*idads, year, mileage, price, make, length, height, nbseats, acceleration, consumption, co2emission*}.

Common sense fuzzy partitions have been defined on the attributes of this relation, which led to a shared vocabulary made of 59 fuzzy predicates. Figure 5 illustrates the way users may employ this vocabulary to construct their fuzzy queries.

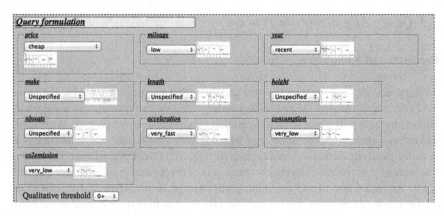

Fig. 5 Query interface relying on the shared vocabulary

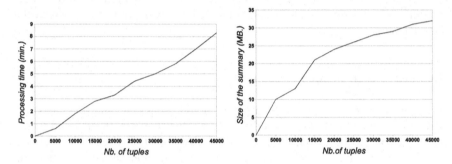

Fig. 6 Evolution of the processing time and space wrt. the number of tuples

Using this predefined vocabulary, we have first evaluated the time needed to compute a complete fuzzy-cardinality-based summary and also its evolution with respect to the size of the database. Figure 6 (left) shows the evolution of the time needed to compute the fuzzy cardinalities for a database whose size varies from 5000 to 45,000. Figure 6 (right) shows the evolution of the memory space needed to store the computed fuzzy-cardinality-based summary. These results have been obtained on a basic computer configuration (Intel Core 2 Duo 2.53GHz with 4Go 1067 MHz of DDR3 ram) and Postgresql as the RDBMS for the storage of the relation *secondHandCars* and its summary.

As expected, the time needed to compute the fuzzy cardinality-based summary linearly increases wrt. the size of the database. The most interesting phenomenon that can be observed in Fig. 6 is that the size of the memory used to store the fuzzy cardinalities is very reasonable and increases in a logarithmic way according to the number of tuples. Indeed, the number of fuzzy cardinalities that have to be stored increases quickly from 0 to 15,000 tuples, then very slowly to 35,000 and is almost stable from 35,000 to 45,000. This phenomenon was predictable and can be explained by the fact that whatever the number of tuples, the possible combinations of

Fig. 7 Evolution of the processing time wrt. the number of predicates

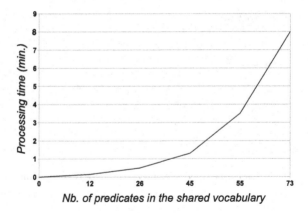

Nb. of predicates in the shared vocabulary

properties to describe them is finite and can quickly be enumerated. As an example, let us consider the failing fuzzy query Q composed of 8 predicates: *"year is recent and mileage is low and price is low and acceleration is very_high and consumption is very_low and co2emission is very_low"*. Whatever the number of tuples in the database, some combinations of properties are not observed, such as: *"year is recent and mileage is low and price is low"*, *"acceleration is very_high and consumption is very_low"*, *"acceleration is very_high and co2emission is very_low"*,etc. So, one can expect that the size of the memory used to store the fuzzy cardinalities will not increase significantly in general, even when the database grows a lot.

To complete these observations, Fig. 7 shows the evolution of the time needed to compute a complete summary of the database (with 46.089 tuples) with respect to the number of predicates in the vocabulary.

This first experimentation clearly shows that this fuzzy-cardinality-based summary can be considered even for large databases as long as the vocabulary contains a reasonably small number of fuzzy predicates. It is worth noticing that this characteristics correspond to most of the applicative contexts, especially for web sites proposing a query interface to their database.

6.2 A Prototype for Explaining Failing Queries

The query interface illustrated in Fig. 5 has been completed with the cooperative approach described in Sect. 4 in order to provide the users with some explanations about the failure of their queries [30]. In the first part of this experimentation, we have used the fuzzy cardinalities precomputed according to the predefined vocabulary and estimated the time needed to generate the explanations of failing queries. For this purpose, we have submitted 50 failing or unsatisfactory queries containing various numbers of predicates, from 1 up to 10. Figure 8 illustrates the explanations of the failing query *"year is vintage and price is low"*, whereas Fig. 9 shows the evolution

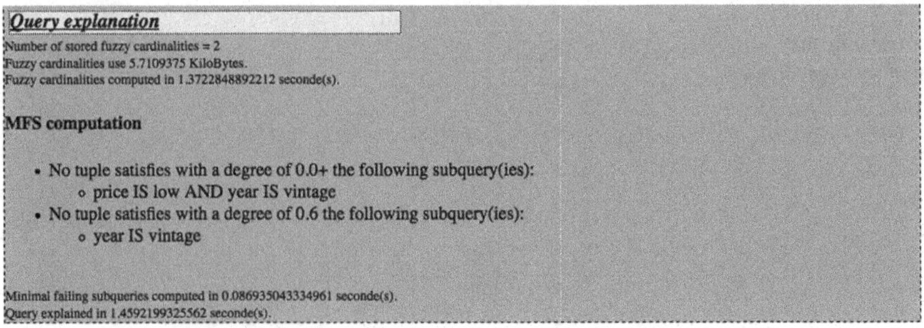

Query explanation

Number of stored fuzzy cardinalities = 2
Fuzzy cardinalities use 5.7109375 KiloBytes.
Fuzzy cardinalities computed in 1.3722848892212 seconde(s).

MFS computation

- No tuple satisfies with a degree of 0.0+ the following subquery(ies):
  - price IS low AND year IS vintage
- No tuple satisfies with a degree of 0.6 the following subquery(ies):
  - year IS vintage

Minimal failing subqueries computed in 0.086935043334961 seconde(s).
Query explained in 1.4592199325562 seconde(s).

Fig. 8 Explanations for the failing query "*year is vintage and price is low*"

of the average time needed to compute these explanations for failing queries whose number of predicates varies from 1 to 10. These results have been observed on three queries containing a single predicate, six containing two predicates and ten for other numbers of predicates. Despite the exponential aspect of the curve, these results show that for a reasonable number of predicates involved in the query, the time needed to compute the MFSs is very limited. Moreover, it is worth noticing that the performances of this explanation process could certainly be improved using parallel programming and a compiled language such as C instead of PHP.

To complete this first experimentation, we have also implemented the "naive" approach studied in [18], which does not make use of a summary but processes every possible subquery. To make the comparison meaningful, we have implemented a version of our approach where a fuzzy-cardinality-based summary is dynamically computed for each submitted query. In this case, the sole predicates involved in the query are concerned by the summarization process. Figure 10 graphically shows the difference in computation time for these two approaches and empirically shows the benefits of a single scan of the database. This comparison is performed for queries with at most six predicates, as the time needed to compute the MFSs for longer queries is prohibitive with the technique proposed in [18].

In Sect. 6, we have seen that the size of a fuzzy-cardinality-based summary is very limited, and that it is not linearly related to the size of the database but rather depends on the applicative context and on the correlations between the attributes. Indeed, Fig. 6 shows that the size of the summary quickly converges as soon as all the "plausible combinations" of predicates have been enumerated.

The experimentations that we carried out show the benefits of an approach whose complexity is not very sensitive to the number of tuples in the database. However, such an approach can only be used when the queries involve a relatively small number of predicates. As said previously, this is not a problem in practice as the number of predicates specified by a user is rather low (≤ 10) in most applicative contexts. To

Fig. 9 MFS computation
time using precomputed fuzzy
cardinalities

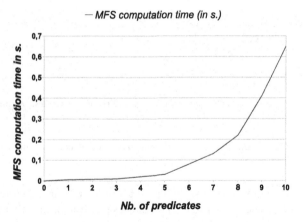

Fig. 10 Naive method versus
dynamic computation of the
summary

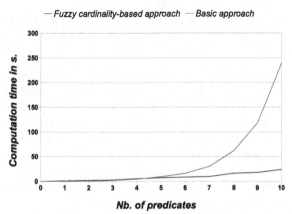

support this assertion, we have analyzed the query interface of 12 web sites[1] proposing an access to ads about second hand cars. The maximum number of constraints
(*i.e.* predicates) a user can specify through these interfaces varies from 5 to 12 with
an average of 8.8 predicates.

MFSs-Based Failing Queries Revision

When faced with a failing query, the explanations given by the layered MFSs help the
user revise his/her initial query. Depending on the nature of the conflicts underlined
in the MFSs, a user may:

- reconsider the qualitative threshold α_u specified in the query,

---

[1] Examples of web portals to databases containing ads about second hand cars: http://
www.annoncesauto.com, http://www.paruvendu.fr, http://www.auroreflex.com, http://www.ebay.
fr, http://www.lacentrale.fr, ...

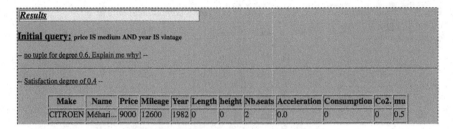

Fig. 11 Example of an MFS-guided revision of an initial failing query

- remove one or several predicates involved in a conflict,
- replace one or several predicates involved in an MFS by predicates from the shared vocabulary that appear less conflicting,
- apply a repair step which aims at relaxing the definition of some predicates [7] or replace the conjunctive query Q by a fuzzy quantified statement of the type $Q^* = most(P_1, P_2, \ldots, P_n)$ [37].

Figure 8 illustrates a failing situation for an initial query "*year is vintage and price is low*" and a user-defined qualitative threshold $\alpha_u = 0.2$. The explanation related to this failure clearly point out that the predicate "*price is low*" is in conflict with the property "*year is vintage*". Guided with this explanation, one may replace the conflicting predicate "*price is low*" by a less demanding one such as "*price is medium*" (Fig. 11).

Thanks to the gradual MFSs, the user knows that it is useless to expect answers with a high level of satisfaction if he/she keeps the predicate "*year is vintage*" which constitutes an atomic MFS for $\alpha = 0.6$.

6.3 A Prototype for Reducing Plethoric Answer Sets

To help users revise their queries when they return a plethoric answer set, we have augmented the query interface illustrated in Fig. 5 with a cooperative functionality that implements the approach described in Sect. 5 [10]. Using this interface, users may define their fuzzy queries and specify a quantitative threshold k corresponding to the number of answers they expect.

A concrete example given below illustrates the relevance of the predicates suggested by the system for augmenting the initial query.

Example 2 Let us consider the following query Q composed of fuzzy predicates chosen among the shared vocabulary (Fig. 5):

$Q = \textbf{select} * \textbf{from } second\ Hand\ Cars \textbf{ where } year\ is\ very\_old \textbf{ with } k = 50.$

Executed on the second hand cars DB, Q returns an answer set whose cardinality is:
$F_Q = \{1/179 + 0.8/179 + 0.6/179 + 0.4/323 + 0.2/323\}$. We are faced with a
PAS problem, which means that the query augmentation process is triggered.

The following candidates are suggested along with the fuzzy cardinality of the
corresponding augmented queries:

1. *mileage is medium* $(\mu_{cor}(Q, P^p_{mileage, medium}) = 0.11)$

$$F_{Q \wedge P^p_{mileage, medium}} = \{1/24 + 0.8/27 + 0.6/28 + 0.4/72 + 0.2/77\}$$

2. *mileage is very high* $(\mu_{cor}(Q, P^p_{mileage, veryhigh}) = 0.19)$

$$F_{Q \wedge P^p_{mileage, very high}} = \{1/7 + 0.8/7 + 0.6/8 + 0.4/18 + 0.2/19\}$$

3. *mileage is high* $(\mu_{cor}(Q, P^p_{mileage, high}) = 0.37)$

$$F_{Q \wedge P^p_{mileage, high}} = \{1/101 + 0.8/106 + 0.6/110 + 0.4/215 + 0.2/223\}.$$

For each candidate query Q', the user may decide to process Q' (i.e. retrieve the
results) or to repeat the augmentation process on Q'. If the latter option is chosen,
the table of fuzzy cardinalities is checked in order to retrieve relevant predicates for
augmenting Q' (i.e. properties correlated to Q') along with their associated fuzzy
cardinalities that are ranked according to k. Let us assume that the user selects

$$Q' = year\ is\ very\_old\ and\ mileage\ is\ medium$$

for a second augmentation step. The following candidates are suggested along with
their fuzzy cardinalities:

1. *price is low* $(\mu_{cor}(Q', P^p_{price, low}) = 0.34)$

$$F_{Q' \wedge P^p_{price, low}} = \{1/18 + 0.8/20 + 0.6/21 + 0.4/46 + 0.2/51\}$$

2. *price is medium* $(\mu_{cor}(Q', P^p_{price, medium}) = 0.15)$

$$F_{Q' \wedge P^p_{price, medium}} = \{1/6 + 0.8/7 + 0.6/7 + 0.4/22 + 0.2/22\}. \diamond$$

From this experimentation on a real-world database, one may observe that query
augmentation based on semantic correlation provides the users with useful information about data distributions and the possible queries that can be formulated in
order to retrieve coherent answer sets. By coherent answer set, we mean a group of
items that share correlated properties and that may correspond to what the user was
looking for without knowing initially how to retrieve them. Moreover, thanks to the
precomputed knowledge tables, it is not necessary to process the candidate queries

to inform the user about the size of their answer sets and the predicates that can be used to augment them.

This experimentation shows that the predicates suggested to augment the queries are meaningful and coherent according to the initial underspecified queries. One can find below some examples of suggested augmented queries Q' starting from underspecified queries Q:

- $Q = year\ is\ old\ and\ mileage\ is\ high\ and\ price\ is\ very\_low$
  intensified after two iterations into:
  $Q' = year\ is\ old\ and\ mileage\ is\ high\ and\ price\ is\ very\_low\ and\ acceleration\ is\ slow\ and\ consumption\ is\ high$
  with $|\Sigma_Q^*| = 63$ and $|\Sigma_{Q'}^*| = 26$.
- $Q = year\ is\ recent$
  intensified after two iterations into:
  $Q' = year\ is\ recent\ and\ mileage\ is\ low\ and\ price\ is\ medium$
  with $|\Sigma_Q^*| = 4.060$ and $|\Sigma_{Q'}^*| = 199$.
- $Q = price\ is\ high$
  intensified after two iterations into:
  $Q' = price\ is\ high\ and\ year\ is\ last\_model\ and\ co2emission\ is\ low$
  with $|\Sigma_Q^*| = 180$ and $|\Sigma_{Q'}^*| = 45$.

7 Related Work

The practical need for endowing intelligent information systems with the ability to exhibit cooperative behavior has been recognized since the early '90s. As pointed out in [13, 17], the main intent of cooperative systems is to provide correct, non-misleading and useful answers, rather than literal answers to user queries. During the last two decades, several cooperative approaches have been proposed for different aspects related to the problems dealt with here. In this section, we first recall the main existing approaches for database summarization. Then, we situate the uniform fuzzy-cardinality-based cooperative approach we propose with respect to work related to failing queries and plethoric answers respectively.

Database Summarization

In [34], Saint-Paul et al. propose an approach to the production of linguistic summaries structured in a hierarchy, i.e., a summarization tree where the tuples from the database are rewritten using the linguistic variables involved in fuzzy partitions of the attribute domains. The deeper the summary in the tree, the finer its granularity. First, the tuples from the database are rewritten using the linguistic variables involved in fuzzy partitions of the attribute domains. Then, each candidate tuple is incorporated into the summarization tree and reaches a leaf node (which can be seen as a classification of the tuple). In the hierarchical structure, a level is associated with the relative proportion of data that is described by the associated summary. However, the relative semantic poorness of these summaries in terms of cardinality-related information makes its interest limited when it comes to helping the user reformulate his/her query in an EAS or PAS situation.

Developed by Rasmussen and Yager, SummarySQL [32] is a fuzzy query language which can evaluate the truth degree of a summary guessed by the user. A summary expresses knowledge about the database in a statement under the form "Q objects in DB are S" or "Q R objects in DB are S" where DB stands for the database, Q is a linguistic quantifier (*almost all, about half*, etc.) and R and S are linguistic terms (*young, well-paid*, and so on). The expression is evaluated for each tuple and the associated truth values are later used to obtain a truth value for the whole summary. A similar type of approach is proposed in [28]. Anyway, this view of database summarization is purely oriented toward knowledge discovery, and does not aim at providing tools to support database querying/browsing.

Failing Queries

We discuss here only some studies that are most related to the approach proposed. For a complete and rich synthesis of works about failing queries, the reader can refer to [6, 7]. Jannach [19] proposes an algorithm which is somewhat similar to ours, but which does not precompute the cardinalities. Instead, it builds a binary matrix containing the satisfaction degrees obtained by each tuple for each atomic predicate, and combines these degrees in order to detect the MFSs. The main problems with this technique are that (i) such a table can be very large to the point of not fitting in memory (cf. the experimental results reported in [30]), and (ii) a query is processed for each atomic predicate on the whole dataset.

The algorithm proposed in [18] *processes* every query corresponding to a candidate MFS, which is obviously quite expensive. Similarly, the approach described in [23, 24], processes every maximally successful subquery of a failing query in order to retrieve what the author calls a *recovery set*. Compared to these works, the major interest of our approach is that the determination of the MFSs does not imply any additional query processing, thanks to the precomputation of fuzzy cardinalities. Thus, the complexity of our algorithm is linear in the size of the data (cf. Sect. 6.2).

Finally, apart from the study done in [7] and to the best of our knowledge, there is no other work that has addressed the problem of MFS detection in the context of preference queries, which covers an application context that goes beyond failing queries *stricto sensu*.

Plethoric Answer Sets

In their probabilistic ranking model, Chaudhuri et al. [11] also propose to use a correlation property between attributes and to take it into account when computing ranking scores. However, correlation links are identified between attributes and not predicates, and the identification of these correlations relies on a workload of past submitted queries.

Su et al. [35] have emphasized the difficulty to manage such a workload of previously submitted queries or users feedbacks. This is why they have proposed to learn attribute importances regarding a *price* attribute and to rank retrieved items according to their commercial interest. Nevertheless, this method is domain-dependent and can only be applied for e-commerce databases.

The approach advocated by Ozawa et al. [26, 27] is also based on the analysis of the database itself, and aims at providing the user with information about the data distributions and the most efficient constraints to add to the initial query in order to reduce the initial set of answers. The approach we propose in this chapter is somewhat close to that introduced in [26], but instead of suggesting an attribute on which the user should specify a new constraint, our method directly suggests a set of fuzzy predicates along with some information about their relative interest with respect to the user needs. The main limitation of the approach advocated in [26] is that the attribute chosen is the one which maximizes the dispersion of the initial set of answers, whereas most of the time, it does not have any semantic link with the predicates that the user specified in his/her initial query. To illustrate this, let us consider again the relation *secondHandCars* introduced in Sect. 3.1. Let Q be a fuzzy query on *secondHandCars*: *"select * from secondhandcars where type = 'estate' and year is recent"* resulting in a PAS problem. In such a situation, Ozawa et al. [26] first apply a fuzzy c-means algorithm [2] to classify the data, and each fuzzy cluster is associated with a predefined linguistic label. After having attributed a weight to each cluster according to its representativity of the initial set of answers, a global dispersion degree is computed for each attribute. The user is then asked to add new predicates on the attribute for which the dispersion of the initial set of answers is maximal. In this example, this approach may have suggested that the user should add a condition on the attributes *mileage* or *brand*, on which the recent estate cars are probably the most dispersed. We claim that it is more relevant to reduce the initial set of answers with additional conditions which are in the semantic scope of the initial query. Here for instance, it would be more judicious to focus on cars with a high level of security and comfort as well as a low mileage, which are features usually related to recent estate cars. This issue has been illustrated in Sect. 6.3.

The problem of plethoric answers to fuzzy queries has been addressed in [6] where a query strengthening mechanism is proposed. Let us consider a fuzzy set $F = (A, B, a, b)$ representing a fuzzy query Q. The authors of [6] define a fuzzy tolerance relation E which can be parameterized by a tolerance indicator Z, where Z is a fuzzy interval centered in 0 that can be represented in terms of a trapezoidal membership function by the quadruplet $Z = (-z, z, \delta, \delta)$. From a fuzzy set $F = (A, B, a, b)$ and a tolerance relation $E(Z)$, the erosion operator builds a set F_Z such that $F_Z \subseteq F$ and $F_Z = F \ominus Z = (A + z, B - z, a - \delta, b - \delta)$. However, such an erosion-based approach can lead to a deep modification of the meaning of the user query, if the erosion process is not correctly controlled.

8 Conclusion

This chapter is a synthesis of several works that we have carried out in the context of cooperative query answering. The main originality of our approach is that it addresses symmetrical problems with a unified framework based the notion of a *database fuzzy summary*. The type of summary we consider is based on fuzzy cardinalities and offers a concise formalism to represent the data distributions over a predefined vocabulary composed of fuzzy partitions. We have empirically shown on a concrete applicative context that this method is efficient and that it provides meaningful information that may help the user retrieve the items he/she is looking for. An important point is that the summarization process has a linear data complexity. On the other hand, this fuzzy-cardinality-based cooperative approach is realistic only when the number of predicates that compose the predefined fuzzy vocabulary is reasonably small. An interesting perspective would be to study the benefits of an incremental computation of the summaries bootstrapped with correlation between attributes or predicates that can be identified in a workload of previously submitted queries.

Concerning the failing query problem, we have proposed an approach that provides informative explanations about the reasons of the failure. A perspective is to define a strategy that automatically repairs the failing queries, the goal being to suggest a relaxed query that returns a non-empty set of answers and, if possible, whose cardinality is as close as possible to the quantitative parameter k. As mentioned above, an interesting solution could be to consider reformulations involving fuzzy quantified statements.

References

1. Agrawal, R., Srikant, R.: Fast algorithms for mining association rules in large databases. In: Bocca, J.B., Jarke, M., Zaniolo, C. (eds.) VLDB, pp. 487–499. Morgan Kaufmann, San Francisco (1994)
2. Bezdek, J.: Pattern Recognition with Fuzzy Objective Function Algorithm. Plenum Press, New York (1981)
3. Bodenhofer, U., Küng, J.: Fuzzy ordering in flexible query answering systems. Soft Comput. **8**, 512–522 (2003)
4. Bosc, P., Buckles, B., Petry, F., Pivert, O.: Fuzzy databases. In: Bezdek, J., Dubois, D., Prade, H. (eds.): Fuzzy Sets in Approximate Reasoning and Information Systems, pp. 403–468. The Handbook of Fuzzy Sets Series. Kluwer Academic Publishers, Dordrecht (1999)
5. Bosc, P., Dubois, D., Pivert, O., Prade, H., de Calmès, M.: Fuzzy summarization of data using fuzzy cardinalities. In: Proceedings of the 9th International Conference on Information Processing and Management of Uncertainty in Knowledge-Based Systems (IPMU'02), pp. 1553–1559, Annecy, France (2002)
6. Bosc, P., Hadjali, A., Pivert, O.: Empty versus overabundant answers to flexible relational queries. Fuzzy Sets Syst. **159**(12), 1450–1467 (2008)
7. Bosc, P., Hadjali, A., Pivert, O.: Incremental controlled relaxation of failing flexible queries. J. Intell. Inform. Syst. **33**(3), 261–283 (2009)
8. Bosc, P., Pivert, O.: SQLf: a relational database language for fuzzy querying. IEEE Trans. Fuzzy Syst. **3**(1), 1–17 (1995)

9. Bosc, P., Pivert, O., Dubois, D., Prade, H.: On fuzzy association rules based on fuzzy cardinalities. In: FUZZ-IEEE, pp. 461–464 (2001)
10. Bosc, P., Pivert, O., Hadjali, A., Smits, G.: Correlation-based query expansion. In: Actes des 26e Journées Bases de Données Avancées (2010)
11. Chaudhuri, S., Das, G., Hristidis, V., Weikum, G.: Probabilistic ranking of database query results. In: Proceedings of VLDB'04, pp. 888–899 (2004)
12. Chomicki, J.: Querying with intrinsic preferences. In: Proceedings of EDBT'02, pp. 34–51 (2002)
13. Corella, F., Lewison, K.: A brief overview of cooperative answering. In: Technical report http://www.pomcor.com/whitepapers/cooperative_responses.pdf (2009)
14. Cuppens, F., Demolombe, R.: Cooperative answering: a methodology to provide intelligent access to databases. In: Proceedings of DEXA'88, pp. 333–353 (1988)
15. Dubois, D., Prade, H.: Fuzzy cardinalities and the modeling of imprecise quantification. Fuzzy Sets Syst. **16**, 199–230 (1985)
16. Dubois, D., Prade, H.: Fundamentals of fuzzy sets, volume 7 of The Handbooks of Fuzzy Sets. Kluwer Academic, The Netherlands (2000)
17. Gaasterland, T., Godfrey, P., Minker, J.: Relaxation as a platform for cooperative answering. J. Intell. Inform. Syst. **1**(3–4), 296–321 (1992)
18. Godfrey, P.: Minimization in cooperative response to failing database queries. Int. J. Cooperative Inform. Syst. **6**(2), 95–149 (1997)
19. Jannach, D.: Techniques for fast query relaxation in content-based recommender systems. In: Proceedings of KI'06, pp. 49–63 (2006)
20. Kaplan, S.-J.: Cooperative responses from a portable natural language query system. Artif. Intell. **19**, 165–187 (1982)
21. Kiessling, W.: Foundations of preferences in database systems. In: Proceedings of VLDB'02 (2002)
22. Zadeh, L.A.: Fuzzy sets. Inform. Control **8**(3), 338–353 (1965)
23. McSherry, D.: Incremental relaxation of unsuccessful queries. In: Proceedings of ECCBR'04, pp. 331–345 (2004)
24. McSherry, D.: Retrieval failure and recovery in recommender systems. Artif. Intell. Rev. **24**(3–4), 319–338 (2005)
25. Motro, A.: Cooperative database system. In: Proceedings of FQAS'94, pp. 1–16 (1994)
26. Ozawa, J., Yamada, K.: Cooperative answering with macro expression of a database. In: Proceedings of IPMU'94, pp. 17–22 (1994)
27. Ozawa, J., Yamada, K.: Discovery of global knowledge in database for cooperative answering. In: Proceedings of Fuzz-IEEE'95, pp. 849–852 (1995)
28. Pilarski, D.: Linguistic summarization of databases with quantirius: a reduction algorithm for generated summaries. Int. J. Uncertainty Fuzziness Knowl. Based Syst. **18**(3), 305–331 (2010)
29. Pivert, O., Bosc, P.: Fuzzy Preference Queries to Relational Databases. Imperial College Press, London (2012)
30. Pivert, O., Smits, G., Hadjali, A., Jaudoin, H.: Efficient detection of minimal failing subqueries in a fuzzy querying context. In: Eder, J., Bieliková, M., Tjoa, A.M. (eds.) ADBIS. Lecture Notes in Computer Science, vol. 6909, pp. 243–256. Springer (2011)
31. Ras, R.-W., Dardzinska, A.: Intelligent query answering. In: Wang, J. (ed.) Encyclopedia of Data Warehousing and Mining, 2nd edn, vol. II, pp. 1073–1078. Idea Group, Inc., Hershey (2008)
32. Rasmussen, D., Yager, R.R.: Summary SQL: a fuzzy tool for data mining. Intell. Data Anal. **1**(1–4), 49–58 (1997)
33. Ruspini, E.: A new approach to clustering. Inform. Control **15**(1), 22–32 (1969)
34. Saint-Paul, R., Raschia, G., Mouaddib, N.: General purpose database summarization. In: Proceedings of VLDB'05, pp. 733–744 (2005)
35. Su, W., Wang, J., Huang, Q., Lochovsky, F.: Query result ranking over e-commerce web databases. In: Proceedings of CIKM'06 (2006)

36. Ughetto, L., Voglozin, W.A., Mouaddib, N.: Database querying with personalized vocabulary using data summaries. Fuzzy Sets Syst. **159**(15), 2030–2046 (2008)
37. Zadeh, L.: A computational approach to fuzzy quantifiers in natural languages. Comput. Math. Appl. **9**, 149–183 (1983)

Chapter 13
Scalability and Fuzzy Systems: What Parallelization Can Do

Malaquias Q. Flores, Federico Del Razo, Anne Laurent and Nicolas Sicard

Abstract (Fuzzy) Database management systems aim to provide tools for data storage and ing. Based on the stored information, systems can offer analytical functionalities in order to deliver decisional database environments. In many application areas, fuzzy systems have proven to be efficient for modeling, reasoning, and predicting with imprecise information. However, expanding the frontiers of such areas or exploring new domains is often limited when facing real world data: as the space to search get bigger, more computation time and memory space are required. In this chapter, we discuss how the parallelization of fuzzy algorithms is crucial to tackle the problem of scalability and optimal performance in the context of fuzzy database mining. More precisely, we present the parallelization of fuzzy database mining algorithms on multi-core architectures of two knowledge discovery paradigms, namely fuzzy gradual pattern mining and fuzzy tree mining (for example in the case of XML databases). We also present a review of other two related problems, namely fuzzy association rule mining and fuzzy clustering.

M. Q. Flores (✉) · A. Laurent
LIRMM, University Montpellier 2 CNRS, 161 rue Ada 34095, Montpellier, France
e-mail: quinterofl@lirmm.fr

A. Laurent
e-mail: laurent@lirmm.fr

M. Q. Flores
Instituto Tecnológico de Apizaco, Tlaxcala, Mexico
e-mail: quinterofl@lirmm.fr

F. Del Razo
Instituto Tecnológico de Toluca, DGEST-SEP Av. S/N, Metepec, Edo de Mexico, Mexico
e-mail: delrazo@ittoluca.edu.mx

N. Sicard
AllianSTIC-EFREI, Paris, 30-32 av. de la République, 94800 Villejuif Cedex, France
e-mail: nicolas.sicard@efrei.fr

O. Pivert and S. Zadrożny (eds.), *Flexible Approaches in Data, Information and Knowledge Management*, Studies in Computational Intelligence 497, DOI: 10.1007/978-3-319-00954-4_13, © Springer International Publishing Switzerland 2014

1 Introduction

In recent years fuzzy set and fuzzy logic theory have found applications in mathematical theory, artificial intelligence, non-linear control, real-time systems, database mining, machine learning, database management systems, decision making, consumer electronics, expert systems, economics, finance, software engineering, among other interesting areas of application [35, 36, 63, 71]. The expression "*fuzzy systems*" is the name commonly used to refer in general to the systems resulting from the different applications of fuzzy logic [36]. Whereas for referring to specific systems, we use expressions such as *fuzzy control, fuzzy database management systems, fuzzy database mining techniques*, and so on.

Fuzzy systems are computer systems inspired from the linguistic processing of information, where representation and processing of imprecise and uncertain data is done through fuzzy set theory and fuzzy logic (fuzzy inference) respectively [36, 63]. Such systems aim at implementing on the machines, models and algorithms related to imprecise information processing by *approximate reasoning*. *Fuzzy database mining techniques* are methods to extract *automatically meaningful knowledge* from complex databases [35, 71].

Informally, *approximate reasoning* is here defined as the process of inferring meaningful conclusions from imprecise antecedents [36]. The *automatic extraction of knowledge from databases*, also known as *Knowledge Discovery in Databases* is defined as a multi-step process of discovering potentially useful information from large and complex databases [54]. In this framework, *fuzzy database management systems* and *fuzzy database mining techniques* have an important role [5, 33, 35, 50].

With the emergence of innovative and accessible models of parallel computation, fuzzy systems can improve their performance by using parallel computing architectures. In this chapter, we discuss how important the parallelization of fuzzy algorithms is to tackle the problem of scalability and optimal performance in the framework of fuzzy database mining. More precisely, we discuss the parallelization of fuzzy database mining algorithms on multi-core architectures of four knowledge discovery paradigms, namely fuzzy association rules, fuzzy clustering, fuzzy gradual dependencies, and fuzzy tree mining (for example in the case for XML databases). In all cases, the role of fuzzy databases is important in the process of extraction of fuzzy patterns.

The outline of this chapter is as follows: In Sect. 2, we present an overview about fuzzy database systems, definitions related to the representation of fuzziness in the relational fuzzy database model, an introduction to fuzzy database mining techniques, and the importance of optimizing the performance of such techniques by parallelization. We present a brief review about parallel programming models, taxonomy of computer architectures and definition of performance measurements of parallel programs in Sect. 3. In Sect. 4, we present our approach of parallelization of a fuzzy gradual patterns mining method, our approach of parallelization of a fuzzy tree mining algorithm on multi-core architectures, we also present a review of two related problems, namely: parallel fuzzy association rule mining and parallel fuzzy clustering. Finally, we conclude and give some suggestions for future research directions in Sect. 5.

2 Fuzzy Databases and (Fuzzy) Database Mining Techniques

In the framework of traditional database management systems, it is common to assume that the data are precise and certain. Unfortunately, real-world data are often uncertain, imprecise, inconsistent, ambiguous or vague due to different reasons such as: human errors, instrument errors, recording errors, noisy data, and so on [67]. *Fuzzy databases* (FDB) aim at providing tools for storage and querying data with the previously mentioned imperfections [64].

In order to represent and manage imperfect data, during the last thirty-five years has been carried out extensive scientific research work aimed at developing different approaches of how to incorporate *fuzziness*[1] at different levels into *FDB* models [47, 53].

The most studied fuzzy database models have been the fuzzy relational database models *(FRDBM)* [1, 3, 10] and fuzzy object-oriented database models *(FOODBM)* [46, 47]. In the next section, we review necessary theoretical background and terminology of *FRDBM*, on which this chapter focuses. For a comprehensive review related to *FOODBM*, refer to [3, 46, 47]. This is because in this chapter, we present an approach to address the problem of optimizing the automatic extraction of gradual patterns from *fuzzy relational databases*.

2.1 Fuzzy Databases

In *FRDBM*, fuzziness is introduced at the tuple level and at the attribute level. At the tuple level, the tuples are described by the membership degrees that indicate the extent to which their characteristics belong to their considered fuzzy relations or a possibility distribution measuring the possibility that the tuples belongs to their fuzzy relations [3, 59]. At the attribute level, the attributes are described through fuzzy linguistic variables or possibility distributions that indicate the extent to which attribute values belong to their fuzzy sets defined on the domains of the attributes [1, 3].

The fuzzy relations that are employed in recent *FRDBM* to introduce fuzziness at the tuple level are: *similarity relations* [3, 11], *proximity relations* [3, 59], or

---

[1] According to OXFORD DICTIONARY. Fuzziness is deterministic uncertainty Fuzziness is concerned with the degree to which events occur rather than the likelihood of their occurrence (probability).

resemblance relation [3, 57], where each pair of values in the attribute domain are mapped, through similarity, proximity or resemblance relation, to interval [0, 1].

Definition 1 Let X and Y be non-empty sets. A fuzzy subset R of the Cartesian product $X \times Y$ is called a *binary fuzzy relation from X to Y*. For $(x, y) \in R$, for some pair (x, y), $R(x, y)$ is the degree to which x is R-*related* to y in the unit interval [0, 1].

If $X = Y$, that is to say if R is a subset of $X \times X$ then R is a *binary fuzzy relation on X* denoted as $R : X^2 \rightarrow [0, 1]$.

Zadeh introduced the notion of *fuzzy similarity relation* [70] in 1971, later generalized by the proximity relation [23] and the resemblance relation [14]. In 1999, Bodenhofer proposed a generalization in the form of the *fuzzy equivalence relation* [9].

Fuzzy equivalence relation is a concept that plays an outstanding role for modeling gradual similarity under fuzzy environment, where information concerning the objects of study and analysis is often expressed in linguistic terms, e.g., very low between very poor and poor, about \$580, approximately between \$6,090 and \$4,700, fair, very high, and so on [38].

Definition 2 A fuzzy relation $E : X^2 \rightarrow [0, 1]$ is called *equivalence relation* on a domain X with respect to a *t-norm T*, for brevity *T-equivalence*, if and only if the following three axioms are fulfilled for all $x, y, z \in X$:

(i) E-reflexivity: $E(x, x) = 1$,

(ii) E-symmetry: $E(x, y) = E(y, x)$, and

(iii) T-transitivity: $T(E(x, y), E(y, z)) \le E(x, z)$.

$E(x, y)$, $E(y, z)$ and $E(x, z)$ are the grade of membership of the ordered pairs (x, y), (y, z), and (x, z) in E, with respect to a triangular *minimum (t-norm) T*. Along the last twenty years, the above concept of *equivalence relation* has been developed and generalized. For instance, alternative interpretations of the property of *T*-transitivity have been proposed, e.g., Bezdek and Harris [8, 9] introduced an interpretation of *T*-transitivity based on the *Lukasiewicz t-norm* $T_L(x, y) = max(0, x + y - 1)$. The interpretation based on the *product t-norm* $T_P(x, y) = (x \cdot y)$ was introduced by Faurous and Fillard [8, 9].

In *FRDBM*, the concepts of *fuzzy linguistic variable* and *possibility distribution* [10] play an important role in representing imprecise data at the attribute level.

Definition 3 A *fuzzy linguistic variable* V is defined as a quadruple of the form $V = (X, D, T, MF)$, where X is the name of V, D is the values domain of V, T represents the set of fuzzy subsets defined in D, and MF represents the membership functions that characterize each *fuzzy subset* $\in T$.

Table 1 Example of a fuzzy database

Attribute	Size				Weight				Sugar rate			
Id	X_1	NX_1	low	high	X_2	NX_2	low	high	X_3	NX_3	low	high
o_1	6	0.00	0.85	0.15	6	0.00	0.80	0.20	5.3	1.00	0.45	0.55
o_2	10	0.24	0.60	0.40	12	0.75	0.30	0.70	5.1	0.50	0.48	0.52
o_3	14	0.47	0.44	0.56	14	1.00	0.20	0.80	4.9	0.00	0.48	0.52
o_4	23	1.00	0.15	0.85	10	0.50	0.45	0.55	4.9	0.00	0.48	0.52
o_5	6	0.00	0.85	0.15	8	0.25	0.55	0.45	5.0	0.25	0.5	0.50
o_6	14	0.47	0.45	0.14	9	0.38	0.5	0.5	4.9	0.00	0.48	0.52

Definition 4 A *fuzzy subset A* defined in a domain D is a set with fuzzy boundaries and therefore totally characterized by a membership function $(A(d))$, which denotes the degree of membership of d in the *fuzzy subset A* $\forall d \in D$.

$$A = \{(d, A(d)) \mid d \in D\} \tag{1}$$

Table 1 illustrates the concept of fuzzy database representing imprecise data at the attribute level, the attributes are defined by *fuzzy linguistic variables*. For example the non normalized values of attribute *Size* are represented by the variable X_1, with their membership degrees in the fuzzy sets *low* and *high* in the interval [0, 1]. The variable NX_1 represents the normalized values of attribute *Size* in the interval [0, 1].

2.2 Fuzzy Database Mining Techniques

The aim of database mining can be defined as finding patterns or rules that describe the meaning of the relationships or dependencies between the data contained in big and complex databases. Database mining is an interdisciplinary field, which combines research from areas such as machine learning, statistics, theory of fuzzy sets and fuzzy logic, neural networks, evolutionary computing, high performance computing, parallel programming, databases, and *FDB* models [25, 29, 35].

In the pattern mining field, an important problem is the extraction of patterns that are intrinsically vague, imprecise, uncertain and that can involve data disturbed by the noise [34]. This problem comes from the fact that real-world data tend to be uncertain due to human errors, instrument errors, recording errors, noisy data, and so on [26, 67]. Fuzzy databases allow a natural and flexible representation of patterns and data with the characteristics mentioned above. In this framework, in recent years, several extensions of database mining techniques have been developed on the basis of fuzzy sets and fuzzy logic theory [35], such extensions are known as fuzzy database mining techniques (FDMT) [5].

Scaling algorithms of FDMT is a challenge [28], because their search spaces, requirements of computation time and memory are larger than of the algorithms used in crisp database mining methods [27]. In this context, we present the parallelization of fuzzy database mining algorithms on multi-core architectures, more precisely, we present parallel gradual pattern mining based on fuzzy orderings and parallel fuzzy tree mining. We also present a review of other two related approaches, namely parallel fuzzy association rule mining and parallel fuzzy clustering. Before stating that, in Sect. 3 some definitions of Parallel programming will be recalled.

3 Parallel Programming Models and Parallel Computers: An Overview

Parallel computing is a viable means to improve performance of algorithms of fuzzy computing [66]. With the emergence of new generations of multi-core processors and the new generations of graphics processing units ($GPUs$) as key components of high performance hardware of a computer system, optimization of fuzzy systems through its parallelization is possible on general purpose computing platforms [15, 24, 39].

Parallelization-based optimizations of algorithms [6, 66] aim to: (i) reduce the execution time, (ii) allow real-time processing, (iii) solve large problems, and (iv) exploit the computing power of the more and more present high-performance systems (e.g., multi-core processors that now even equip mobile phones and tablets).

Below we present the taxonomy of parallel computers, concepts, and terminology about parallel programming models, that is used through subsequent sections.

3.1 Taxonomy of Computer Architecture

According to instruction and data streams, Flynn in 1966 [22, 58] defined a taxonomy of computer architecture, as is presented in Table 2, where instruction streams are the operations to be performed by the processors/cores, the data streams are sequences of data to be processed and that circulate between memory and the processors/cores.

- **A SISD system** is the standard architecture of uniprocessor von Neumann computers.

Table 2 Flynn's taxonomy of computer architecture	Single-data stream	Multiple-data stream
Single-instruction stream	SISD	SIMD
Multiple-instruction stream	MISD	MIMD

- **In SIMD architecture,** same instruction stream is executed in all processors with different data streams in a synchronized fashion.
- **In a MISD machine**, different instruction streams on the same stream of data, this type of machine has never been used in practice.
- **In the MIMD category**, each processor has its own control unit and uses its own stream of data and executes its own stream of instructions (or part of the program), this type of architecture is considered the more flexible.

3.2 Parallel Programming Models

The parallel programming models are divided into three categories [22, 58, 66]:

- Distributed memory systems, any cluster and/or single symmetric multiprocessors (SMP), each processor has its own system memory that cannot be accessed by other processors, the shared data are transferred usually by message passing, e.g., sockets and message passing interface (MPI).
- Shared memory, SMP only, the processors share the global memory, the processors have direct access to the entire set of data. Access to the same data need synchronization and sequential memory access, e.g., Posix threads, OpenMP, and automatic parallelization.
- Hierarchical systems, is a combination of shared and distributed models. They are composed by multiprocessor nodes in which memory is shared by intra-node processors and distributed over inter-node processors. Hierarchical models are implemented on fast networks and share disk drives.

3.3 Process and Thread

In multicore architectures, a parallel program is executed by the processors through one or multiple control flows referred to as processes or threads [31, 66]. A process can consist of several threads that share a common address space whereas each process works on a different address space [31]. In order to achieve efficiency, the multicore *CPUs* can use only a few threads, while *GPUs* may use thousands [16].

In a multiprocessor/multicore system SIMD, data parallelism (loop-level parallelism) takes place when different threads execute the same code or task on different data streams. Task parallelism (control parallelism) take place when each processor/core executes a different thread on the same or different data stream.

In the massively multi-threaded SIMD architecture provided by *GPUs*, threads are extremely lightweight and grouped into *thread block* [16]. Threads within the same *thread block* are divided into SIMD groups, called *warps*, each one of them contain 32 threads [16, 24].

The parallel portions of an application are executed on the device GPU as *kernels*, one *kernel* is executed at a time by an array of threads, where all treads run the same code and each thread has an ID that it uses to compute memory addresses and make control decisions [16, 24].

3.4 About Speedup and Scaleup of Parallel Programs

The *speedup* of a parallel program expresses the relative diminution of response time that can be obtained by using a parallel execution on p processors or cores compared to the best sequential implementation of that program.

The *speedup* ($Speedup(p)$) of a parallel program with parallel execution time $T(p)$ is defined as

$$Speedup(p) = \frac{T(1)}{T(p)} \tag{2}$$

where:

- p is the number of processors/cores or threads;
- T(1) is the execution time of the sequential program (with one thread or core);
- $T(p)$ is the execution time of the parallel program with p processors, cores, or threads.

Scale Up ($Scaleup(p)$) evaluates throughput of a parallel implementation and can be expressed as:

$$Scaleup(p) = \frac{T(1, D)}{T(p, pD)} \tag{3}$$

where $T(1, D)$ is the execution time of the sequential program on 1 core with data size of D, $T(p, pD)$ is the execution time of the parallel program on p cores and p times D.

4 Parallel Fuzzy Database Mining

Parallelizing fuzzy database mining algorithms is a viable means to improve their performance and for making feasible fuzzy database mining to large-scale [29].

Within the framework of multiprocessor/multicore architectures of share or distributed memory, parallel fuzzy database mining as well as parallel database mining follow two approaches of parallelization: *task parallelization* and *data parallelization* [24, 25, 60]. In *task parallelization* the processors/cores execute a different task on the (fuzzy) database. In *data parallelization* the (fuzzy) database is partitioned among the processors/cores and all execute the same task.

4.1 Parallel Mining of Gradual Patterns

In *fuzzy gradual pattern mining*, the aim is to find dependencies between the *variation* and *direction of change* of attribute values of gradual patterns in the fuzzy database instead of finding the degree of presence or absence of attributes in a transaction [32, 40]. Gradual patterns allow describing complex interactions in the behaviour of the attribute values of a DB, interactions represented as: $\{(pollution\_IO, + \mid -),$ $(measurement\ error, + \mid -)\}$ interpreted as "*The higher/lower light pollution, the higher/lower measurement error*".

Given a fuzzy database DB (as that shown in Table 1) consisting of N transactions/objects $O = \{o_1, \ldots, o_N\}$, m attributes $X = \{X_1, \ldots, X_m\}$ corresponding to fuzzy linguistic variables of the form $V_j = \{X_j, \{A_j, B_j, \ldots\}, \{A(o_i) \in [0, 1], B(o_i) \in [0, 1], \ldots\}, D_j$ where $\{A_j, B_j, \ldots\}$ are the fuzzy sets defined on the domain of values D_j of the attribute X_j, $\{A(o_i) \in [0, 1], B(o_i) \in [0, 1], \ldots\}$ are membership functions denoting the degree to which the value $x_i \in D_j$ for object o_i belongs to the fuzzy sets $\{A, B, \ldots\}$, for $i = 1, 2, \ldots, N$ and $j = 1, 2, \ldots, m$.

In this framework, we present the definitions of *gradual item*, *gradual pattern*, of *concordant couple*, and *support of a GP*, since they are basic concepts in the complex task of *gradual pattern mining*.

Definition 5 A *gradual item* is defined as a tuple of the form (X_l, d), where X_l is a attribute in DB, $d \in \{+ \mid -\}$ denotes the direction of change (tendency) in the attribute values X_l. For instance $(X_l, +)$ represents the fact that the X_l values have the tendency to increase or to decrease in case $(X_l, -)$.

Definition 6 A *gradual pattern (GP)* is defined as a combination of two or more *gradual items*, semantically interpreted as their conjunction denoted as $GP = \{(X_l, d_l), l = 1, \ldots k \mid 2 \le k \le m\}$. For instance $GP = \{(X_1, +), (X_2, +), (X_3, -)\}$ is interpreted as $(X_1, more), (X_2, more), (X_3, less)$, i.e., $\{The\ higher X_1, the\ higher X_2, the\ lower X_3\}$.

Definition 7 A *concordant couple (cc)* is an index pair (i, j), where the objects (o_i, o_j) satisfy all the variations d expressed by the *gradual items* in a GP of size k, e.g., let $GP = \{(X_1, -), (X_2, -)\}$ with size $k = 2$, an index pair $cc(i, j)$ is a *concordant couple* if $((x_i^1 > x_j^1)\ implies\ (x_i^2 > x_j^2))$ then $cc(i, j) = 1$ else $cc(i, j) = 0$, where $i = (x_i^1, x_i^2)$ and $j = (x_j^1, x_j^2)$, for $i, j \in \{1, 2, 3, \ldots, n\}$ and $i \ne j$.

Definition 8 An index pair (i, j) has a *fuzzy concordance degree (c̃c)* in the interval $[0, 1]$, if the objects (o_i, o_j) satisfy in a degree given in the interval $[0, 1]$ all the variation constraints d expressed by the k *gradual items* contained in a GP, e.g., let $GP = \{(X_1, +), (X_2, +)\}$ size $k = 2$, assuming that X_1 and X_2 are two equivalence relations $E_{X_1} : X_1^2 \to [0, 1]$, $E_{X_2} : X_2^2 \to [0, 1]$, we can define a *strict* $T_L - E_{X_1}$ ordering on X_1 as in (4, 5) and a *strict* $T_L - E_{X_2}$ ordering on X_2 as in (6, 7), we compute $\tilde{c}c$ for an index pair (i, j) as is done by Eq. (8) where $\tilde{T}$ is a Lukasiewicz *t-norm*.

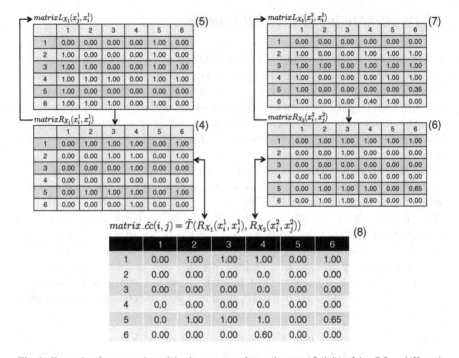

matrix $L_{X_1}(x_j^1, x_i^1)$ (5)

	1	2	3	4	5	6
1	0.00	0.00	0.00	0.00	1.00	0.00
2	1.00	0.00	0.00	0.00	1.00	0.00
3	1.00	1.00	0.00	0.00	1.00	1.00
4	1.00	1.00	1.00	0.00	1.00	1.00
5	1.00	0.00	0.00	0.00	0.00	0.00
6	1.00	1.00	1.00	0.00	1.00	0.00

matrix $L_{X_2}(x_j^2, x_i^2)$ (7)

	1	2	3	4	5	6
1	0.00	0.00	0.00	0.00	0.00	0.00
2	1.00	0.00	0.00	1.00	1.00	0.00
3	1.00	1.00	0.00	1.00	1.00	1.00
4	1.00	0.00	0.00	0.00	1.00	1.00
5	1.00	0.00	0.00	0.00	0.00	0.35
6	1.00	0.00	0.00	0.40	1.00	0.00

matrix $R_{X_1}(x_i^1, x_j^1)$ (4)

	1	2	3	4	5	6
1	0.00	1.00	1.00	1.00	0.00	1.00
2	0.00	0.00	1.00	1.00	0.00	1.00
3	0.00	0.00	0.00	1.00	0.00	0.00
4	0.00	0.00	0.00	0.00	0.00	0.00
5	0.00	1.00	1.00	1.00	0.00	1.00
6	0.00	0.00	0.00	1.00	0.00	0.00

matrix $R_{X_2}(x_i^2, x_j^2)$ (6)

	1	2	3	4	5	6
1	0.00	1.00	1.00	1.00	1.00	1.00
2	0.00	0.00	1.00	0.00	0.00	0.00
3	0.00	0.00	0.00	0.00	0.00	0.00
4	0.00	1.00	1.00	0.00	0.00	0.00
5	0.00	1.00	1.00	1.00	0.00	0.65
6	0.00	1.00	1.00	0.60	0.00	0.00

matrix $\tilde{cc}(i, j) = \tilde{T}(R_{X_1}(x_i^1, x_j^1), R_{X_2}(x_i^2, x_j^2))$ (8)

	1	2	3	4	5	6
1	0.00	1.00	1.00	1.00	0.00	1.00
2	0.00	0.00	0.00	0.0	0.00	0.00
3	0.00	0.00	0.00	0.0	0.00	0.00
4	0.0	0.00	0.00	0.0	0.00	0.00
5	0.0	1.00	1.00	1.0	0.00	0.65
6	0.00	0.00	0.00	0.60	0.00	0.00

Fig. 1 Example of computation of the *fuzzy concordance degrees* ($\tilde{cc}(i, j)$) of the $GP = \{(X_1, +),$ $(X_2, +)\}$ with the normalized values of attributes *Size* and *Weight* of Table 1

$$R_{X_1}(x_i^1, x_j^1) = 1 - L_{X_1}\left(x_j^1, x_i^1\right) \qquad (4)$$

$$L_{X_1}(a, b) = min\left(1, max\left(0, 1 - \frac{1}{r} * (a - b)\right)\right); r > 0 \qquad (5)$$

$$R_{X_2}\left(x_i^2, x_j^2\right) = 1 - L_{X_2}\left(x_j^2, x_i^2\right) \qquad (6)$$

$$L_{X_2}(a, b) = min\left(1, max\left(0, 1 - \frac{1}{r} * (a - b)\right)\right); r > 0 \qquad (7)$$

$$\tilde{cc}(i, j) = \tilde{T}(R_{X_1}(x_i^1, x_j^1), R_{X_2}(x_i^2, x_j^2)) \qquad (8)$$

Figure 1 shows the computation of the *fuzzy concordance degrees* ($\tilde{cc}(i, j)$) of the $GP = \{(X_1, +), (X_2, +)\}$ with the normalized values of attributes *Size* and *Weight* of Table 1, where for example given the pair of objects (o_5, o_6), $o_5 = (0.00, 0.25)$, $o_6 = (0.47, 0.38)$, the *concordant couple*$((x_5^1, x_5^2), (x_6^1, x_6^2))$ is set as $x_5^1 = 0.00$, $x_6^1 = 0.47$, $x_5^2 = 0.25$, and $x_6^2 = 0.38$. Then the results of Eqs. (5), (4), (7), (6) and

(8) are $L_{X_1}(x_6^1, x_5^1) = 0.00$, $R_{X_1}(x_5^1, x_6^1) = 1.00$, $L_{X_2}(x_6^2, x_5^2) = 0.35$, $R_{X_2}(x_5^2, x_6^2) = 0.65$, and finally the *fuzzy concordance degree* $\tilde{c}c(5, 6) = 0.65$.

Definition 9 The *support of a GP* $(supp(GP))$ is an indicator of reliability of the occurrence of *GP* in *DB*, which is calculated in absolute and relative terms. In absolute terms the support of a *GP* is defined as the number of objects in *DB* that respect the variation d described by the *gradual items* in *GP*. In relative terms the support of a *GP* is defined as the fraction (frequency) of the absolute support of *GP* with respect to the total number of objects in *DB*.

4.1.1 State of the Art

(a) Sequential extraction of gradual patterns: related work

In order to measure the strength of the dependency or correlation between the variation and direction of change of attribute values of a gradual pattern/dependency, there are various approaches and each has its own method to compute the support (see [40, 42, 49] for more details), a brief description is given below.

- Numerical approach: such as analysis of contingency diagrams by means of techniques from statistical regression analysis, suggested in [32], the validity of the gradual tendency is evaluated from the quality of the regression, measured by the normalized mean squared error R^2, together with the slope of the regression line.
- Qualitative alternative: count the number of pairs of points (x_i^1, x_i^2) and (x_j^1, x_j^2), where $x_i^1, x_j^1 \in X_1$ and $x_i^2, x_j^2 \in X_2$ for which $(x_i^1 < x_j^1)$ and $(x_i^2 < x_j^2)$, association rules in [48], and fuzzy association rules in [49] are used in order to mine gradual dependencies type { *the more X_1, the more X_2* }. Other methods and algorithms of this category are: approach based on conflict sets [42], approach based on the precedence graph [43, 44], and approach based on rank correlation measures (GRAANK) [42].
- Numerical-qualitative approaches: this kind of techniques combines properties of both approaches, the numerical and the qualitative one, in order to measure not only the existence of a tendency, but its strength in terms of a *fuzzy rank correlation* measure [40, 56], or terms of fuzzy association rules and fuzzy gradual dependence [48, 49].

(b) Parallel extraction of gradual patterns: related work

Recently, in [43] and [44], Laurent et al. have presented PGP-mc a multicore parallel approach for mining gradual patterns where the evaluation of the correlation and support is based on conflict sets and precedence graph approaches [42]. In this approach, new tasks are dynamically assigned to a pool of threads on a *"first come, first serve"* basis.

PGP-mc was implemented on two MIMD multi-core computer employing the parallel programming model of Posix threads. Experiments were led on synthetic databases automatically generated by a tool based on adapted version of IBM Synthetic Data Generation Code for Associations and Sequential Patterns. For example, the sequential processing of the 350 attributes database took more than five hours while it spend approximatively 13 min using 24 threads. Furthermore, speed-up results are particularly stable from one architecture to another (for 24 to 32 cores).[2]

An efficient parallel mining of *closed frequent gradual patterns*, named PGLCM has been proposed by Do et al. [61]. This approach is based on the principle of the LCM algorithm for mining *closed frequent patterns*, an adaptation of LCM named GLCM in order to mine *closed frequent gradual patterns*, and parallelization of the GLCM algorithm named PGLCM based on the Melinda parallelism environment. It consists of shared memory space, called *TupleSpace*, where all the threads can either deposit or retrieve a data unit called *Tuple*, via two primitives *get*(*Tuple*) and *put*(*Tuple*). All the synchronizations for accessing the *TupleSpace* are handled by the Melinda framework.

The comparative experiment is based on synthetic databases produced with the same modified version of IBM Synthetic Data Generator for Association and Sequential Patterns. All the experiments were conduced on a MIMD multi-core computer in two stage, the first one to evaluate the performance of the sequential version of GLCM and PGP-mc (known as Grite), the second one to evaluate the scaling capacities of PGLCM and PGP-mc (known as Grite-MT). Evaluation criteria were the execution time and memory consumption. Where GLCM/PGLCM compute only the *closed frequent gradual patterns*, whereas Grite/PGP-mc compute all the *frequent gradual patterns* (see [43] and [44] for more details).

4.1.2 Parallel Fuzzy Orderings for Fuzzy Gradual Pattern Mining

In this section, we address the problem of automatically finding correlations between positive and/or negative small variations in the values of attributes affected by noise and non-linear nature. Consequently, we implemented a method of extraction of gradual patterns based on the concepts of *fuzzy concordance degree* and *fuzzy strict ordering* presented in Definition 8.

In our approach we propose to compute the support of a *GP* as

$$supp(GP) = \frac{\sum_{i=1}^{N} \sum_{j \neq i} \tilde{c}c(i, j)}{N(N-1)} \qquad (9)$$

where each $\tilde{c}c(i, j)$ is computed using (8) and stored in a matrix named *matrix of fuzzy concordance degree* ($m\tilde{c}c(i, j)$) (see Fig. 2).

More precisely, our sequential approach of fuzzy gradual pattern mining is shown in Fig. 3, where the algorithm works as follows:

- Step 1. For each attribute $X_l \in DB$, build their gradual items $\{(X_l, +), (X_l, -)\}$.
- Step 2. Initialization ($k = 2$): (a) with the gradual items of step 1 generate all gradual pattern candidates of size k, (b) computing their $m\tilde{c}c(i, j)$ according to

---

[2] Detailed results are available on-line at http://www.lirmm.fr/~laurent/.

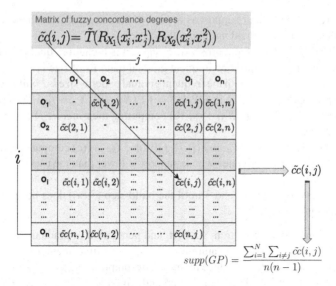

Fig. 2 Illustration of the structure of a *matrix of fuzzy concordance degrees* $m\tilde{c}c(i, j)$

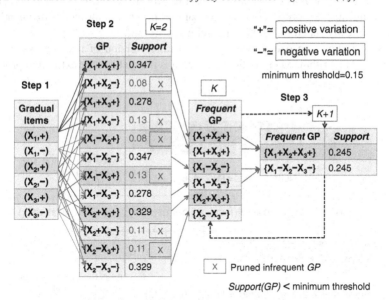

Fig. 3 Sequential fuzzy gradual pattern mining (*fuzzyMGP*)

Eq. (8), (c) compute their support as in (9), (d) prune candidates whose support is lower than the user-defined *minimum threshold*,

- Step 2.1 classify gradual patterns whose support is higher or equal than the user-defined *minimum threshold* as frequent gradual patterns of size k,
- Step 3. Set $k = k+1$: (a) with the frequent gradual patterns size k-1 build all gradual pattern candidates of size k, e.g., $GP_c\{(X_1, +), (X_2, +), (X_3, +)\} \leftarrow \{GP_a\{(X_1, +), (X_2, +)\}, GP_b\{(X_1, +), (X_3, +)\}\}$ (b) computing their $m\tilde{c}c(i, j)$ according to a *t-norm* T, e.g.,
  $GP_c.m\tilde{c}c(i, j) \leftarrow \min(GP_a.m\tilde{c}c(i, j), GP_b.m\tilde{c}c(i, j))$ (c) compute their support as in (9), (d) prune candidates whose support is lower than the user-defined *minimum threshold*,
- Step 4. Iterate on step 2.1 and 3 until the build step (3 (a)) does not provide any new candidate.

The evaluation of the correlation, support, and generation of gradual pattern candidates are tasks that require huge amounts of processing time, memory consumption, and load balance. In order to reduce memory consumption, each matrix of *fuzzy concordance degrees* $m\tilde{c}c(i, j)$ is represented and stored according to the *Yale Sparse Matrix Format*, such as only non-zero coefficients are retained. In order to reduce processing time we propose to use the parallel programming model of OpenMP, which is ideally suited for multi-core architectures [58].

Figure 4 shows an overall view of the parallel version of two regions of our fuzzyMGP algorithm, where in the first region is parallelized the extraction process

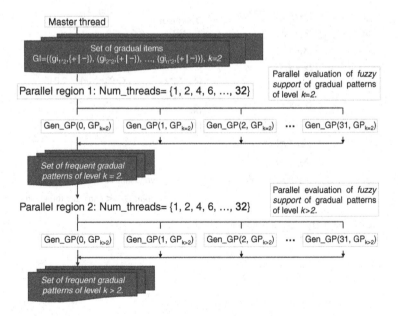

Fig. 4 Parallel extraction of gradual patterns (parfuzzyMGP)

of gradual patterns of size $k = 2$. In the second region is parallelized the extraction cycle of gradual patterns of size $k > 2$. To implement our parallel fuzzyGPM algorithm we choose OpenMP for the following reasons.

Open Multi-Processing (OpenMP) is a shared memory architecture API, that supports multi-platform for writing shared memory parallel applications in C, C++, and Fortran on many architectures, including Linux, OS X, UNIX and Microsoft Windows platforms. It consists of a set compiler directives, runtime routines, and environment variables that influence runtime behaviour [65, 69].

In OpenMP, a parallel region is a block of code executed by a team of threads simultaneously with shared or private data. A team threads is defined by a master thread and a set of N worker threads, which is defined by the environment variable OMP_NUM_THREADS [58, 65, 69].

4.1.3 Experiments and Results

We present an experimental study of the scaling capacities of our approach on several cores, for the database C500A50 with 500 records and 50 attributes, and database C500A100 with 500 records and 100 attributes, which were used in [43, 61] and produced with the IBM Synthetic Data Generator for Association and Sequential Patterns.

Our experiments were performed on a MIMD computer with up to 32 processing cores and 64 GB of RAM with Linux Centos 5.1 and GCC OpenMP 3.1.

The first experiment involves a database with 500 records and 50 attributes, Table 3 shows the results regarding memory consumption and number of gradual patterns (NGP) found for minimum thresholds of 0.30 and 0.35, with uncompressed and compressed matrices of concordance degrees. Figures 10 and 11 in Appendix 1 depict the execution time and speedup reached for 1 to 32 threads.

It should be noted that these results are highly sensitive to the thresholds that are set, which a common drawback of data mining techniques. Some works have tried to study the automatic setting of the threshold, for instance by using genetic programming. However, no general and good result has been shown, neither from the theoretical studies nor from the experimental approaches. In the cases where a threshold is difficult to set, the users can consider giving the set of patterns they would like to retrieve by using $top - k$ approaches [52].

The second experiment involves a database with 500 records and 100 attributes, Table 4 shows the results regarding memory consumption and number of gradual patterns (NGP) found for minimum thresholds of 0.375 and 0.38, with uncompressed and compressed matrices of concordance degrees. Figures 12, 13, 14 and 15 in Appendix 1 depict the execution time and speedup reached for 1 to 32 threads.

4.2 Parallel Mining of Fuzzy Trees

With the development of Internet and Web, frequent pattern mining has been extended to more complex patterns like tree mining, graph mining, and fuzzy tree mining. Such applications arise in complex domains like bioinformatics, Web mining, banking, marketing, biology, health, *etc.* especially to handle complex databases such as semi-structured data or tree databases (for example in the case of XML databases) [19, 20].

Definition 10 A *tree* is a direct, acyclic, connected graph, and rooted labelled tree of the form $T=(V, E)$, with $V = \{0, 1, \ldots, n\}$ as the set of nodes, $E = \{(x, y) \mid x, y \in V\}$ as the set of edges. There is a special vertex $r \in V$ called the root of T and $\forall x \in V$, there is a unique path from r to x. Then y is a *descendant* of x if $x, y \in V$ and there is a path (L) from x to y. If $\mid L \mid$ from two vertices x, y is reduced to one, then the *descendant* relationship is considered as a *children* relationship. We assume that the children $\{y_1, y_2, \ldots, y_n\}$ $(n \geq 0)$ of a node $x \in V$ are ordered from left to right [19, 20, 60].

Tree mining consists in discovering all the frequent subtrees F_S from a database of trees (D) [20], as shown in Fig. 5. The frequency is computed using the notion of support: Given a database D and a tree S, the *support* of a tree S is the proportion of trees T from D where S is included:

$$Support(S) = \frac{\# \, of \, trees \, T \, where \, S \, is \, embedded}{\# \, of \, trees \, in \, D}$$

S is said to be frequent if $Support(S) \geq \sigma$ where σ is a user-defined minimal support threshold. There are two types of strict inclusion: induced inclusion and embedded inclusion, see Fig. 6, where a tree S is included in another tree $T \in D$, if all nodes in S are included in T.

Fuzzy approaches have been proposed in order to soften the constraint on the patterns (*frequent subtrees*) found by the algorithms. In fuzzy tree mining a tree S is included in another tree $T \in D$, with a degree of inclusion $\tau(S, T)$. Four types of fuzzy inclusion have been proposed: *ancestor-descendant degree*, *sibling ordering degree*, *partial inclusion*, and *node similarity*. A detailed treatment of these approaches is

Table 3 Results of experiment with data set C500-A50-50 of 50 attributes and 500 records

Type of matrix	Minimum threshold	Elapsed time	Speedup	Memory consumption (%)	NGP found
Uncompressed	0.30	Fig. 10	Fig. 11	12.6	59810
	0.35	Fig. 10	Fig. 11	0.6	2758
Compressed	0.30	–	–	3.5	59810
	0.35	–	–	0.3	2758

Table 4 Results of experiment with data set C500-A100-50 of 100 attributes and 500 records

Type of matrix	Minimum threshold	Elapsed time	Speedup	Memory consumption (%)	NGP found
Uncompressed	0.375	Fig. 12	Fig. 13	36.2	186994
	0.38	Fig. 12	Fig. 13	24.7	121154
Compressed	0.375	Fig. 14	Fig. 15	14.4	186994
	0.38	Fig. 14	Fig. 15	10.3	121154

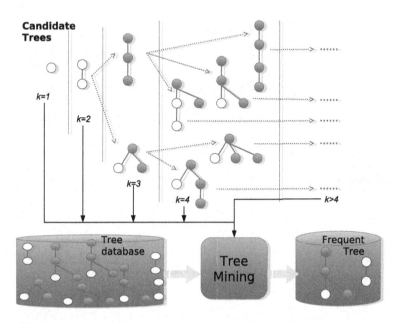

Fig. 5 Scheme of process of tree mining

given in [20, 41]. Figure 7 shows partial inclusion of trees and ancestor-descendant relationship.

In our approach, we consider the fuzzy inclusion *ancestor-descendant degree*, to which we propose to give a scope considering the *number of nodes* between ancestor and descendant nodes. We propose the fuzzy membership function in Eq. (10) for characterizing the *ancestor-descendant relationship* depending on the *number of nodes* separating the two nodes being considered. The frequency is computed using the notion of *fuzzy support*: given a database D and a tree S, the *support* of S in D is given by:

$$Support(S) = \frac{\sum_{T \in D}(\tau(S, T))}{\# of\ trees\ in\ D}$$

Fig. 6 Induced inclusion and embedded inclusion of trees

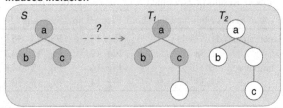

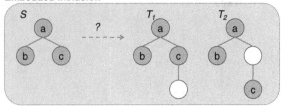

Fig. 7 Partial inclusion of trees and ancestor-descendant relationship

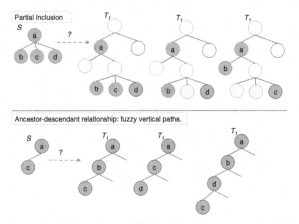

$$\tau(S, T) = \begin{cases} 1 & No\ more\ than\ 5\ nodes \\ 0.5 & if\ Number\ of\ nodes = 6 \\ 0 & if\ Number\ of\ nodes \geq 7 \end{cases} \qquad (10)$$

The core of the process for fuzzy tree mining is briefly described in the algorithm 1. Several methods have been proposed for generating candidates from frequent subtrees [2, 12, 13, 72]. Most of the methods rely on the construction of candidates by using an extension on the right most branch. The trees are numbered in a depth-first enumeration from the root to the right most leaf. Then, for every frequent tree of size k (containing k nodes), candidates are generated by adding a node on the right after considering all the possible 2-trees whose first node corresponds to the anchoring node.

Algorithm 1 Fuzzy Frequent SubTrees Mining

Data: Tree Database D

Result: Fuzzy Frequent Subtrees $\mathscr{F}$

1 $\mathscr{F} \leftarrow \emptyset$

  $k \leftarrow 1$

  repeat

2     $\mathscr{S}_k \leftarrow Gen\_Cand(k)$

    foreach $s \in \mathscr{S}_k$ **do**

3        $Support(s) \leftarrow 0$

      foreach $T \in D$ **do**

4   /* If degree of fuzzy inclusion is relevant, computing Support(s) */

        if $Fuzzy\_inclusion\_degree\ \tau(s,T)$ **then**

5            $Support(s) = Agg_{T \in D}(\tau(s, T))$;

6   /* minSupp stands for a user-specified minimum support value */

      if $Support(s) \geq minSupp$ **then**

7          $\mathscr{F} \leftarrow \mathscr{F} \cup \{s\}$;

8   k++

9 **until** $\mathscr{F}$ does not grow any more;

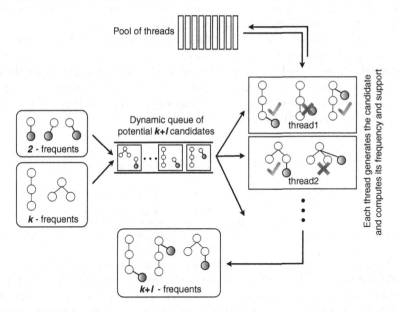

Fig. 8 Parallel fuzzy tree mining

Recently, [60] have developed *PaFUTM: Parallel Fuzzy Tree Mining* a parallel version of Algorithm 1. Figure 8 illustrates the general structure of *PaFUTM* where the computation of the fuzzy support is parallelized using a pool of 1 to 32 threads and a dynamic queue of tasks type *"first come, first served"*. For each level k, potential $k+1$ candidates are queued. Then each idle thread extract a non-processed candidate, calculates its frequency and fuzzy support according to a fuzzy inclusion $\tau(S, T)$ type *Ancestor-descendant degree*. This fuzzy inclusion is defined by a discrete fuzzy set interpreted as a fuzzy scope for the ancestor-descendant relationship *"scope no more than 5 nodes"*.

PatFUTM was implemented using the g++ 3.4.6 and evaluated with POSIX threads, on a 32-core machine, with 8 AMD Opteron 852 processors (each with 4 cores), 64 GB of RAM with Linux Centos 5.1, g++ 3.4.6. and evaluated with two types of datasets: *B* datasets (*BA, BB* and *BC*) that contain lots of relatively small trees and *C* datasets (*CG, CH, CJ*) that contain a smaller amount of larger trees.

4.3 Related Problems

4.3.1 Parallel Mining of Fuzzy Association Rules

Fuzzy association rule mining is an important fuzzy database mining task which is defined as a process to find out the fuzzy patterns or fuzzy attributes which frequently occur together from a fuzzy database [18, 45, 50].

According to the notation of fuzzy database introduced in Sect. 4.1, given two attributes X_1, X_2, a fuzzy set $A \in t X_1$ and a fuzzy set $B \in Y_2$, then a *fuzzy association rule* is defined as an implication of the form, $A \rightarrow B$, where A and B are considered as two *fuzzy patterns*, the problem in fuzzy association rules mining is to find all rules $A \rightarrow B$ that hold in O with the fuzzy support $(fsupp(\{A, B\}))$ and the fuzzy confidence $(fconf(A \rightarrow B))$, defined as

$$fsupp(\{A, B\}) = \sum_{i=1}^{N} min(A(o_i), B(o_i)) \tag{11}$$

$$fconf(A \rightarrow B) = \frac{\sum_{i=1}^{N} min(A(o_i), B(o_i))}{\sum_{i=1}^{N} A(o_i)} \tag{12}$$

The fuzzy association rules with at least a minimum support and a minimum confidence respectively are extracted and considered as interesting [18, 45, 50].

Bao-wen et al. [4], and Jian-jian et al. [37], presented the adaptation of the Count Distribution Parallel Algorithm to design the parallel algorithm for mining fuzzy association rules. Quantitative attributes are partitioned into several fuzzy sets by the parallel fuzzy c-means algorithm (PFCM) [62]. The parallel algorithm for mining Boolean association rules is improved to extract frequent fuzzy patterns. Finally, the fuzzy association rules with at least fuzzy confidence are generated on all processors. The parallel algorithm based on MPI was implemented on the distributed linked PC/workstation of six computers. The results of experimental work showed that the parallel mining algorithm had an excellent *scaleup* and *speedup*.

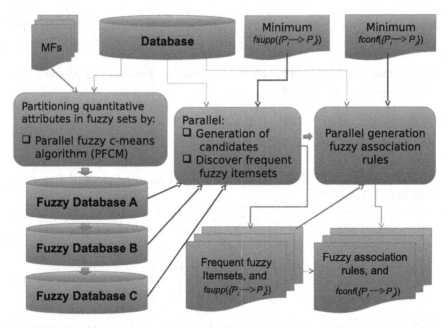

Fig. 9 Scheme of parallel mining fuzzy association rules

In another approach, in order to extract both association rules and membership functions from quantitative attributes, Hong et al. [30], propose a parallel genetic-fuzzy mining algorithm based on the master-slave architecture. Where the master processor uses a single population as a simple genetic algorithm, and distributes the tasks of fitness evaluation to slave processors. The crossover, mutation and production are performed by the master processor. The results showed that the *speedup* can increase nearly linear along with the number of individuals to be evaluated.

Figure 9 shows the general structure of our interpretation of the parallel process of extracting fuzzy association rules.

4.3.2 Parallel Fuzzy Clustering: c-Means

Cluster analysis is defined as the process of grouping a data set, where the similarity between data within a cluster is maximised while the similarity between data of different clusters is minimized [55, 62]. In *classical (hard) cluster analysis* it is considered that each point of the data set belongs to only one cluster [68]. Whereas in *fuzzy cluster analysis* each point of the data set may belong to more than one cluster, according to a set of membership degrees [62, 68].

The most widely used fuzzy clustering algorithm is the *Fuzzy c-Means* (FCM) algorithm proposed by Dunn [21] and generalized by Bezdek [7], FCM is a method based on an objective function of the form:

$$J = \sum_{i=1}^{n} \sum_{j=1}^{c} (u_{ji})^m \left\| x_i - c_j \right\|^2 \tag{13}$$

The clustering is achieved by an iterative optimisation process that minimises the objective function (13) subject to:

$$\sum_{j=1}^{c} u_{ji} = 1 \tag{14}$$

FCM achieves the optimisation of J by the iterative calculations of c_j and u_{ji} using Eqs. (15) and (16).

$$c_j = \frac{\sum_{i=1}^{n} (u_{ji})^m x_i}{\sum_{i=1}^{n} (u_{ji})^m} \tag{15}$$

$$u_{ji} = \left(\sum_{k=1}^{c} \left(\frac{\left\| x_i - c_j \right\|}{\left\| x_i - c_k \right\|} \right)^{\frac{2}{m-1}} \right)^{-1} \tag{16}$$

The process stops when the condition in Eq. (17) is met for successive iterations $t, t+1$, ε is the minimum permissible error and the weighting exponent m is often set to the value 2 [51, 55, 62, 68].

$$Max \left\{ \left\| u_{ji}^{t+1} - u_{ji}^{t} \right\| \right\} < \varepsilon \forall j, i \tag{17}$$

Given n data points $\{x_1, x_2, \ldots, x_n\} \in \mathbf{R}^d$ and assuming that C clusters are to be generated, $c_j \in \mathbf{R}^d$ is the centroid of the cluster j in C, u_{ji} is the matrix of the membership degrees (in [0, 1]) of each x_i in each cluster j in C, the $\sum_{j=1}^{C} u_{ji} = 1$ $\forall i$, for $j = 1, \ldots, C; i = 1, \ldots, n$.

In fuzzy database mining, the FCM method is used to partition the quantitative attributes of crisp database into several fuzzy sets [62]. As the database size becomes larger and larger, this usually requires a high volume of computations, and considerable amount of memory which may lead to frequent disk access, making the process inefficient. With the development of high performance parallel systems, parallel fuzzy clustering may be used to improve performance and efficiency of fuzzy clustering algorithm [51, 62].

Terence et al. [62], present a parallel version of FCM algorithm, where P process are generated and assigned to P processors, the set of data points is divided into equal number of data points, so that each process computes with its n/P data points loaded into its own local memory. The processes can exchange data through of calls to the MPI library. The parallelization of FCM algorithm takes place in two stages, Eq. (15) is evaluated in parallel in the first stage by each cluster, Eq. (16) in the same way is evaluated in the second stage. In their experimental work, their approach of parallel

FCM algorithm demonstrated to reach almost ideal *speedups* and excellent *scaleup* for larger data sets, and it performs equally well when more clusters are requested.

5 Conclusion

In this chapter, we discuss the importance of the scalability of fuzzy systems in general and the scalability of fuzzy database mining algorithms in particular. We analyzed the possibilities of scalability offered by architectures of multi-core processors and its potential for parallel processing. We presented a study of parallel programming of fuzzy database mining algorithms based on Multithreading.

In this chapter, we have discussed how the parallelization of fuzzy algorithms is crucial to tackle the problem of scalability and optimal performance in the context of fuzzy database mining. More precisely, we presented the parallelization of fuzzy database mining algorithms on multi-core architectures of two knowledge discovery paradigms, namely fuzzy gradual pattern mining and fuzzy tree mining (for example in the case of XML databases), so also we presented a review of other two related problems, namely fuzzy association rule mining, fuzzy clustering.

Parallel programming models we are interested in exploring are: Task Parallelism, Data parallelism, and Task-Data parallelism.

The obtained results show the interest of parallel approaches in the fuzzy data mining context. The feasibility having been done, future work will include the integration of more complex optimisations that will enable to tackle with larger databases such as the ones encountered on the Web. For this purpose, depth-first approaches will be explored,together with distributed implementations of the algorithms using GPU processors and the MPI paradigm.

Another perspective of our work is to consider fuzzy data.When considering fuzzy databases where each data is a fuzzy set, fuzzy orderings may then be defined over fuzzy subsets. The computation will be more time and memory consuming, thus requiring more work on the clever parallel implementation. In particular, the data structures will have to be redefined.

Acknowledgments This work was realized with the support of HPC@LR, a Center of Competence in High-Performance Computing from the Languedoc-Roussillon region, funded by the Languedoc-Roussillon region, the Europe and the Universit Montpellier 2 Sciences et Techniques. The HPC@LR Center is equipped with an IBM hybrid Supercomputer.

Appendix 1: Results of Parallel Gradual Pattern Mining

Graphic representation of the Speedup obtained in the experimental work with the parallel fuzzyGPM algorithm (Figs. 10, 11, 12, 13, 14, 15).

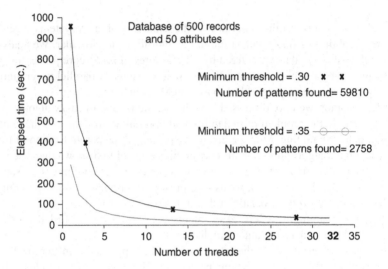

Fig. 10 Threads versus elapsed time with a database of 500 × 50 and minSupp = .30 and .35, using uncompressed binary matrices of concordance degrees

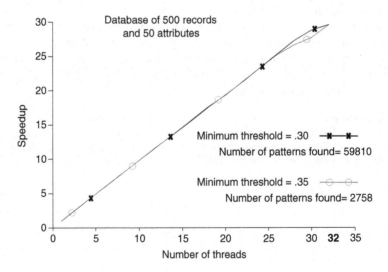

Fig. 11 Speedup with a database of 500 × 50 and minSupp = .30 and .35, using uncompressed binary matrices of concordance degrees

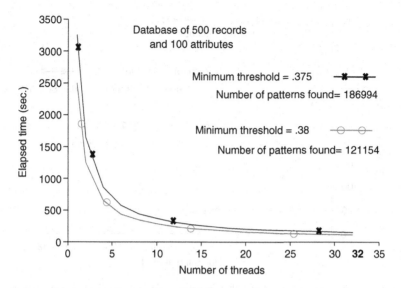

Fig. 12 Threads versus elapsed time with a database of 500×100 and minSupp = .375 and .38, using uncompressed binary matrices of concordance degrees

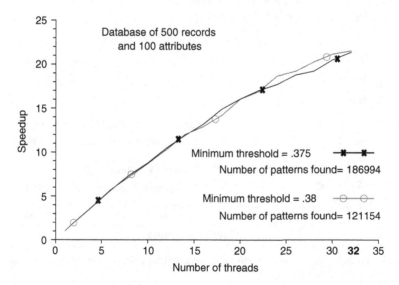

Fig. 13 Speedup with a database of 500×100 and minSupp = .375 and .38, using uncompressed binary matrices of concordance degrees

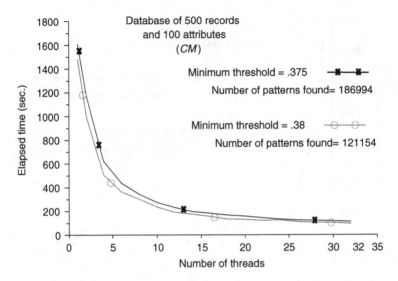

Fig. 14 Threads versus elapsed time with a database of 500 × 100 and minSupp = .375 and .38, using compressed matrices of concordance degrees

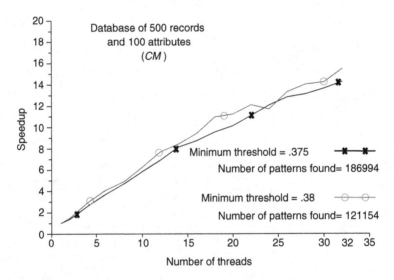

Fig. 15 Speedup with a database of 500 × 100 and minSupp = .375 and .38, using compressed matrices of concordance degrees

References

1. Angryk, R.A., Petry, E.F.: Discovery of abstract knowledge from non-atomic attribute values in fuzzy relational databases. In: Bouchon-Meunier, B., Goletti, G., Yager, R.R. (eds.) Modern Information Processing: From Theory to Applications, pp. 1–11. Elsevier, Amsterdam (2005)
2. Asai, T., Arimura, H., Uno, T., Nakano, S.: Discovering frequent substructures in large unordered trees. In: Proceedings of the 6th International Conference on Discovery Science (2003)
3. Bahri, A., Chakhar S., Yosr, N., Bouaziz R.: Implementing imperfect information in fuzzy databases. In: International Syposium on Computational Intelligence and Intelligent Informatics, October 14–16, Hammamet, Tunisia, pp. 1–8 (2005)
4. Bao-wen, X., Jian-jiang, L., Yingz-hou, Z., Lei, X., Huowang, C., Hong-ji, Y.: Parallel algorithm for mining fuzzy association rules. In: Proceedings of International Conference on Cyberworlds, IEEE (2003)
5. Basterretxea, K., Del Campo, I.: Electronic hardware for fuzzy computing. In: Laurent, A., Lesot, M.-J. (eds.) Scalable Fuzzy Algorithms for Data Management and Analysis: Methods and Design, pp. 1–30. Information Science Reference (2010)
6. Barney, B.: Introduction to Parallel Computing. Lawrence Livermore National Laboratory. http://computing.llnl.gov/tutorials/parallel_comp/#ModelsData. Cited 29 September (2012)
7. Bezdek, J.C.: Pattern Recognition with Fuzzy Objective Function Algorithms. Plenum Press, New York (1981)
8. Bodenhofer, U., Klawonn, F.: Robust rank correlation coefficients on the basis of fuzzy orderings: initial steps. Mathware Soft Comput. **15**, 5–20 (2008)
9. Bodenhofer, U.: A similarity-Based Generalization of Fuzzy Orderings. Johannes-Kepler-Universitat Linz, Linz (1999)
10. Bosc, P., Prade, H.: An introduction to fuzzy set and possibility theory based approaches to the treatment of uncertainty and imprecision in database management systems. In: Proceedings of UMIS'94: From Needs and Solutions, Catalina, CA, USA (1994)
11. Buckles, W.P., Petry, F.E.: Fuzzy representation of data for relational databases. Fuzzy Sets Syst. **7**, 213–226 (1982)
12. Chi, Y., Nijssen, J., Muntz, R., Kok, J.: Frequent subtree mining: an overview. Fundam. Inform. **66**(1–2), 161–198 (2005)
13. Chi, Y., Xia, Y., Yang, Y., Muntz, R.: Mining closed and maximal frequent subtrees from databases of labeled rooted trees. IEEE Trans. Knowl. Data Eng. **17**(2), 190–202 (2005)
14. Cubero, J.C., Medina, J.M., Pons, O., Vila, M.A.: Extensions of resemblance relation. ELSEVIER Fuzzy Sets Syst. **86**, 197–212 (1997)
15. CUDA-NVIDIA: What is GPU computing? In GPU Computing Solutions. http://www.nvidia.com/object/GPU_Computing.html. Cited 29 September, (2012)
16. CUDA Training: Cuda Parallel Programming Model Overview. In Downloadable CUDA Training Podcast. http://www.developer.nvidia.com/cuda-training, Cited 29 September (2012)
17. Data Mining, Analytics, and Databases. In: GPU Computing Solutions. http://www.nvidia.com/object/data_mining_analytics_database.html, (2011)
18. Delgado, M., Marin, N., Martín-Bautista, M., J., Sánchez, D., Vila, M.-A.: Mining fuzzy association rules: an overview. In: Soft Computing for Information Processing and Analysis: Studies in Fuzziness and Soft Computing, vol. 11/2005-vol. 276, Springer (2005)
19. Del Razo, F., Laurent, A., Poncelet, P., Teisseire, M.: Fuzzy tree mining: go soft on your nodes. In: Foundations of Fuzzy Logic and Soft Computing, 12th International Fuzzy Systems Association World Congress IFSA, pp. 145–154. Lecture Notes in Computer Science. Springer, Heidelberg (2007)
20. Del Razo, F., Laurent, A., Poncelet, P., Teisseire, M.: FTMnodes:Fuzzy tree mining based on partial inclusion. Elsevier, ScienceDirect Fuzzy sets and systems (2009)
21. Dunn, J.C.: A fuzzy relative of the ISODATA process and its use in detecting compact well-separated clusters. J. Cybernet. **3**, 32–57 (1973)

22. El-Rewini, H., Abd-el-Barr, M.: Advanced Computer Architecture and Parallel Processing. Wiley, New York (2005)
23. Fan, J., Xie, W.: Some notes on similarity measure and proximity measure. ELSEVIER Fuzzy Sets Syst. **101**, 403–412 (1999)
24. Fang, W.-F., Lu, M., Xiao X., He, B., Luo, Q.: Frequent itemset mining on graphics processors. In: Proceedings of the Fifth International Workshop on Data Management on New Hardware (DaMoN 209), ACM (2009)
25. Freitas, A.A.: A survey of parallel data mining. In: 2nd International Conference on the Practical Applications of Knowledge Discovery and Data Mining, pp. 287–300 (1998)
26. Golkar, C.: Predictive in-database analytics bringing analytics to the data. In: Fuzzy Logix. www.fuzzyl.com/products/in-database-analytics/ (2011)
27. Golkar, C.: Fuzzy Logix Unveils NVIDIA GPU-Based Analytics Appliance The Tanay ZXnW Series. www.fuzzyl.com/press-releases/fuzzy-logix-uneils-nvidia-gpu-based-analytics-appliance/ (2011)
28. Hall, L., O., Goldgof, D., B., Canul-Reich, J., Hore, P., Cheng W., Shoemaker, L.: Scaling fuzzy models. In: Laurent, A., Lesot, M.-J. (eds.) Scalable Fuzzy Algorithms for Data Management and Analysis: Methods and Design, pp. 31–53. Information Science Reference (2010)
29. Hirota, K., Pedrycz, W.: Fuzzy computing for data mining. In: Proceedings of the IEEE, vol. 87, no. 9 (September 1999)
30. Hong, T.P., Lee, Y.C., Wu, M.T.: Using the master-slave parallel architecture for genetic-fuzzy data mining. In: Proceedings of IEEE International Conference on Systems, Man and, Cybernetics (2005)
31. Hughes, C., Hughes, T.: Professional Multicore Programming: Design and Implementation for C++ Developers. Wrox & Wiley Publishing, Inc., Hoboken (2008)
32. Hüllermeier, E.: Association rules for expressing gradual dependencies. In: PKDD, LNAI 2431. Springer, Berlin (2002)
33. Hüllermeier, E.: Fuzzy methods in machine learning and data mining: status and prospects. Fuzzy Sets Syst. **156**(3), 387–407 (2005)
34. Hüllermeier, E.: Why fuzzy set theory is useful in data mining. In Successes and New Directions in Data Mining, IGI Global (2008)
35. Hüllermeier, E.: Fuzzy sets in machine learning and data mining. Appl. Soft Comput. J. **11**, 1493–1505 (2011)
36. Jang, J.-S. R., Sun, C.-T., Mizutani, E.: Neuro-Fuzzy and Soft Computing: A Computational Approach to Learning and Machine Intelligence. Prentice Hall Engineering Science Mathematics, New Jersey (1997)
37. Jian-jiang, L., Bao-wen, X., Xiao-feng, Z., Da-zhou, K., Yan-hui, L., Jin Z.: Parallel mining and application of fuzzy association rules. In: Higher Education Press and Springer-Verlag (2006)
38. Julian-Iranzo, P.: A procedure for the construction of a similarity relation. In: Proceedings of IPMU'2008, Terremolinos (Malaga), pp. 489–496 (2008)
39. Kim, S.: A GPU based parallel hierarchical fuzzy art clustering. In: Proceedings of the International Joint Conference on Neural Networks (IJCNN), IEEE-Computational Intelligence Society (2011)
40. Koh, H., W., Hüllermeier, E.: Mining gradual dependencies based on fuzzy rank correlation. In: Proceedings of SMPS 2010, 5th International Conferebce on Soft Methods in Probability and Statistics. Oviedo/Mieres (Asturias), Spain, October (2010)
41. Laurent A., Poncelet, P., Teisseire, M.: Fuzzy data mining for the semantic web: building XML mediator schemas. In: Fuzzy Logic and the Semantic Web, pp. 249–265. Elsevier, Amsterdam (2006)
42. Laurent, A., Lesot, M., J., Fifqi, M., GRAANK: Exploiting rank correlations for extracting gradual itemsets. In: FQAS 2009, LNAI 5822. Springer, Berlin (2009)
43. Laurent, A., Negrevergne, B., Sicard, N., Termier, A.: PGP-mc: towards a multi-core parallel approach for mining gradual patterns. In: Proceedings of DASFAA (2010)

44. Laurent, A., Negrevergne, B., Sicard, N., Termier, A.: Efficient parallel mining of gradual patterns on multi-core processors. In: AKDM-2, Advances in Knowledge Discovery and Management, vol. 2. Springer (2010)
45. Lin, N., P., Chueh H., E.: Fuzzy correlation rules mining. In: Proceedings of the 6th WSEAS International conference on Applied Computer Science, Hangzhou, China (2007)
46. Ma, Z., M.: Advances in Fuzzy Object-Oriented Databases: Modeling and Applications. Idea Group Publishing, Hershey (2004)
47. Ma, Z.M., Yan, L.: A literature overview of fuzzy database models. J. Inform. Sci. Eng. **26**(2), 427–441 (2008)
48. Molina, C., Serrano, J.M., Sánchez, D., Vila, M.A.: Measuring variation strength in gradual dependencies. In: Proceedings of the 5th EUSFLAT Conference Contents of Volume I, New Dimensions in Fuzzy Logic and Related Technologies (2007)
49. Molina, C., Serrano, J.M., Sánchez, D., Vila, M.A.: Mining gradual dependencies with variation strength. In: Mathware& Soft Computing, vol. 15 (2008)
50. Martin, T., Shen, Y.: Fuzzy association rules to summarise multiple taxonomies in large databases. In: Laurent, A., Lesot, M.-J. (eds.) Scalable Fuzzy Algorithms for Data Management and Analysis: Methods and Design, pp. 273–301. Information Science Reference (2010)
51. Murugavalli, S., Rajamani, V.: A high speed parallel fuzzy C-mean algorithm for brain tumor segmentation. BIME J. **6**(1) (2006)
52. Ngan, S.C., Lam, T., Wong, R., Wai-Chee Fu, A.: Mining N-most interesting itemsets without support threshold by the COFI-tree. Int. J. Bus. Intell. Data Mining **1**(1) (2005)
53. Petry, F., Bosc, P.: Fuzzy Databases: Principles and Applications. Kluwer Academic Publishers, Boston (1996)
54. Piatetsky-Shapiro, G., Frawley, W.J.: Knowledge Discovery in Databases. AAAI Press/The MIT Press (1991)
55. Polimi, D.: A tutorial on clustering algorithms: introduction, k-means, and fuzzy c-means clustering. In: home.dei.polimi.it/matteucc/Clustering/tutorial-html/cmeans.html. Cited 29 September (2012)
56. Quintero, M., Laurent, A., Poncelet, P.: Fuzzy ordering for fuzzy gradual patterns. In: FQAS 2011, LNAI 7022. Springer, Berlin (2011)
57. Rundensteiner, E.A., Hawkes, L.W., Bandler, W.: On nearness measures in fuzzy relational data models. Int. J. Approx. Reason. **3**, 267–298 (1989)
58. Rauber, T., Rünger, G.: Parallel Programming: for Multicore and Cluster Systems. Springer, Berlin (2010)
59. Shenoi, S., Melton, A.: Proximity relations in the fuzzy relational database model. ELSEVIER Fuzzy Sets Syst. (Supplement) **100**, 51–62 (1999)
60. Sicard, N., Laurent, A., Del Razo, F., Quintero Flores, P.M.: Towards multi-core parallel fuzzy tree mining. In: FUZZ-IEEE'2010, IEEE World Congress on Computational Intelligence, IEEE Computational Intelligence Society (2010)
61. Thac Do, T.D., Laurent, A., Termier, A.: PGLCM: efficient parallel mining of closed frequent gradual itemsets. In: Proceedings of International Conference on Data Mining (ICDM) (2010)
62. Terence, K., Kate, A., S., Sebastian, L., David, T.: Parallel fuzzy c-means clustering for large data sets. In: Proceedings of the 8th International Euro-Par Conference on Parallel Processing, pp. 365–374 (2002)
63. Timothy, J.R.: Fuzzy Logic with Engineering Applications. John Wiley & Sons, West Sussex (2010)
64. Touzi, A.G., Ben Hassine, A.B.: New architecture of fuzzy database management systems. Int. Arab J. Inform. Technol. **6**(3), 213–220 (2009)
65. Van der Pas, R.: An overview of OpenMP. In: OpenMP the OpenMP API specification for parallel programming. http://openmp.org/wp/resources/#Tutorials (2011)
66. Van der Pas, R.: Basic concepts in parallelization. In OpenMP the OpenMP API specification for parallel programming. http://openmp.org/wp/resources/#Tutorials. Cited 29 September (2011)
67. Wen, C.H., Chen Y.L.: Mining fuzzy association rules from uncertain data. Knowl. Inform. Syst. *23*(2), Springer (2010)

68. Yang, M.-S.: A survey of fuzzy clustering. Mathl. Comput. Modeling **18**(11), 1–16 (1993)
69. Yang, C.-T., Huang, C.-L., Lin C.-F.: Hybrid CUDA, OpenMP, and MPI parallel programming on multicore GPU clusters. Computer Physics Communications, ELSEVIER, Volume (182), pp. 266–269 (2011)
70. Zadeh, L.A.: Similarity relations and fuzzy orderings. Inform. Sci. ELSEVIER **3**(2), 177–200 (1971)
71. Zadeh, L.A., Hirota, K., Klir, G.J., Sanchez, E., Wang, P.-Z., Yager, R.R.: Advances in Fuzzy Systems: Applications and Theory. World Scientific, Singapore (2011)
72. Zaki, M.J.: Efficiently mining frequent trees in a forest: algorithms and applications. IEEE Trans. Knowl. Data Eng. **17**(8), 1021–1035 (2005)

Printed in the United States
By Bookmasters